Recent Developments in Electrochemical Sensing

Recent Developments in Electrochemical Sensing

Editors

Iulia Gabriela David
Dana Elena Popa

MDPI

Basel • Beijing • Wuhan • Barcelona • Belgrade • Novi Sad • Cluj • Manchester

Editors
Iulia Gabriela David
Department of Analytical
Chemistry and Physical
Chemistry
University of Bucharest
Bucharest
Romania

Dana Elena Popa
Department of Analytical
Chemistry and Physical
Chemistry
University of Bucharest
Bucharest
Romania

Editorial Office
MDPI AG
Grosspeteranlage 5
4052 Basel, Switzerland

This is a reprint of articles from the Special Issue published online in the open access journal *Chemosensors* (ISSN 2227-9040) (available at: https://www.mdpi.com/journal/chemosensors/special_issues/C92WDT1EFS).

For citation purposes, cite each article independently as indicated on the article page online and as indicated below:

Lastname, A.A.; Lastname, B.B. Article Title. *Journal Name* **Year**, *Volume Number*, Page Range.

ISBN 978-3-7258-2179-2 (Hbk)
ISBN 978-3-7258-2180-8 (PDF)
doi.org/10.3390/books978-3-7258-2180-8

Contents

chemosensors

MDPI

Article

Bioengineered Flagellin–TiO_2 Nanoparticle-Based Modified Glassy Carbon Electrodes as a Highly Selective Platform for the Determination of Diclofenac Sodium

Juan Santiago Hidalgo [1,2], Éva Tóth [3], Hajnalka Jankovics [3,†], Carmen Ioana Fort [2], Graziella Liana Turdean [2,*], Etelka Tombacz [3,†] and Ildiko Galambos [1]

1 Soós Ernő Research and Development Center, University of Pannonia, Zrínyi Miklós St. 18, H-8800 Nagykanizsa, Hungary; juansannin2595@gmail.com (J.S.H.); galambos.ildiko@pen.uni-pannon.hu (I.G.)
2 Research Center of Electrochemistry and Non-Conventional Materials, Department of Chemical Engineering, Faculty of Chemistry and Chemical Engineering, "Babes-Bolyai" University, Arany Janos St. 11, 400028 Cluj-Napoca, Romania; iladiu@chem.ubbcluj.ro
3 Bio-Nanosystems Laboratory, Research Institute of Biomolecular and Chemical Engineering, University of Pannonia, Egyetem St. 10, H-8200 Veszprém, Hungary; toth.eva@mk.uni-pannon.hu (É.T.); jankovics.hajnalka@mk.uni-pannon.hu (H.J.); tombacz.etelka@pen.uni-pannon.hu (E.T.)
* Correspondence: graziella.turdean@ubbcluj.ro
† These authors contributed equally to this work.

Abstract: This study describes the incorporation of bioengineered flagellin (4HIS) protein in conjunction with TiO_2 anatase nanoparticles into a chitosan (Chit) polymeric matrix as a highly sensitive electrode modifier for the determination of diclofenac sodium (DS) in wastewater. Two types of electrodes were prepared using a simple drop-casting method. The inner structure of the obtained modified electrode was characterized by scanning electron microscopy (SEM) in combination with energy-dispersive X-ray spectroscopy and isothermal titration calorimetry (ITC). The electrochemical and electroanalytical parameters of DS oxidation at the nanostructured interface of the modified electrode were obtained via cyclic voltammetry and square-wave voltammetry. The analytical parameters for diclofenac electro-detection showed a 50% decrease in LOD and LOQ at Chit + TiO_2 + 4HIS/GCE-modified electrode compared with the Chit + 4HIS/GCE-modified electrode. The obtained tools were successfully used for DS detection in drug tablets and wastewater samples. Thus, it was demonstrated that in the presence of a histidine-containing flagellin variant, the electrode has DS recognition capacity which increases in the presence of TiO_2 nanoparticles, and both induce excellent performances of the prepared tools, either in synthetic solution or in real samples.

Keywords: biosensor; flagellin; TiO_2 nanoparticles; cyclic voltammetry; square-wave voltammetry; electrochemical impedance spectroscopy; diclofenac sodium; wastewater

Citation: Hidalgo, J.S.; Tóth, É.; Jankovics, H.; Fort, C.I.; Turdean, G.L.; Tombacz, E.; Galambos, I. Bioengineered Flagellin–TiO_2 Nanoparticle-Based Modified Glassy Carbon Electrodes as a Highly Selective Platform for the Determination of Diclofenac Sodium. *Chemosensors* **2023**, *11*, 576. https://doi.org/10.3390/chemosensors11120576

Academic Editor: Núria Serrano

Received: 31 October 2023
Revised: 22 November 2023
Accepted: 28 November 2023
Published: 7 December 2023

1. Introduction

Human and veterinary pharmaceutical residue, especially diclofenac sodium (DS), often arrives in the aquatic ecosystem, leading to direct toxicological consequences on humans, the long-term (chronic) exposure being reported to have effects worldwide [1]. In this context, diclofenac was included in the first Watch List of the Water Framework Directive with a predicted no-effect concentration of 50 ng/L (i.e., 0.17 nM) [2,3].

Diclofenac sodium (DS, 2-(2-((2,6-dichlorophenyl)amino)phenyl)acetic acid) is a non-steroidal anti-inflammatory drug (NSAID), having antipyretic and analgesic properties, often used to reduce inflammation and certain types of pain.

Given the importance of DS in groundwater and the European Union legislation, numerous techniques have been employed to determine DS, including chromatography (i.e., gas chromatography–mass spectrometry [4,5], liquid chromatography [6–8], liquid chromatography–mass spectrometry [9], thin layer chromatography [10], spectrophotometry [11–15], colorimetry [16], spectrofluorimetry [17], and capillary zone electrophoresis (CZE) [18].

These methods are time-consuming and/or expensive (e.g., specialized equipment, highly skilled operators, and the use of high-purity reagents and standards) and often need the sample pretreatment step. To overcome these drawbacks, electrochemical methods based on chemically modified electrodes [18–22] are used extensively for their properties such as selectivity, reproducibility, low cost, and simplicity [18–23]. DS determination via electrochemical methods was reported in the literature, using either unmodified electrodes (e.g., edge plane pyrolytic graphite electrode [24]), or modified electrodes, using as modifiers metal or metal oxide nanomaterials [25,26], carbon nanomaterials (e.g., carbon nanotubes [27], graphene [28]), conducting polymers [29], ionic liquids (ILs), inorganic complexes (e.g., nickel hydroxide-modified nickel electrode [30]), etc.

In order to enhance the selectivity of the modified electrode, engineered biological macromolecules immobilized on a modified electrode surface can be an efficient alternative for drug determination. Proteins of large size and varied composition are particularly suitable for developing highly specific sensing materials into biosensors due to their complex structure and interaction pattern. Thus, for example, flagellin-based proteins able to build long (up to 10 μm), highly stable filaments from thousands of monomer units by self-assembly can be modified to have a great specificity for binding a certain contaminant [31]. The redox process facilitated by the presence of a designed protein immobilized at the electrode surface is often mediated by the presence of nanomaterials, resulting in improved electroanalytical performance of the device. [32].

Nanotechnology and nanomaterials have received a lot of attention in electrode development in recent years due to their positive impacts on analytical performances such as increased sensitivity, selectivity, and repeatability, which are due to the low density, high porosity, specific surface area, surface-to-volume ratio, reactivity, biocompatibility, and adsorption capacity of used nanocomposites [33,34]. TiO_2 nanoparticles are catalytically active electrode materials [35], having a porous structure that preserves the nano-architecture of an interconnected nanoparticle porous network [36]. Also, as a result of its other unique properties (i.e., good conductivity, high surface area, high porosity, strong adsorptive capability, chemical inertness, optical transparency, chemical/thermal stability, non-toxicity, and good biocompatibility), it has a high potential for use as modifying agents for electrode materials. [36,37].

Furthermore, when compared to solid graphite or noble metal electrodes, TiO_2 nanoparticle-based electrodes have a reduced background current. Other advantages include a larger potential window, improved stability, and increased repeatability. Consequently, TiO_2 nanoparticles can be combined with other compounds (such as graphite, carbon nanotubes, etc.), and the synergistic effect leads to a significant improvement in electrode performance (due to an increase in both surface electroactive area and electron transfer rate between the electrode and the target molecule) [35].

Based on these advantages, the aim of this paper was to combine bioengineered flagellin (4HIS) protein with TiO_2 nanoparticles immobilized via an eco-friendly polymer (chitosan) in order to create, for the first time, to our knowledge, a novel modified electrode useful for a more specific determination of DS. The modified electrode prepared via the drop-casting method was electrochemically characterized by cyclic voltammetry, square-wave voltammetry, and electrochemical impedance spectroscopy. The developed modified electrode demonstrated a good detection limit, linear range, and selectivity, being successfully applied for the determination of DS in drug tablets and wastewater samples.

2. Materials and Methods

2.1. Materials and Reagents

The following reagent were used: acetic acid (0.1 molar, $C_2H_4O_2$, "Reactivul" Bucuresti), chitosan ($C_{12}H_{24}N_2O_9$, from crab shells, Sigma-Aldrich, St. Louis, MO, USA), disodium phosphate ($Na_2HPO_4 \cdot 12H_2O$, 99% pure, "Reactivul" Bucuresti), potassium phosphate monobasic (KH_2PO_4, 99.5% pure, Sigma Aldrich), phosphoric acid (H_3PO_4, 0.1 M, Sigma-Aldrich), potassium hexacyanoferrate(III) ($K_3[Fe(CN)_6]$, 99% pure, Feinbiochemica, Bucuresti), potassium ferrocyanide ($K_4[Fe(CN)_6 \cdot H_2O$, 99% pure, Chimopat, Bucuresti), diclofenac sodium (DS, $C_{14}H_{10}Cl_2NNaO_2$, 99% pure, Sigma-Aldrich), glucose monohydrate ($C_6H_{12}O_6 \cdot H_2O$, 99% pure, Merck, Darmstadt, Germany), citric acid monohydrate ($C_6H_8O_7 \cdot H_2O$, 99% pure, Sigma Aldrich), fructose ($C_6H_{12}O_6$, 99% pure, Sigma Aldrich), Ibuprofen ($C_{13}H_{18}O_2$, Sun Pharma, Cluj-Napoca, Romania), paracetamol ($C_8H_9NO_2$, Sun Pharma, Cluj-Napoca), dopamine hydrochloride ($C_8H_{11}NO_2$, 99%, Alfa Aesar, Ward Hill, MA, USA), sodium hydroxide (NaOH, 85% pure, Merck), potassium chloride (KCl, 99% pure, "Reactivul" Bucuresti), titanium oxide anatase (TiO_2, anatase, 99.5%, particle size 10–30 nm, Sigma Aldrich), ethanol (98% pure, "Reactivul", Bucuresti). Diclofenac tablets (50 mg, from Rambaxy-Terapia, Cluj-Napoca, Romania) were purchased from a local pharmacy.

A suspension of a 4HIS flagellin variant in polymeric form was prepared in the laboratory of the Research Institute of Biomolecular and Chemical Engineering (Veszprem, Hungary) according to [37], except that polymer formation was performed in 20 mM 4-(2-hydroxyethyl)-1-piperazine ethanesulfonic acid (HEPES) (Sigma-Aldrich), 150 mM NaCl, and pH 7.0 buffer, via the addition of ammonium sulfate in a final concentration of 0.6 M.

The phosphate buffer solution (PB, 0.025 M, pH 4.6) was prepared by dissolving the appropriate amount of $Na_2HPO_4 \times 12\ H_2O$ and KH_2PO_4 in distilled water [38]. A 0.1 M DS stock solution was prepared in buffer solution and stored at 4 °C in the refrigerator.

Diluted H_3PO_4 and NaOH solutions were used for adjusting the pH of the buffer solutions. All chemicals having analytical purity were used as received, without any purification.

2.2. Characterization Technique

The morphological structure of the prepared biosensor was carried out using a scanning electron microscope (VEGAS 3 SBU, Tescan, Brno–Kohoutovice, Czech Republic), which was equipped with a primary and secondary electron detector, to observe the inner part of the sample. Furthermore, the scanning electron microscopy (SEM) was equipped with spectra of energy-dispersive X-ray spectroscopy (EDXS), and chemical maps for the elements were acquired using a Dual EDX System (Bruker, Karlsruhe, Germany).

Isothermal titration calorimetry (ITC) experiments were carried out at 25 ± 0.2 °C by a MicroCal PEAQ-ITC calorimeter. Due to the low solubility of DS at pH 4.6, ITC measurements were performed at pH 7.0. A 5 mg/mL solution of the 4HIS flagellin variant in monomeric form was prepared in 100 mM HEPES and 150 mM NaCl (pH 7.0). The diclofenac Na salt solution was prepared in final volumes of 10 mL (1.35 mM) and 5 mL (6.475 mM) via direct volumetric adjustment with the same buffer to volume. In all cases, the diclofenac solution was titrated into the protein at the end point of titration in molar excesses of 2.2 (monomeric only) and 11-fold (monomeric and polymeric protein). The ITC data were fitted to a one-binding-site model with the MicroCal PEAQ-ITC Analysis Software package provided by MicroCal, using a non-linear least-squares algorithm.

2.3. Preparation of the Modified Electrode

The glassy carbon electrode (inner diameter of 3 mm, from ALS Co., Tokyo, Japan) was polished on felt material with alumina slurry (0.3 µm, Buehler, Crissier, Switzerland). After that, it was placed in distilled water in ultrasound equipment for 10 min and then was thoroughly rinsed with ethanol and distilled water.

The 0.1 % chitosan solution was prepared via sonication for 30 min, with 10 mg chitosan in 10 mL of 0.1 M acetic acid. A suspension containing 1 µL 4HIS (corresponding to 2 µg flagellin) and 1 mg TiO_2 anatase in 1 mL chitosan solution was sonicated for 3 h.

Electrode modification via the drop-casting method consists of depositing three times a volume of 3 µL of the above-prepared suspension onto the clean GCE surface and drying it at room temperature via slow solvent evaporation for 2 h. Thus, the obtained electrodes were symbolized as Chit/GCE, Chit + 4HIS/GCE, and Chit + TiO_2 + 4HIS/GCE.

2.4. Electrode Characterization

The electrochemical characterization of the prepared modified electrodes was performed using a computer-controlled AutoLab potentiostat (PGSTAT302N and PGSTAT 12, EcoChemie, Utrecht, The Netherlands) operated by GPES 4.7 software for cyclic voltammetry (CV) and square-wave voltammetry (SWV) measurements. A conventional three-electrodes cell, equipped with a working GC-modified electrode, an Ag/AgCl, KCl_{sat} reference electrode, and a platinum plate auxiliary electrode, was connected to the potentiostat.

2.5. Analysis of the Water Sample

To demonstrate the real application perspective of the performed modified electrode, a real water sample was collected from the city's wastewater treatment plant (Cluj Napoca, Romania) and analyzed using the standard addition method. The wastewater sample was used as it was obtained, without any pretreatment. Also, an HPLC-MS analysis was performed on the wastewater samples (SOOS laboratory, Nagykanizsa, Hungary) to compare the sensor's results.

3. Results

3.1. Morpho Structural Characterization of GCE-Modified Electrodes

The morphological characterization of the films deposited on the GC surface was performed using SEM and is presented in Figure 1A,B. The Chi + 4HIS coating displays a slightly rough surface (Figure 1A) without noticeable cracks along the electrode surface.

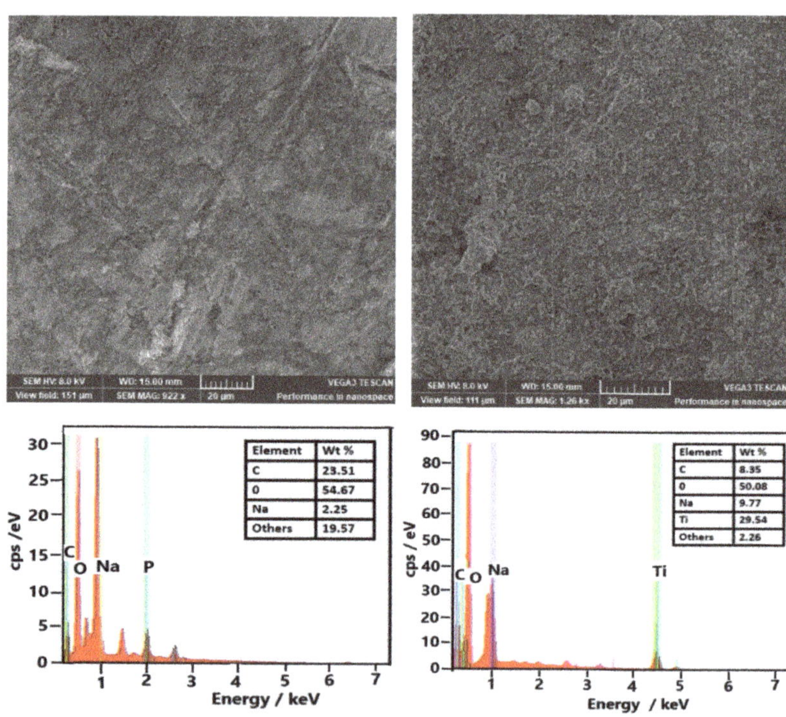

(A) (B)

Figure 1. SEM images of Chit + 4HIS (**A**) and Chit + TiO_2 + 4HIS (**B**) samples and the corresponding EDX spectra indicating the film composition.

The Chi + TiO_2 + 4HIS coating is much thicker, and a dense arrangement of submicron-sized clumps appears to uniformly cover the entire electrode surface (Figure 1B). In contrast to a uniform planar structure of the Chit, some agglomerations can be seen in the presence of 4HIS and TiO_2 surfaces. From EDX spectra, the presence of Ti is confirmed in the samples containing TiO_2, proving its immobilization which occurs during the preparation of the modifier film.

3.2. Electrochemical Behavior of GCE-Modified Electrodes for DS Detection

The electrochemical characteristics of the prepared modified electrode were evaluated by cyclic voltammetry. At 50 $mV \cdot s^{-1}$ (Figure 2), the DS oxidation peak is not visible at the Chit/GCE-modified electrode, while the DS well-defined irreversible oxidation peak is placed at 0.763 V vs. Ag/AgCl, KCl_{sat} at the Chit + 4HIS/GCE-modified electrode and is shifted in the presence of TiO_2 nanoparticles to a lower value of 0.698 V vs. Ag/AgCl, KCl_{sat} at Chit + TiO_2 + 4HIS/GCE-modified electrodes. DS oxidation is an irreversible process, as no cathodic peak is recorded. Similar values have been reported in the literature [39–42]. Also, the oxidation peak current intensity increases from 4.11×10^{-6} A at Chit + 4HIS/GCE to 7.01×10^{-6} A at Chit + TiO_2 + 4HIS/GCE, respectively. This means that there is an increase in current intensities of 70% in the presence of TiO_2 in the modified matrix of the electrode, demonstrating the electrocatalytic performance effect of the TiO_2 anatase nanoparticles.

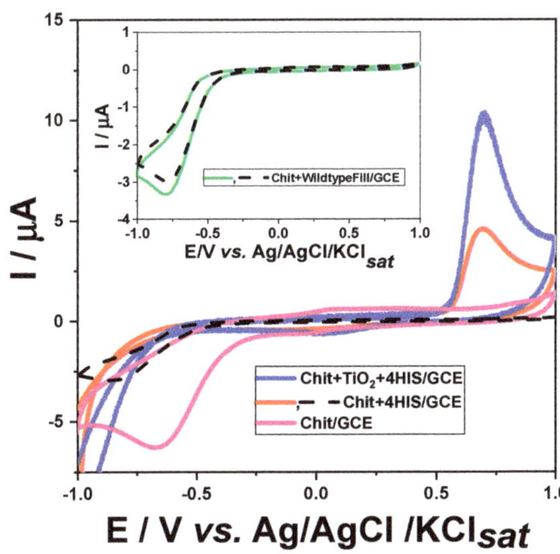

Figure 2. Cyclic voltammograms in the absence (black dash line) and in the presence of 10^{-3} M diclofenac in 0.025 M PB (pH 4.6) at different modified electrodes (see inset, colored solid line). Experimental conditions: scan rate, 50 $mV \cdot s^{-1}$; starting potential, −1 V vs. Ag/AgCl, KCl_{sat}.

Based on the above, the detection of DS is clearly associated with the presence of 4HIS flagellin among the electrode surface components. To understand this more thoroughly, isothermal titration calorimetry was used. ITC measurements of the polymer and monomer 4HIS at pH 7 (due to solubility problems of DS at a lower pH) showed no interaction with DS even at 10-fold molar excess. This is in agreement with our previous observation that the 4HIS flagellin variant specifically binds Ni(II) at a neutral pH [43]. In addition, it was observed that the wild-type flagellin-containing electrode (Chit + wild-type Fill/GCE) did not exhibit an oxidation peak, indicating that the non-modified filaments were unable to detect diclofenac sodium. The 4HIS flagellin was constructed from the wild-type with the modification of L209H-V235H-K241H-S264H, suggesting that the role of the protein in DS detection can be reduced to its histidine side chains. ITC titrations suggest that when the protein is properly folded, the buried binding pocket formed by the four histidine side chains is not accessible to DS. However, the electrode was prepared in 0.1 M acetic acid solution (pH~2.9) by stirring for 3 h, which is likely to lead to the decomposition of the starting polymer and unfolding of the monomers. With the unfolding of the 4HIS, the binding pocket is eliminated, and the histidine side chains can be positioned on the protein surface, accessible to DS. As well, the behavior of the Chit + TiO_2 + 4HIS/GCE electrode could be explained by the synergetic effect of the biochemical arrangement of 4HIS in recognizing DS in the solution and of the spatial distribution of the TiO_2 nanoparticles in the modified electrode matrix, which increases the electrochemically active surface area of the electrode matrix.

In order to estimate the electrochemical active area, CVs were performed in a scan rate range of 0.010–1 V s^{-1} in a solution of 0.1 M KCl containing 0.5 mM of $K_3[Fe(CN)_6]/K_4[Fe(CN)_6$. For both studied modified electrodes, an increase in the anodic peak current intensities ($I_{p,a}$) and the scan rate (ν) was observed (Figure 3A,B).

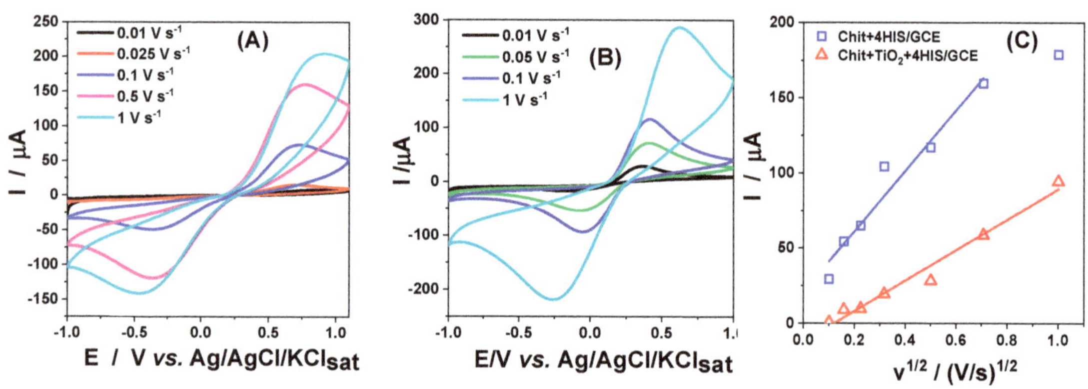

Figure 3. Cyclic voltammograms at Chit + 4HIS/GCE (**A**) and Chit + TiO_2 + 4HIS/GCE (**B**)-modified electrodes in 0.1 M KCl containing 0.5 mM $K_3[Fe(CN)_6]/K_4[Fe(CN)_6$ solution and the corresponding I versus $v^{1/2}$ dependencies (**C**). Experimental conditions: scan rates, see inset; starting potential, −1 V vs. Ag/AgCl, KCl_{sat}.

The electrode process, involving the redox couple $[Fe(CN)_6]^{3-}/[Fe(CN)_6]^{4-}$, is a quasi-reversible one-electron transfer process which obeys the Randles–Ševčik Equation (1) [44]:

$$I_{p,a} = (2.69 \times 10^5)\, n^{3/2} \cdot A \cdot C_0 \cdot D^{1/2} \cdot \nu^{1/2} \quad (1)$$

where $I_{p,a}$ is the anodic peak current (in A), n is the number of electrons transferred (in our case, one electron for the $[Fe(CN)_6]^{3-}/[Fe(CN)_6]^{4-}$ redox couple), A is the active surface area of the electrode (in cm^2), C_0 is the concentration of the solution (in mol/cm^3; in our case, 0.5×10^{-6} mol/cm^3), D is the diffusion coefficient ($cm^2 \cdot s^{-1}$; in our case, 7.60×10^{-6} $cm^2 \cdot s^{-1}$ [45]), and ν is the scan rate (in V/s).

The obtained linear dependences between I versus $v^{1/2}$ are as follows (Figure 3C):

$I/A = (-1.21 \times 10^{-5} \pm 3.81 \times 10^{-6}) + (1.01 \times 10^{-4} \pm 7.24 \times 10^{-6})v^{1/2}/(V/s)^{1/2}$, R/no. points = 0.9875/7, for Chit + 4HIS/GCE;

$I/A = (2.08 \times 10^{-5} \pm 9.08 \times 10^{-6}) + (2.02 \times 10^{-4} \pm 2.3 \times 10^{-5})v^{1/2}/(V/s)^{1/2}$, R/no. points = 0.9750/6, for Chit + TiO_2 + 4HIS/GCE.

Using the slope value of I versus $v^{1/2}$ dependencies and Equation (1), for the active surface area, the values obtained were 0.136 cm^2 for Chit + 4HIS/GCE and 0.272 cm^2 for Chit + TiO_2 + 4HIS/GCE-modified electrodes, respectively. Compared with the geometric area of the electrode of 0.07065 cm^2, the active surface area of the modified electrodes is greater. Also, it must be pointed out that the presence of TiO_2 increases the active surface area of the electrode twice, as expected.

3.2.1. Influence of Scan Rate on DS Oxidation

Figure 4A shows the influence of the scan rate on DS oxidation at the Chit + 4HIS + TiO_2/GCE-modified electrode. As expected, the peak potential shifted to a more positive value of potential with the increase in the scan rate. The following linear log I versus log v dependencies were obtained:

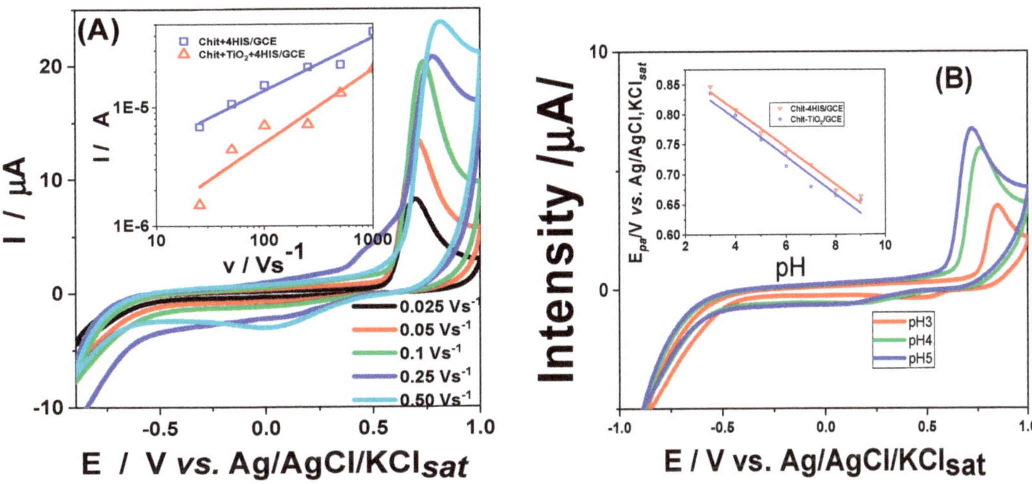

Figure 4. Influence of the scan rate on the cyclic voltammograms of 10^{-3} M DS at Chit + 4HIS + TiO_2/GCE-modified electrode. Inset: log I vs. log v dependencies for different modified electrodes (see inset) (**A**). Influence of pH on the peak potential of 10^{-3} M DS oxidation (**B**). Experimental conditions: electrolyte, 0.025 M PB (pH 4.6); starting potential, −1 V vs. Ag/AgCl, KCl_{sat}.

log I/A = (−6.53·± 0.218) + (0.618 ± 0.095) log v/(V/s), R/no. points = 0.9550/6, for Chit + 4HIS/GCE;

log I/A = (−5.76 ± 0.103) + (0.447 ± 0.046) log v/(V/s), R/no. points = 0.9799/6, for Chit + TiO_2 + 4HIS/GCE.

The slopes of log I vs. log v dependencies (Figure 4A inset) with values close to 0.5 are characteristic of redox processes controlled by diffusion, while values close to 1 indicate processes controlled by adsorption [44]. From the data in the previous equations, it can be concluded that the redox process at the electrode occurred under diffusion control in both cases of modified electrodes.

3.2.2. Influence of the pH on DS Oxidation

It is worth mentioning that the pH of the electrolyte could have an important impact on the catalytic properties of the modifier and thus on its behavior, the reaction mechanism at the electrode, and the analytical parameters of the modified electrode. In this context, the effect of the pH of the supporting electrolyte in the range of 3.0–8.0 on the electrooxidation of 10^{-3} M DS at different modified GCEs was studied using cyclic voltammetry at a scan rate of 50 mV s^{-1} (Figure 4B). It can be observed that the anodic peak potential values gradually shift to more negative potential values as the pH increases. This behavior suggests that protons are involved in the oxidation reaction at the electrode interface [45,46].

The linear dependencies of the anodic peak potential on pH have the following slopes: −0.033 ± 0.003 (R/n = 0.9935/5) at Chit + 4HIS/GCE and −0.040 ± 0.001 (R/n = 0.9993/5) at Chit + TiO_2 + 4HIS/GCE, respectively. Those values show a slight sub-Nernstian behavior compared to the theoretical Nernstian slope of 0.059 V/pH. This action is probably due to the bioengineered flagellin (4HIS) protein component of the electrode matrix, whose behavior is influenced by the variation in pH.

3.3. Analytical Characterization

For the analytical detection of diclofenac, the square-wave voltammetry (SWV) method was used due to its high sensitivity, enhanced peak resolution, and ability to discriminate the capacitive current [47] or to have low background current [48]. Figure 5A presents, as an example, the voltammograms recorded at the Chit + TiO_2 + 4HIS/GCE-modified electrode, and the calibration curves for all modified electrodes are presented in Figure 5B.

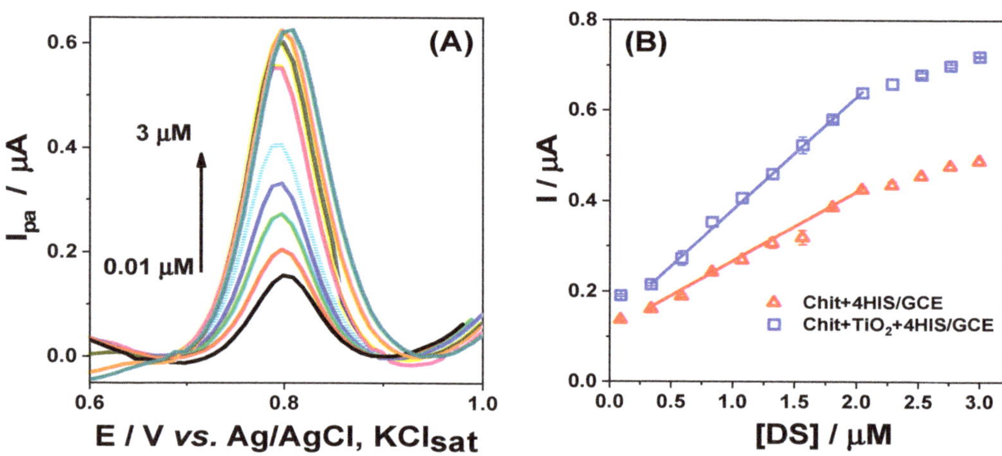

Figure 5. SWVs recorded at Chit + TiO_2 + 4HIS/GCE (**A**) and the corresponding calibration curves for different modified electrodes (see inset) (**B**). Experimental conditions: electrolyte, 0.025 M PB (pH 4.6); stock solution, 10^{-4} M DS; frequency, 10 Hz; step potential, 0.01 V; amplitude, 0.01 V; starting potential, 0.6 V vs. Ag/AgCl, KCl_{sat}.

Linear dependencies between the anodic peak current intensities and DS concentrations were obtained in the concentration ranges from 0.25×10^{-6} to 2×10^{-6} M DS and 0.11×10^{-6} to 2×10^{-6} M DS at Chit + 4HIS/GCE and Chit + TiO_2 + 4HIS/GCE-modified electrodes, respectively, as described by the following regression equations:

$I_{p,a}$/A = $(1.13 \cdot \times 10^{-7} \pm 3.87 \cdot \times 10^{-9}) + (0.153 \pm 0.004)$ [DS]/M, R = 0.9976 n = 8 points, at Chit + 4HIS/GCE;

$I_{p,a}$/A = $(1.29 \times 10^{-7} \pm 2.42 \cdot \times 10^{-9}) + (0.250 \pm 0.003)$ [DS]/M, R = 0.9997, n = 8 points, at the Chit + TiO_2 + 4HIS/GCE-modified electrode, respectively.

When compared to the Chit+4HIS/GCE-modified electrode, the sensitivity of the Chit + TiO_2 + 4HIS/GCE-modified electrode is about twice as high. The limits of detection (LOD) and the limit of quantification (LOQ) were calculated as $3 \times s_b/m$, and $10 \times s_b/m$, where s_b is the standard deviation blank analyte signal and m is the slope of the calibration equation [49]). The obtained values of LOD were 0.066 µM and 0.033 µM DS, and 0.252 µM and 0.111 µM DS for LOQ, at Chit + 4HIS/GCE and Chit + TiO_2 + 4HIS/GCE-modified electrodes, respectively. As seen, a 50% lower LOD and LOQ were obtained for the Chit + TiO_2 + 4HIS/GCE-modified electrode in comparison with the Chit + 4HIS/GCE-modified electrode, proving the beneficial effect of the presence of TiO_2 in the modifier matrix of the electrode. The obtained LOD values are comparable [29] or even better [20,25,26,39,40,50] than those obtained for quasi-similar sensors reported in the literature (Table 1).

Table 1. Comparison of the electroanalytical performances of the different modified electrodes for the determination of DS.

Electrode	LOD (μM)	Linear Range (μM)	Real Sample	References
GO-COOH/GCE	0.09	1.2–400	/	[20]
Au–Pt NPs/f-MWCNTs/Au	0.30	0.5–1000	/	[25]
CeO_2NPs/SPCE	0.40	0.1–26	Tablets	[26]
nanoTiO_2/PEDOT/GCE	0.03	4–15	/	[29]
Amino-labeled	0.27	5–1000	Blood serum	[39]
aptamer/Fe_3O_4/AuNP/CNT/GCE	20	10^{-5}–$1.3 \cdot \times 10^{-3}$	Tap and surface water	[40]
NiNPs/erGO/GCE	0.09	0.25–125	/	[50]
Chit + 4HIS/GCE	0.066	0.25–2	Wastewater	This work
Chit + TiO_2 + 4HIS/GCE	0.033	0.11–2	Wastewater	This work

SPCE = graphite-based screen-printed carbon electrodes; AuNPs = gold nanoparticles (AuNPs); CNT = carbon nanotubes; GO-COOH/GCE = carboxyl-functionalized graphene oxide; Au-Pt NPs/f-MWCNTs/Au = Au-Pt nanoparticles (Au-Pt NPs) and functionalized multiwalled carbon nanotubes (f-MWCNTs); PEDOT = poly (3,4-ethylenedioxythiophene; erGO = electrochemically reduced graphene oxide.

The repeatability and reproducibility of the modified electrodes were investigated by recording the square-wave voltammograms for 3 μM DS in 0.025 M PB solution (pH 4.6). The repeatability was examined using the same electrode for six consecutive measurements, obtaining a relative standard deviation (RSD) of the peak current intensities of 2.75% (Chit + TiO_2 + 4HIS/GCE) and 4.30% (Chit + 4HIS/GCE), respectively, showing very good repeatability of responses at each modified electrode.

The reproducibility was examined by comparing the responses of six different modified electrodes prepared following the same experimental procedure. The relative standard deviations of the obtained current peak intensities were 2.78% at Chit + TiO_2 + 4HIS/GCE and 3.5% at Chit + 4HIS/GCE-modified electrodes, respectively.

The short-time stability of the studied modified electrodes was estimated by performing cyclic voltammograms during 25 cycles in 10^{-3} M DS, with a scan rate of 50 mV s^{-1}. The intensity of the peak was similar during the first 25 cycles, with a slight decrease in the anodic peak intensity of about 5% for all prepared electrodes.

The obtained results show that the developed nanocomposite-based modified electrodes had good repeatability, reproducibility, and stability toward the determination of DS.

3.4. Interferences Study

The ability of the prepared modified electrode to be selective for DS determination was evaluated by adding different concentrations of potential interfering compounds that might be present in real biological samples together with the analyte. Thus, 10^{-3} M glucose, citric acid, fructose, ibuprofen, paracetamol, or dopamine were added to a 10^{-5} M DS prepared in 0.025 M PB (pH 4.6), and the peak current intensities in the presence and absence of interference compounds were compared. The signal change, expressed in percent, was calculated as the ratio $(I_{int+DS} - I_{DS})/I_{DS}$, where I_{DS} is the peak current intensity for DS and I_{int+DS} is the peak current intensity for DS in the presence of the interfering compound.

As seen in Figure 6, for all prepared modified electrodes, the signal change of less than 5% is probably due to the selectivity caused by the histidine side chains of the unfolded flagellin contained in all electrode matrices. However, the greatest values for the signal changes were recorded in the case of citric acid and glucose for all prepared electrodes, especially at the Chit + TiO_2 + 4HIS/GCE-modified electrode. This behavior is most likely caused by the presence of TiO_2 anatase nanoparticles in the sensitive electrode matrix, which offers a high active surface area.

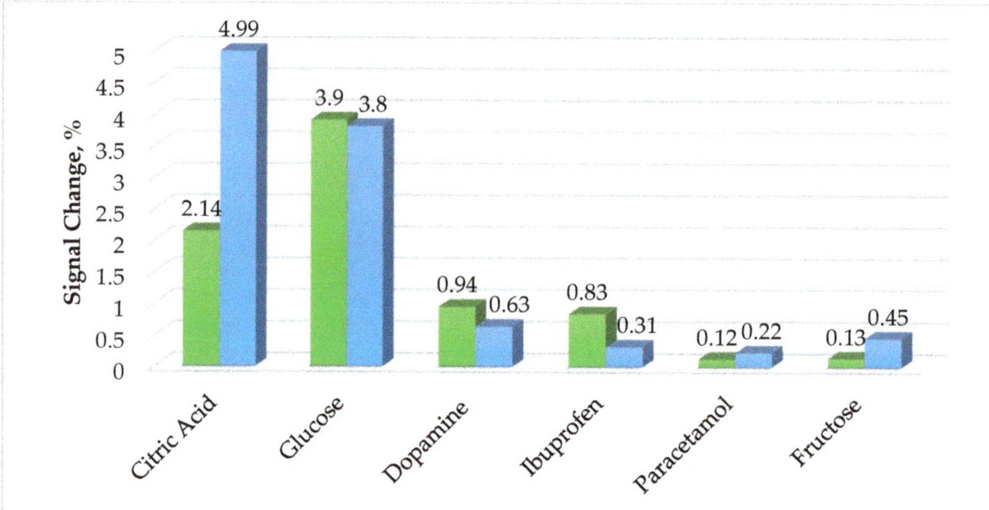

Figure 6. Interference studies at Chit + 4HIS/GCE (green) and Chit + TiO_2 + 4HIS/GCE (blue)-modified electrodes in the presence of 10^{-5} M DS in 0.025 M PB (pH 4.6) and 10^{-3} M of different contaminants.

These results suggest a good selectivity of the prepared electrodes, proving that the modified electrodes are a suitable technique for DS detection.

3.5. Determination of DS in the Pharmaceutical Sample and in the Wastewater Sample

The applicability of the developed modified electrodes was investigated for the determination of DS from pharmaceutical tablets containing 50 mg DS into an inert matrix of lactose monohydrate, povidone, corn starch, colloidal anhydrous silicon dioxide, magnesium stearate, and talc. In order to overcome the possible matrix effects, the standard addition method was performed. For this purpose, the tablet was crushed in an agate mortar, and a weighted quantity of powder was diluted in 100 mL 0.025 M PB (pH 4.6) to obtain a 1.57 μM DS final concentration and was sonicated for 10 min. A total of 10 μL of the drug solution was added to 10 mL of 0.025 M PB (pH 4.6), then additions of 10 and 25 μL of standard 10^{-4} M DS were performed and SWVs were recorded after each addition.

Figure 7A and Table 2 summarize the obtained results. It can be observed that the recovery was between 96–98.7% for all prepared modified electrodes, with the best results for the Chit + TiO_2 + 4HIS/GCE-modified electrode, proving that the presence of TiO_2 nanoparticles induces an enhanced behavior of the modifier matrix of the electrode. The obtained values are in good agreement with those indicated by the pharmaceutical producer. Due to this behavior, both Chit + 4HIS/GCE and Chit + TiO_2 + 4HIS/GCE-modified electrodes are sufficiently accurate and precise to be used for the determination of DS in biological samples.

Table 2. Comparative determination of DS in pharmaceutical products. Experimental conditions: see Figure 7A. Mean ± standard deviation of 3 measurements with different electrodes.

Type of Electrode	[DS]/μM Added	[DS]/μM Found	Recovery (%)	R/n
Chit + 4HIS/GCE	1.57	1.55 ± 0.63	98.72 ± 0.58	0.9964/6
Chit + TiO_2 + 4HIS/GCE	1.57	1.51 ± 0.77	96.18 ± 0.4	0.9991/6

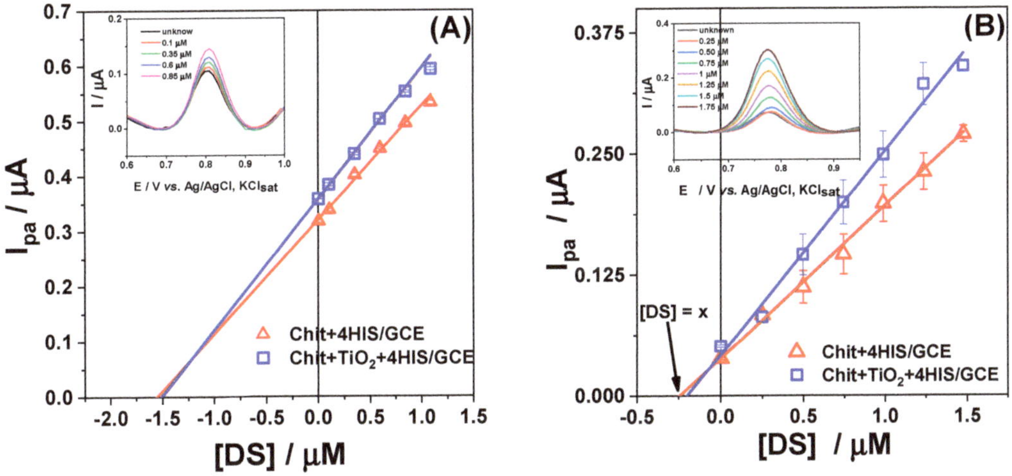

Figure 7. Standard addition plot obtained with a Chit + 4HIS/GCE (red) and Chit + TiO_2 + 4HIS/GCE (blue)-modified electrode, and the corresponding SWVs recorded at Chit + TiO_2 + 4HIS/GCE-modified electrode (inset) for DS determination in pharmaceutical product (**A**) and in wastewater (**B**) samples. Experimental conditions: frequency, 10 Hz; step potential, 0.01 V; amplitude, 0.01 V; starting potential, −0.4 V vs. Ag/AgCl, KCl_{sat}.

Another potential applicability of the prepared modified electrodes is the determination of DS concentration in wastewater samples, using the same standard addition method. No pretreatment of the wastewater samples was required. As previously described, the intensity of the peak increased linearly after each addition of aliquots of standard DS solution. The unknown concentration of DS was estimated by extrapolating the calibration curves (intersection with 0Y = 0 axis) from Figure 7B, and the obtained data are summarized in Table 3.

Table 3. Comparative determination of DS in wastewater sample. Experimental conditions: see Figure 7B. Mean ± standard deviation of 3 measurements with different electrodes.

Type of Electrode	SWV/ μM	HPLC/ μM	Relative Error/ %	RSD/ %
Chit + 4HIS/GCE	0.24 ± 0.026	0.19 ± 0.0018	20.83	3.53
Chit + TiO_2 + 4HIS/GCE	0.20 ± 0.09		5	0.70

4. Conclusions

A newly modified electrode based on a matrix containing a novel bioengineered histidine-containing flagellin (4HIS) and TiO_2 nanoparticles immobilized with chitosan (Chit) polymer on the surface of GCE was prepared (Chit + TiO_2 + 4HIS/GCE) and characterized by electrochemical investigation techniques. TiO_2 nanoparticles in the polymeric matrix led to obtaining an active surface area that is double that of Chit + 4HIS/GCE. For diclofenac sodium detection, the Chit + TiO_2 + 4HIS/GCE-modified electrode had a LOD of less than 50% than the Chit + 4HIS/GCE-modified electrode, proving the beneficial effect of the presence of the TiO_2 anatase structure in the modifier matrix. The device was used for the determination of DF either from pharmaceutical or wastewater real samples, with good recoveries and results when compared with the standard methods of analysis. Compared to wild-type (histidine-free) flagellin-containing matrices, it can be concluded that the interaction of DS with the aromatic side chains of histidine plays an important role in the specific detection of DS.

Thus, based on the advantages of using TiO_2 in the electrode matrix, our work combines for the first time, to our knowledge, TiO_2 with bioengineered flagellin (4HIS) protein and improves the GCE-modified electrode performances (detection limit, linear range, selectivity, stability, etc.) for DS detection. Moreover, the obtained results, which are sufficiently accurate and precise, proved that the obtained modified electrodes can be used for the determination of DS in real samples.

Author Contributions: Conceptualization, J.S.H., G.L.T., C.I.F., I.G., H.J. and E.T.; methodology, J.S.H., G.L.T., C.I.F. and H.J.; software, J.S.H., G.L.T., C.I.F. and H.J.; validation, J.S.H., G.L.T. and C.I.F.; formal analysis, J.S.H., G.L.T., C.I.F., H.J., E.T. and É.T.; investigation, J.S.H. and É.T.; data curation, J.S.H., G.L.T., I.G. and J.S.H.; writing—original draft preparation, J.S.H., G.L.T., C.I.F., I.G., H.J. and E.T.; writing—review and editing, J.S.H., G.L.T., C.I.F., I.G., H.J. and E.T.; visualization, J.S.H., G.L.T., C.I.F. and H.J.; supervision, G.L.T., C.I.F. and I.G. All authors have read and agreed to the published version of the manuscript.

Funding: This research received no external funding.

Institutional Review Board Statement: Not applicable.

Informed Consent Statement: Not applicable.

Data Availability Statement: Data are contained within the article.

Acknowledgments: Juan Hidalgo acknowledges the Erasmus fellowship from Pannonia University (Veszprem, Hungary) for financing the research stay at "Babes-Bolyai" University (Cluj-Napoca, Romania). The authors thank Dorina Simedru for providing SEM measurements. Also, this work has been partially implemented by the TKP2021-NKTA-21 project with the support provided by the Ministry of Culture and Innovation of Hungary from the National Research, Development and Innovation Fund, financed under the 2021 Thematic Excellence Programme funding scheme.

Conflicts of Interest: The authors declare no conflict of interest.

References

1. Oaks, J.L.; Gilbert, M.; Virani, M.Z.; Watson, R.T.; Meteyer, C.U.; Rideout, B.; Shivaprasad, H.L.; Ahmed, S.; Jamshed, M.; Arshad, M.; et al. Diclofenac residues as the cause of vulture population decline in Pakistan. *Nature* **2014**, *427*, 630–633. [CrossRef] [PubMed]
2. EU. *Commission Implementing Decision (EU) 2015/495 of 20 March 2015 Establishing a Watch List of Substances for Union-Wide Monitoring in the Field of Water Policy Pursuant to Directive 2008/105/EC of the European Parliament and of the Council (Notified under Document C(2015) 1756)*; European Commission, Ed.; European Commission: Brussels, Belgium, 2015.
3. Simon, E.; Duffek, A.; Stahl, C.; Frey, M.; Scheurer, M.; Tuerk, J.; Gehrmann, L.; Konemann, S.; Swart, K.; Behnisch, P.; et al. Biological effect and chemical monitoring of Watch List substances in European surface waters: Steroidal estrogens and diclofenac—Effect-based methods for monitoring frameworks. *Environ. Int.* **2022**, *159*, 107033. [CrossRef] [PubMed]
4. Yilmaz, B. GC–MS determination of diclofenac in human plasma. *J. Chromatogr.* **2010**, *71*, 549–551. [CrossRef]
5. Borenstein, M.; Xue, S.; Cooper, S.; Tzeng, T. Sensitive capillary gas chromatographic-mass spectrometric selected ion monitoring method for the determination of diclofenac concentrations in human plasma. *J. Chromatogr. B* **1996**, *685*, 59–66. [CrossRef] [PubMed]
6. Elkady, E.F. Simultaneous determination of diclofenac potassium and methocarbamol in ternary mixture with guaifenesin by reversed-phase liquid chromatography. *Talanta* **2010**, *82*, 1604–1607. [CrossRef]
7. Bhupendra, L.K.; Kaphalia, S.; Kumar, S.; Kanz, M.; Treinen-Moslen, M. Efficient high-performance liquid chromatography/ultraviolet method for determination of diclofenac and 4′-hydroxy diclofenac in rat serum. *J. Chromatogr. B* **2006**, *830*, 231–237. [CrossRef]
8. Lee, H.S.; Jeong, C.K.; Choi, S.J.; Kim, S.B.; Lee, M.H.; Ko, G.; Sohn, D.H. Simultaneous determination of aceclofenac and diclofenac in human plasma by narrow bore HPLC using column-switching. *J. Pharm. Biomed. Anal.* **2020**, *23*, 775–781. [CrossRef]
9. Sparidans, R.W.; Lagas, J.S.; Schinkel, A.H.; Schellens, J.H.; Beijnen, J. Liquid chromatography–tandem mass spectrometric assay for diclofenac and three primary metabolites in mouse plasma. *J. Chromatogr. B* **2008**, *872*, 77–82. [CrossRef]
10. Bhushan, R.; Gupta, D.; Mukherjee, A. Liquid chromatographic analysis of certain commercial formulations for non-opioid analgesics. *Biomed. Chromatogr.* **2007**, *21*, 1284–1290. [CrossRef]
11. Botello, J.C.; Caballero, G.P. Spectrophotometric determination of diclofenac sodium with methylene blue. *Talanta* **1995**, *42*, 105–108. [CrossRef]
12. de Souza, R.L.; Tubino, M. Spectrophotometric determination of diclofenac in pharmaceutical preparations. *J. Braz. Chem. Soc.* **2005**, *16*, 1068. [CrossRef]

13. Garcia, Z.M.; Albero, M.I. Flow-injection spectrophotometric determination of diclofenac sodium in pharmaceuticals and urine samples. *J. Pharm. Biomed. Anal.* **1998**, *17*, 267–273. [CrossRef] [PubMed]
14. El-Didamony, A.M.; Amin, A.S. Adaptation of a color reaction for spectrophotometric determination of diclofenac sodium and piroxicam in pure form and in pharmaceutical formulations. *Anal. Lett.* **2004**, *37*, 1151–1162. [CrossRef]
15. de Micalizzi, Y.C.; Pappano, N.B.; Debattista, N.B. First and second-order derivative spectrophotometric determination of benzyl alcohol and diclofenac in pharmaceutical forms. *Talanta* **1998**, *47*, 525–530. [CrossRef] [PubMed]
16. Sastry, S.; Mohana Rao, A.R.; Prasad, N.V. Spectrophotometric analysis of diclofenac sodium and piroxicam and their pharmaceutical preparations. *Anal. Lett.* **1987**, *2*, 75–80. [CrossRef]
17. Arancibia, J.A.; Escandar, G.M. Complexation study of diclofenac with b-cyclodextrin and spectrofluorimetric determination. *Analyst* **1999**, *124*, 1833–1838. [CrossRef] [PubMed]
18. Jin, W.; Zhang, J. Determination of diclofenac sodium by capillary zone electrophoresis with electrochemical detection. *J. Chromatogr. A* **2000**, *868*, 101–107. [CrossRef] [PubMed]
19. Sarhangzadeh, K.; Khatami, A.A.; Jabbari, M.; Bahari, S. Simultaneous determination of diclofenac and indomethacin using a sensitive electrochemical sensor based on multiwalled carbon nanotube and ionic liquid nanocomposite. *J. Appl. Electrochem.* **2013**, *43*, 1217–1224. [CrossRef]
20. Karuppiah, C.; Cheemalapati, S.M.; Chen, S.M.; Palanisamy, S. Carboxyl-functionalized graphene oxide-modified electrode for the electrochemical determination of nonsteroidal anti-inflammatory drug diclofenac. *Ionics* **2014**, *21*, 231–238. [CrossRef]
21. Baghayeri, M.; Maleki, B.; Zarghani, R. Voltammetric behavior of tiopronin on carbon paste electrode modified with nanocrystalline $Fe_{50}Ni_{50}$ alloys. *Mater. Sci. Eng. C* **2014**, *44*, 175–182. [CrossRef]
22. Baghayeri, M.; Nazarzadeh Zare, E.; Mansour Lakouraj, M. A simple hydrogen peroxide biosensor based on a novel electromagnetic poly(p-phenylenediamine) @Fe_3O_4 nanocomposite. *Biosens. Bioelectron.* **2014**, *55*, 259–265. [CrossRef] [PubMed]
23. Vahedi, J.; Karimi-Maleh, H.; Baghayeri, M.; Sanati, A.L.; Khalilzadeh, M.A.; Bahrami, M. A fast and sensitive nanosensor based on MgO nanoparticle room-temperature ionic liquid carbon paste electrode for determination of methyldopa in pharmaceutical and patient human urine samples. *Ionics* **2013**, *19*, 1907–1914. [CrossRef]
24. Goyal, R.N.; Chatterjee, S.; Agrawal, B. Electrochemical investigations of diclofenac at edge plane pyrolytic graphite electrode and its determination in human urine. *Sens. Actuat. B Chem.* **2010**, *145*, 743–748. [CrossRef]
25. Eteya, M.; Rounaghi, M.; Deiminiat, B. Fabrication of a new electrochemical sensor based on Au-Pt bimetallic nanoparticles decorated multi-walled carbon nanotubes for determination of diclofenac. *Microchem. J.* **2019**, *144*, 254–260. [CrossRef]
26. de Carvalho, R.C.; Bettsa, A.J.; Cassidya, J.F. Diclofenac determination using CeO_2 nanoparticle modified screen-printed electrodes—A study of background correction. *Microchem. J.* **2020**, *158*, 105258. [CrossRef]
27. Salahuddin, M.; Akhter, S.; Basirun, W.J.; Akhtaruzzaman, M.; Mohammed, M.A.; Rahman, N.M.M.A.; Salleh, N.M. Biosynthesized copper nanoparticle decorated multiwall carbon nanotube-nano cellulose nanocomposite: An electrochemical sensor for the simultaneous detection of acetaminophen and diclofenac sodium. *Surf. Interfaces* **2022**, *34*, 102385. [CrossRef]
28. Naz, S.; Nisar, A.; Qian, L.; Hussain, S.; Karim, S.; Hussain, S.Z.; Liu, Y.; Sun, H.; Ur-Rahman, A.; Ahmad, M. Graphene Oxide Functionalized with Silver Nanoparticles and ZnO Synergic Nanocomposite as an Efficient Electrochemical Sensor for Diclofenac Sodium. *Nano Brief Rep. Rev.* **2021**, *16*, 2150139. [CrossRef]
29. Thiagarajan, S.; Rajkumar, M.; Chen, S. Nano TiO_2-PEDOT Film for the Simultaneous Detection of Ascorbic Acid and Diclofenac. *Int. J. Electrochem. Sc.* **2012**, *7*, 2109–2122. [CrossRef]
30. Hajjizadeh, A.; Jabbari, H.; Heli, A.; Movahedi, A.M.; Haghgoo, S. Evaluation of the electrocatalytical properties of $NiCo(OH)_2$ composite modified electrodes. *Electrochim. Acta* **2007**, *53*, 1766–1774. [CrossRef]
31. Jankovics, H.; Szekér, P.; Tóth, E.; Balázs, K.; Lábadi, Z.; Saftics, A.; Kalas, B.; Fried, M.; Petrik, P.; Vonderviszt, F. Flagellin-based electrochemical sensing layer for arsenic detection in water. *Sci. Rep.* **2021**, *11*, 3497. [CrossRef]
32. Boumya, W.; Taoufik, N.; Achak, M.; Bessbousse, H.; Elhalil, A.; Barka, N. Electrochemical sensors and biosensors for the determination of diclofenac in pharmaceutical, biological and water samples. *Talanta* **2021**, *3*, 100026. [CrossRef]
33. Shrivastava, S.; Jadon, N.; Jain, R. Next-generation polymer nanocomposite-based electrochemical sensors and biosensors: A review. *Trends. Anal. Chem.* **2016**, *82*, 55–67. [CrossRef]
34. John, B. Polymer Nanocomposite-Based Electrochemical Sensors, and Biosensors. In *Nanorods and Nanocomposites*; InTech Open: London, UK, 2020. [CrossRef]
35. Postolović, K.Ș.; Stanić, Z. Chitosan/TiO_2 nanoparticles modified carbon paste electrode as a sensitive volt-ammetric sensor for the determination of diclofenac sodium as an anti-inflammatory drug. *Mater. Today Commun.* **2023**, *37*, 107416. [CrossRef]
36. Dias, G.V.; Joseane, C.; Bernardes, B.N.; Wesling, D.M.; Larissa, S.; Marques, C.R. Preparation, and electrochemical capacitance of high surface area TiO_2–RuO_2 aerogels. *Open Ceram.* **2021**, *8*, 100196. [CrossRef]
37. Huang, X.J.; Choi, Y.K. Chemical sensors based on nanostructured materials. *Sens. Actuat. B* **2007**, *122*, 659–671. [CrossRef]
38. Seracu, D. *Indreptar de Chimie Analitica (Tabele, Diagrame, Programme)*; Editura Tehnica: Bucuresti, Romania, 1989; p. 116, ISBN 973-3100-935.
39. Kashefi-Kheyrabadi, L.; Mehrgardi, M.A. Design, and construction of a label-free aptasensor for electrochemical detection of sodium diclofenac. *Biosens. Bioelectron.* **2012**, *33*, 184–189. [CrossRef]
40. Deng, A.P.; Himmelsbach, M.; Zhu, Q.S.; Frey, S.; Sengl, M.; Buchberger, W.; Niessner, R.; Knopp, D. Residue analysis of the pharmaceutical diclofenac in different water types using ELISA and GCMS. *Environ. Sci. Technol.* **2003**, *37*, 3422–3429. [CrossRef]

41. Mohamed, M.; El-Wekil, A.; Saad, A.; Alkahtani, H.H.; Ali, R.H.; Mahmoud, A.M. Advanced sensing nanomaterials-based carbon paste electrode for simultaneous electrochemical measurement of esomeprazole and diclofenac sodium in human serum and urine samples. *J. Mol. Liq.* **2018**, *262*, 495–503. [CrossRef]
42. Abbas, A.; Bahiraei, A.; Madrakian, T. Gold nanoparticle/multi-walled carbon nanotube modified glassy carbon electrode as a sensitive voltammetric sensor for the determination of diclofenac sodium. *Mat. Sci. Eng. C* **2016**, *59*, 168–176. [CrossRef]
43. Labadi, Z.; Kalas, B.; Saftics, A.; Illes, L.I.; Jankovics, H.; Bereczk-Tompa, E.; Sebestyén, A.; Tóth, E.; Kakasi, B.; Moldovan, C.; et al. Sensing layer for Ni Detection in Water Created by Immobilization of Bioengineered Flagellar Nanotubes on Gold Surfaces. *ACS Biomater. Sci. Eng.* **2020**, *6*, 3811–3820. [CrossRef]
44. Bard, A.J.; Faulkner, L.R. *Electrochemical Methods: Fundamentals and Applications*, 2nd ed.; Wiley: New York, NY, USA, 2022; pp. 231, 591.
45. Killedar, L.; Ilager, D.; Shetti, N.P.; Aminabhavi, T.M.; Reddy, K.R. Synthesis of ruthenium doped titanium dioxide nanoparticles for the electrochemical detection of diclofenac sodium. *J. Mol. Liq.* **2021**, *340*, 116891. [CrossRef]
46. Shah, A.; Ullah, A.; Rauf, A.; Rehman, Z.; Shujah, S.; Shah, S.M. Detailed electrochemical probing of a biologically active isoquinoline. *J. Electrochem. Soc.* **2013**, *160*, H597. [CrossRef]
47. Honakeri, N.C.; Malode, S.J.; Kulkarni, R.M.; Shetti, N.P. Electrochemical behavior of diclofenac sodium at core-shell nanostructure modified electrode and its analysis in human urine and pharmaceutical samples. *Sens. Int.* **2020**, *1*, 100002. [CrossRef]
48. Nasraoui, S.; Ameur, S.; Al-Hamry, A.; Mounir, B.; Olfa Kanoun, A. Development of an Efficient Voltammetric Sensor for the Monitoring of 4-Aminophenol Based on Flexible Laser-Induced Graphene Electrodes Modified with MWCNT-PANI. *Sensors* **2022**, *22*, 833. [CrossRef]
49. Harvey, D. *Modern Analytical Chemistry*, 1st ed.; McGraw-Hill: Boston, MA, USA, 2000; p. 95.
50. Daizong, J.; Zhouyuanjing, S.; Zixiang, L.; Sze, S.L.; Jingwen, Z.; Tingkai, Z.; Zetao, C.; Xiongjie, Y.; Yanli, L.; Di, L.; et al. Smartphone-based square wave voltammetry system with screen-printed graphene electrodes for norepinephrine detection. *Smart Mater. Med.* **2020**, *1*, 1–9. [CrossRef]

chemosensors

MDPI

Article

Disposable Stochastic Platform for the Simultaneous Determination of Calcipotriol and Betamethasone in Pharmaceutical and Surface Water Samples

Bianca-Maria Tuchiu [1,2,*], Raluca-Ioana Stefan-van Staden [1,2,*], Jacobus (Koos) Frederick van Staden [1] and Hassan Y. Aboul-Enein [3]

1 Laboratory of Electrochemistry and PATLAB, National Institute of Research for Electrochemistry and Condensed Matter, 202 Splaiul Independentei Str., 060021 Bucharest, Romania; koosvanstaden2012@gmail.com

2 Faculty of Chemical Engineering and Biotechnologies, Politehnica University of Bucharest, 060021 Bucharest, Romania

3 Pharmaceutical and Medicinal Chemistry Department, The Pharmaceutical and Drug Industries Research Division, National Research Centre, Dokki, Cairo 12311, Egypt; haboulenein@yahoo.com

* Correspondence: bianca.tuchiu@gmail.com (B.-M.T.); ralucavanstaden@gmail.com (R.-I.S.-v.S.); Tel.: +40-75-150-7779 (R.-I.S.-v.S.)

Abstract: A disposable stochastic platform based on calix [6]arene modified multi-walled carbon nanotubes-gold nanoparticles screen-printed electrode has been developed for the simultaneous determination of calcipotriol and betamethasone. For both analytes, very wide linear concentration ranges and extremely low limits of quantification (LOQ) were obtained: from 1.0×10^{-15} to 1.0×10^{-3} mol L^{-1} and with a 1.0×10^{-15} mol L^{-1} LOQ for calcipotriol, and from 1.0×10^{-16} to 1.0×10^{-2} mol L^{-1} with a 1.0×10^{-16} mol L^{-1} LOQ for betamethasone. The applicability of the sensing platform was successfully tested in commercially available topical pharmaceutical gel and surface water samples, obtaining recovery values ranging from 99.10 to 99.99% and relative standard deviation values under 0.05%. The obtained results render the proposed platform a viable, robust, selective, and sensitive tool that can be employed for the determination of the analytes in on-site routine quality control of pharmaceuticals and water quality monitoring.

Keywords: stochastic platform; disposable sensor; simultaneous determination; calcipotriol; betamethasone

Citation: Tuchiu, B.-M.; Stefan-van Staden, R.-I.; van Staden, J.F.; Aboul-Enein, H.Y. Disposable Stochastic Platform for the Simultaneous Determination of Calcipotriol and Betamethasone in Pharmaceutical and Surface Water Samples. *Chemosensors* **2023**, *11*, 446. https://doi.org/10.3390/chemosensors11080446

Academic Editor: Alina Vasilescu

Received: 18 July 2023
Revised: 29 July 2023
Accepted: 8 August 2023
Published: 10 August 2023

1. Introduction

Calcipotriol and betamethasone (Figure 1) are two active pharmaceutical ingredients used in combination in various formulations for the topical treatment of mild to moderate plaque psoriasis (psoriasis vulgaris). Psoriasis vulgaris is a chronic inflammatory immune-mediated cutaneous condition characterized by scaly, red, and itchy lesions on the limbs and trunk that are delimited from uninvolved skin by a distinct line of demarcation [1]. The lesions are caused by overactive T-helper lymphocytes which trigger an overgrowth of undifferentiated keratinocytes. The severity of psoriasis vulgaris can be intensified by various factors such as cutaneous injuries, sunburn, stress, and certain medications such as beta-blockers, ACE inhibitors, and lithium. The calcipotriol/betamethasone combination has proven effective due to the different mechanisms of action of the components. Betamethasone is a fluorinated corticosteroid that inhibits inflammation and epidermal hyperproliferation by acting on the glucocorticoid receptors while calcipotriol (or calcipotriene) is a vitamin D analog that has the effect of regulating cell proliferation and differentiation by acting via a vitamin D_3 receptor [2]. The combination of the two has proven to be more effective than calcipotriol and betamethasone in monotherapy [3]. Furthermore, based on pharmacoeconomic evaluations, this combination is more financially beneficial than other topical treatments [4].

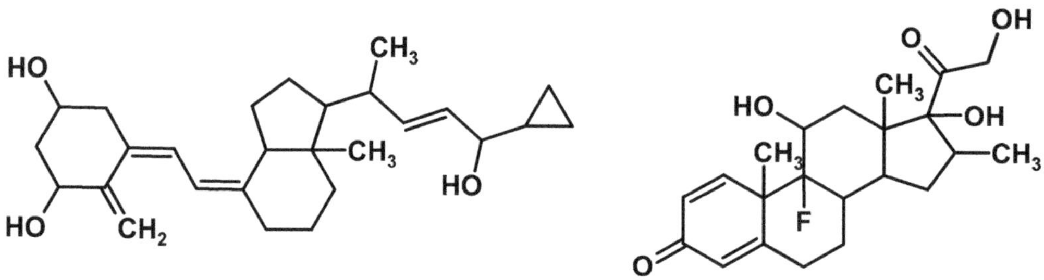

(a) (b)

Figure 1. Chemical structure of: (**a**) calcipotriol; (**b**) betamethasone.

The potential negative impacts of both calcipotriol and betamethasone necessitate careful management of their usage to prevent significant damage to human bodily functions. Prolonged utilization of topical corticosteroids has raised some concerns, as their administration may result in cutaneous side effects such as cutaneous atrophy, the formation of striae, and tachyphylaxis. Extended usage of large quantities may even result in hypothalamic-pituitary-adrenal axis suppression, although it occurs in very rare cases. Adverse reactions such as irritant contact dermatitis have also been reported with topical administration of vitamin D analogs more frequently than with corticosteroids [3].

Like other corticosteroids, betamethasone risks eliciting adverse reactions on the physiological functions and behavior of aquatic organisms once released into aquatic ecosystems. Corticosteroids have been observed to disrupt osmoregulation, metabolism, the immune system, and proper functioning of the musculoskeletal and cardiovascular systems, as well as reproductive functions [5]. For this reason, monitoring the betamethasone levels in water samples to identify sources of pollution and develop effective mitigation strategies could reduce the deleterious effects caused by their presence.

Whether calcipotriol or other vitamin D analogs pose environmental risks has not been identified.

The compendial method provided in the European Pharmacopoeia for the determination of calcipotriol is liquid chromatography whereas for the determination of betamethasone it is UV-Vis spectroscopy [6]. Numerous other methods have been reported for the quantification of calcipotriol and betamethasone up until the present times. These include HPLC with UV detection, TLC-densitometry, UPLC with UV detection, and UV spectrophotometry with two chemometrics methods: fuzzy interference system (FIS) and continuous wavelet transform (CWT) [7–10]. These methods have several drawbacks, such as pricey equipment that needs to be handled by qualified personnel, consumption of large quantities of expensive reagents, and complex and lengthy sample processing, even though they are reproducible, accurate, and sensitive. Electrochemical techniques possess desirable attributes such as sensitivity, selectivity, ease of use, affordability, and basic sample processing rendering them a viable substitute for the simultaneous determination of calcipotriol and betamethasone.

The cornerstone of a new generation of electrochemical sensors, called stochastic sensors, was laid out by Bayley and Cremer in 2001 when they presented membrane-bound channels or pores capable of distinguishing multiple molecules of interest. First, natural nanopores were used, for example, staphylococcal α-hemolysin [11]. Afterwards, the utilization of artificial nanopores embedded within solid membranes demonstrated significant versatility in the assay of individual molecules, as their integration in the design of stochastic sensors presents a compelling viewpoint regarding the identification and determination of molecular interactions. The interactions between the nanopores and individual molecules represent a random and reversible process that modulates the electrical current passing through the nanopore. The current is measured as a function of time and is generated by applying a constant electrical potential. As the analyte molecule present in the solution flows through the nanopore, it generates a current blockade for a specific duration. Artificial nanopores represent advantageous tools as they are capable of interacting with multiple molecules but at distinct time intervals, hence they can be employed for the simultaneous analysis of multiple molecules. When analyzing a mixture of analytes, the signals of each analyte are distinguished based on the amount of time it takes to block the nanopore and the amount of time it takes to bind to the nanopore wall [12].

The utilization of the stochastic approach in electrochemistry enables qualitative and quantitative analysis. To date, numerous stochastic sensors have been developed for screening specific diseases, determining some pharmaceutically active compounds, evaluating water quality and food control, and successfully applying to many types of sample matrices [13–16]. Moreover, the stochastic method also allows for enantio-analysis [17].

Commercially available screen-printed electrodes (SPEs) are versatile tools with numerous applications in electrochemical analysis. This is mainly due to the variety of benefits they offer, such as cost-effectiveness, reproducibility, facile and convenient manufacturing and utilization processes, versatile design and customization options for diverse applications, compatibility with portable devices, and the ability to detect a wide range of compounds [18].

Carbon nanotubes exhibit exceptional mechanical, chemical, and thermal characteristics due to their nanostructure and aspect ratio. These include remarkable electrical conductivity, Young's modulus, tensile strength, flexibility, favorable chemical inertness, and elevated electrical conductivity. Their unique physical and catalytic features render them highly suitable for electrochemical sensors. There are two categories of carbon nanotubes, namely single-walled carbon nanotubes (SWCNTs) which exhibit sp2 hybridized carbon atoms displayed in a hexagonal honeycomb structure subsequently rolled into a tubular shape, and multi-walled carbon nanotubes (MWCNTs) which consist of multiple concentric tubes that encircle one another [19].

Gold nanoparticles (AuNPs) possess several advantageous properties in terms of surface area, conductivity, electro-catalytic characteristics, and stability. Consequently, they enhance the sensitivity and improve the limit of detection of the sensor by increasing the active area of the electrode [20–22].

Calixarenes are flexible macrocyclic compounds presenting hollow cavity-like structures with a conical shape. Calix [6]arene consists of six phenolic units linked by methylene groups. The selection of calix [6]arene as a constituent material for the design of the sensor was based on its 5.0 Å diameter cavity being suitable for stochastic detection [23].

The utilization of techniques that enable the simultaneous determination of multiple analytes offers the benefit of saving analysis time and diminishing the associated costs. Unlike classical methods, electrochemical methods used for simultaneous determination avoid the step of separating the analytes, thus simplifying the whole process.

This present study proposes a disposable stochastic platform based on a calix [6]arene modified multi-walled carbon nanotubes-gold nanoparticles screen printed electrode for the simultaneous determination of calcipotriol and betamethasone in topical pharmaceutical gel and surface water samples. To our knowledge, there have not yet been any electrochemical methods described for the simultaneous determination of calcipotriol and betamethasone. Only a limited number of electrochemical sensors have been proposed for the individual determination of betamethasone, while for calcipotriol there were no reports identified in the literature [24–26]. On site analysis is of high request in pharmaceutical industry (where the quality of the samples must be analyzed in order to produce high quality pharmaceutical formulations), as well as in water analysis, because the quality of water is a very important issue for the health of population. Developing such robust platforms for on site analysis, connected by wireless to mobile instrumentation (e.g., laptop/tablet/smart phone) and further to databases and agencies for the quality of environment facilitate the real-time monitoring of calcipotriol and betamethasone concentrations in different samples. Therefore, the novelty of this paper is given by the simultaneous assay of calcipotriol and betamethasone in pharmaceutical and water samples using newly designed screen-printed electrodes integrated into a platform able to connect to a mobile device which is performing data acquisition, data processing, and data transmission to a central point that requires monitorization of, e.g., the water quality. Compared with the standard HPLC methods recommended by national and international pharmacopoeias, who largely recommend the determination of one of the active compounds (either calcipotriol, either betamethasone), methods which are not cost-effective, given the costs of gas, column, and sample-processing for each of the active compounds, the proposed platform is cost effective as it can be used for more than 100 measurements, no sample preparation is needed, and the cost of the platform is far lower than the cost of one-run chromatographic method used singly for either calcipotriol or betamethasone determination.

2. Materials and Methods

2.1. Materials and Reagents

Calcipotriol, betamethasone, calix [6]arene, monosodium phosphate, disodium phosphate, ethanol, and dimethyl sulfoxide were procured from Sigma Aldrich (Milwaukee, Brookfield, WI, USA). The multi-walled carbon nanotubes-gold nanoparticles modified screen-printed carbon electrodes (ref. DRP-110CNT-GNP) were procured from Metrohm. The electrode has the following measurements: 33.0 mm length, 10.0 mm width, and 0.5 mm height, with the diameter of the working electrode being 4.0 mm. The reference electrode is silver based and the auxiliary electrode is carbon based. The electrode is constructed on a ceramic substrate.

Phosphate buffer solution (PBS, 0.1 mol L^{-1}) was obtained by mixing aqueous monosodium phosphate and disodium phosphate. Then, the pH was adjusted to the desired 5.0 pH using a 0.1 mol L^{-1} HCl solution.

Calcipotriol and betamethasone were dissolved in dimethyl sulfoxide to prepare the stock solutions (1.0×10^{-3} mol L^{-1} calcipotriol and 1.0×10^{-2} mol L^{-1} betamethasone). The solutions used for platform calibration were prepared using the successive dilution method by buffering with PBS pH 5.0.

2.2. Apparatus and Methods

All the stochastic measurements were conducted on an EmStat Pico mini potentiostat linked to a smartphone running the version 2.7 PStouch mobile application (PalmSens BV, Houten, The Netherlands). The pH adjustments were done employing a Mettler Toledo pH meter. Deionized water was acquired using a Direct-Q 3 Water Purification System (Molsheim, France) to prepare the solutions.

The experiments were carried out at ambient temperature.

2.3. Design of the C6A/MWCNT-GNP SCPE Disposable Platform

The disposable sensing platform was constructed by chemical immobilization of calix [6]arene (C6A) on multi-walled carbon nanotubes-gold nanoparticles (MWCNT-AuNPs) substrate by drop casting technique. Firstly, 3.18 mg of C6A was mixed with ethanol to create a dispersion. Then, 1.0 μL of C6A dispersion was drop-casted on the surface of MWCNT-AuNPs SPE, obtaining the disposable stochastic platform denoted C6A/MWCNT-AuNPs SPE. The platform (Scheme 1) was rapidly prepared for use within a matter of seconds due to the utilization of ethanol as a dispersion medium, which exhibits a rapid evaporation rate at room temperature.

The modified platforms were stored in a dry place, at room temperature, and away from direct sunlight.

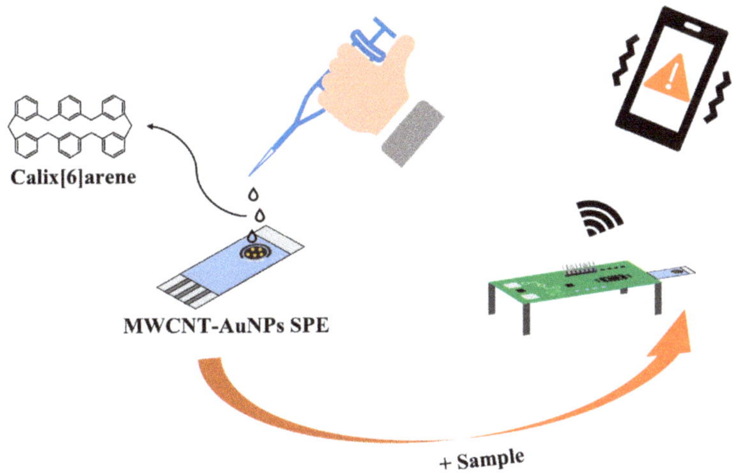

Scheme 1. Design of the platform used for the assay of calcipotriol and betamethasone.

2.4. Stochastic Mode

The measurements were carried out in stochastic mode using the chronoamperometry technique. The current was recorded at a constant potential of 200.0 mV vs. Ag/AgCl. This potential value was optimized by screening from 10 to 10 mV starting with a potential of 10 mV, so that the value of t_{off} could be read reliably using both automatic and manual reading of the diagrams (values lower than 1 ms are not reliably read manually, and values higher than 10 s will enlarge the analysis time while quantitative parameters (sensitivity LOQ_ remain the same. Two parameters, t_{on} and t_{off}, are identified in the obtained diagrams. The t_{on} parameter is used for quantitative analysis, as it describes the frequency of analyte-channel interactions, while the t_{off} parameter represents the signature of the analyte and is used for qualitative analysis, providing information about the duration and amplitude of these interactions. For the calibration of the proposed disposable platform, standard solutions with varying concentrations of calcipotriol and betamethasone, respectively, were utilized. The following calibration equations were established based on the linear regression approach:

$$1/t_{on} = a + b \times C_{calcipotriol} \quad (1)$$

$$1/t_{on} = a + b \times C_{betamethasone} \quad (2)$$

where a—intercept, and b—slope/sensitivity.

Based on these equations, the unknown concentrations of calcipotriol and betamethasone contained in the samples were calculated.

2.5. Samples

The proposed disposable platform was applied for the simultaneous determination of calcipotriol and betamethasone from real samples, namely pharmaceutical gel and surface water samples.

The pharmaceutical gel was acquired from a local drugstore. It contained 50.0 μg per gram of calcipotriol and 0.5 mg per gram of betamethasone as active pharmaceutical ingredients. The other components of the gel are liquid paraffine, polyoxypropylene stearyl ether, hydrogenated castor oil, butylhydroxytoluene (E321), and racemic α-tocopherol. The pharmaceutical sample was not subjected to any preliminary processing.

The samples of surface water were collected from a nearby river and stored in the refrigerator before the analysis. The samples were buffered in a 1:1 (*v*/*v*) ratio with pH 5.0 PBS. The absence of calcipotriol and betamethasone signatures in the water samples indicated the non-existence of these particular molecules in the samples. Subsequently, the water samples were spiked with various concentrations of calcipotriol and betamethasone.

3. Results

3.1. Response Characteristics of the C6A/MWCNT-AuNPs SPE Disposable Platform in the Stochastic Mode

The stochastic mode is an approach that relies on the interactions of the target analyte with a conductive channel. The process of stochastic sensing involves two distinct steps. During the initial stage, a constant potential is applied, resulting in a current flowing through the channel. Then, the analyte is extracted from the solution at the membrane-solution interface. The current subsequently drops to 0 when the analyte enters the channel and blocks the current flow. This process, known as the pattern recognition phase, is useful in qualitative analysis because the time duration in which it takes place is specific to each analyte. Because the t_{off} parameter is closely related to the size and shape of the analyte molecule, it is often referred to as the analyte signature. The second phase occurs when the analyte flows through the pore, where it binds to its wall, and redox processes occur. The time in which this step takes place is the t_{on} parameter used in the quantitative analysis. This stage is called the binding phase and is defined by the following equilibrium reactions:·

$$Ch_{(i)} + calcipotriol_{(i)} \Leftrightarrow Ch \bullet calcipotriol_{(i)} \quad (3)$$

$$Ch_{(i)} + betamethasone_{(i)} \Leftrightarrow Ch \bullet betamethasone_{(i)} \quad (4)$$

where Ch is the channel and i is the interface.

Calibration graphs obtained using the stochastic mode are shown in Figure 2.

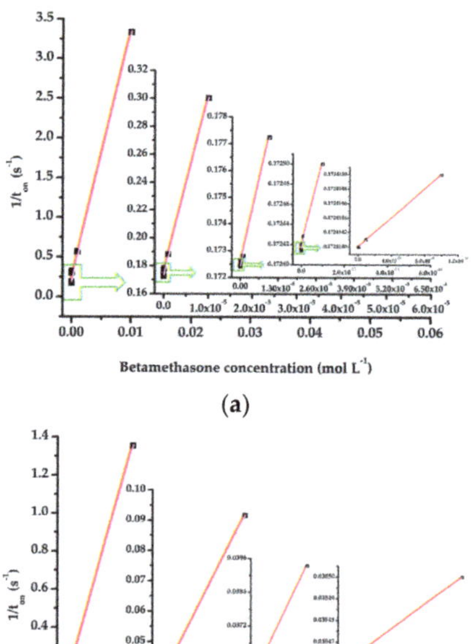

(**a**)

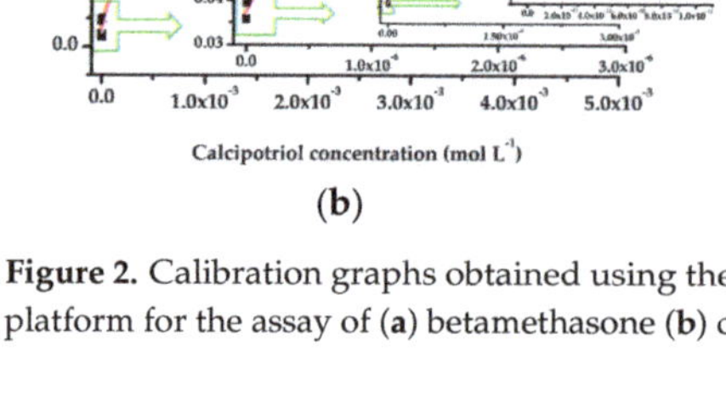

(**b**)

Figure 2. Calibration graphs obtained using the C6A/MWCNT-AuNPs SPE stochastic disposable platform for the assay of (**a**) betamethasone (**b**) calcipotriol.

Table 1 displays the response characteristics of the C6A/MWCNT-AuNPs SPE stochastic disposable platform, as determined from the t_{on} parameter values. The obtained results suggest that the proposed platform represents an accurate and reliable approach for determining calcipotriol and betamethasone simultaneously in real samples with minimal to no pre-treatment. This conclusion is based on the correlation between the very wide linear concentration range, remarkably low limit of quantification (LOQ), and significant sensitivity achieved. Thus, the proposed platform has the potential to be employed for quality control in the manufacturing of topical dosage forms but also in water quality monitoring.

Table 1. Response characteristics of the C6A/MWCNT-AuNPs SPE disposable stochastic platform.

Calibration Equation; Correlation Coefficient (r)	Linear Concentration Range (mol L^{-1})	t_{off} (s)	Sensitivity (mol L^{-1} s^{-1})	LOQ (mol L^{-1})
	Calcipotriol			
$1/t_{on} = 0.04\ (\pm 0.01) + 5.86\ (\pm 0.03) \times 10^8 \times C_{calcipotriol}$ r = 0.9999	1.0×10^{-15}–1.0×10^{-3}	2.2 ± 0.1	$5.86\ (\pm 0.03) \times 10^8$	1.0×10^{-15}
	Betamethasone			
$1/t_{on} = 0.07\ (\pm 0.01) + 3.25\ (\pm 0.02) \times 10^9 \times C_{betamethasone}$ r = 0.9992	1.0×10^{-16}–1.0×10^{-2}	0.7 ± 0.1	$3.25\ (\pm 0.02) \times 10^9$	1.0×10^{-16}

The proposed stochastic platform exhibited a significantly low LOQ (of fg mL^{-1} order of magnitude) for both calcipotriol and betamethasone, as well as a broader linear concentration range in comparison to previously described methods developed for the simultaneous determination of calcipotriol and betamethasone. The LOQ is given by the lowest concentration found in the linear concentration range according to the new IUPAC recommendation (paragraph 3.36, Note 3) [27]. The comparison is depicted in Table 2.

Table 2. Comparison of various proposed methods used for the simultaneous determination of calcipotriol and betamethasone.

Method	Analyte	Linear Concentration Range (mol L^{-1})	LOQ (mol L^{-1})	Ref.
HPLC with UV detection	Calcipotriol Betamethasone	2.42×10^{-6}–4.85×10^{-5} 9.91×10^{-7}–3.96×10^{-4}	2.9×10^{-7} 7.96×10^{-7}	[7]
UPLC with UV detection	Calcipotriol Betamethasone	3.03×10^{-5}–1.82×10^{-4} 3.15×10^{-4}–1.91×10^{-3}	2.37×10^{-5} 9.64×10^{-5}	[9]
UV spectrophotometry with FIS and CWT	Calcipotriol Betamethasone	2.42×10^{-6}–2.42×10^{-5} 1.98×10^{-6}–1.98×10^{-5}	5.45×10^{-8} 5.49×10^{-8}	[10]
Stochastic using C6A/MWCNT-AuNPs SPE disposable platform	Calcipotriol Betamethasone	1.0×10^{-15}–1.0×10^{-3} 1.0×10^{-16}–1.0×10^{-2}	1.0×10^{-15} 1.0×10^{-16}	This work

The results obtained using the proposed stochastic platform were compared in terms of betamethasone determination with other proposed electrochemical sensors in Table 3.

For the assay of both calcipotriol and betamethasone, far lower limits of quantification were obtained when the proposed platform was used; furthermore, wider linear concentration ranges were obtained compared to previously developed methods (Tables 2 and 3). Additionally, the proposed platform and method is able to determine simultaneously, in the same run the two active compounds, both calcipotriol, and betamethasone, making it faster compared to the other methods, as well as also cost-effective. The wide linear concentration range also facilitates the determination of calcipotriol and betamethasone in water samples, where there is a need for fast, reliable, cost-effective, on-site determination, in order to avoid accumulation of calcipotriol and betamethasone reaching values that are dangerous for the health of the population.

Table 3. Comparison of various proposed electrochemical sensors used for the determination of betamethasone.

Electrochemical Method	Sensor	Linear Concentration Range (mol L^{-1})	LOQ (mol L^{-1})	Ref.
SWV	SWNT/EPPGE	1.0×10^{-9}–2.5×10^{-8}	1.0×10^{-9}	[24]
SWV	SWNTs–CTAB/EPPGE	0.5×10^{-9}–1.0×10^{-7}	0.86×10^{-9}	[25]
DPV	Hg(Ag)FE	5.0×10^{-9}–0.8×10^{-6}	5.0×10^{-9}	[26]
Stochastic	C6A/MWCNT-AuNPs SPE disposable platform	1.0×10^{-16}–1.0×10^{-2}	1.0×10^{-16}	This work

SWV = square wave voltammetry; DPV = differential pulse voltammetry; SWNT/EPPGE = single wall carbon nanotube modified edge plane pyrolytic graphite electrode; SWNTs–CTAB/EPPGE = single wall carbon nanotubes–cetyltrimethylammonium bromide nanocomposite film modified edge plane pyrolytic graphite electrode; Hg(Ag)FE = silver-based amalgam film electrode.

The possibility of simultaneous determination of calcipotriol and betamethasone in one run facilitates the on-site uniformity content test of the pharmaceutical formulations containing both active compounds (calcipotriol, and betamethasone). This is a valuable advantage for pharmaceutical industry because the quantitative determinations of calcipotriol, and betamethasone can be performed in real time, with low cost, also facilitating the adjustment of the quantities of calcipotriol and betamethasone if the on-site analyses show that they are not in accordance with the recommendation of national or international pharmacopoeias.

3.2. Selectivity of C6A/MWCNT-AuNPs SPE Disposable Stochastic Platform

The selectivity of stochastic sensors relies on the distinct signature associated with each analyzed molecule. In the case of calcipotriol and betamethasone, the experimental results demonstrate that different values of the t_{off} parameter were obtained, suggesting that the proposed platform is selective towards the two analytes. The selectivity was also checked versus other components of the pharmaceutical formulation: polyoxypropylene stearyl ether, butylhydroxytoluene, and α-tocopherol wherein t_{off} values higher than 2.7 s were recorded: 3.5 s for polyoxypropylene stearyl ether, 3.1 s for α-tocopherol, and 2.8 s for butylhydroxytoluene. Accordingly, the proposed platform is selective versus the tested compounds.

3.3. Reproducibility and Stability of C6A/MWCNT-AuNPs SPE Disposable Stochastic Platform

The reproducibility and stability of the proposed stochastic platform were tested by preparing 10 identical platforms. For the reproducibility studies, the sensitivity of each platform was determined and then the relative standard deviation (RSD%) value was calculated based on the variation of the sensitivities, obtaining a value of 0.09%. For the stability studies, the 10 platforms were stored for 2 months prior to being utilized for the simultaneous determination of the two analytes. The obtained results exhibited consistency over the duration of the study, with no significant variations observed in the sensitivity values of the platforms. These variations were found to be less than 0.15%. These findings indicate that the proposed platform exhibits high reproducibility and long-term stability.

3.4. Simultaneous Determination of Calcipotriol and Betamethasone from Real Samples Using C6A/MWCNT-AuNPs SPE Disposable Stochastic Platform

The applicability of the proposed stochastic disposable platform for the simultaneous determination of calcipotriol and betamethasone was tested on real samples (pharmaceutical gel and surface water). The recorded diagrams were analyzed and the t_{off} and t_{on} parameters were determined for both the analytes, as illustrated in Figure 3a,b. Based on the values of t_{off} (signatures) recorded, the active components: calcipotriol and betamethasone were determined, and the t_{on} value read just after the signature (in between two t_{off} values) was used to create the calibration curves reported in Table 1 to determine the concentrations of calcipotriol and betamethasone in the samples (see also the paragraph related to the stochastic mode). Table 4 shows the computed recovery and relative standard deviation (RSD) values.

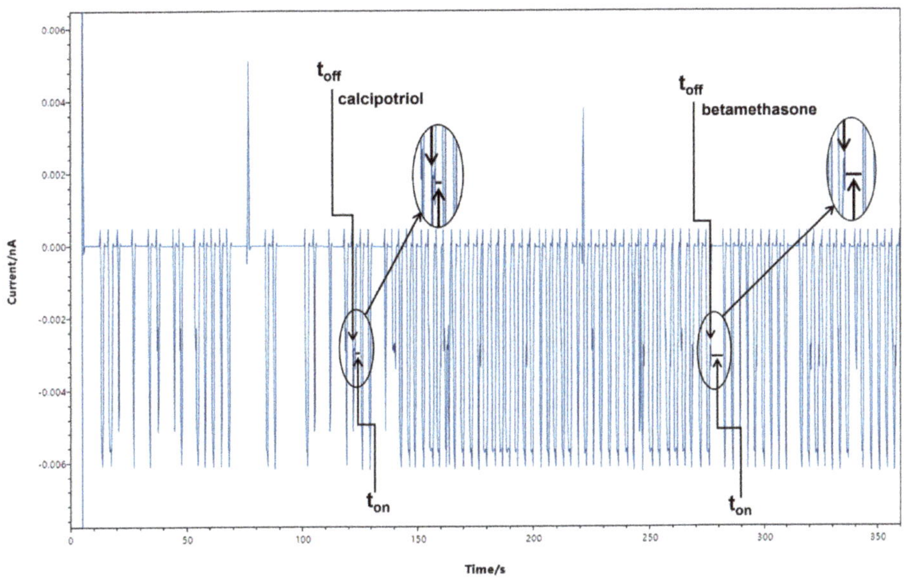

(**a**)

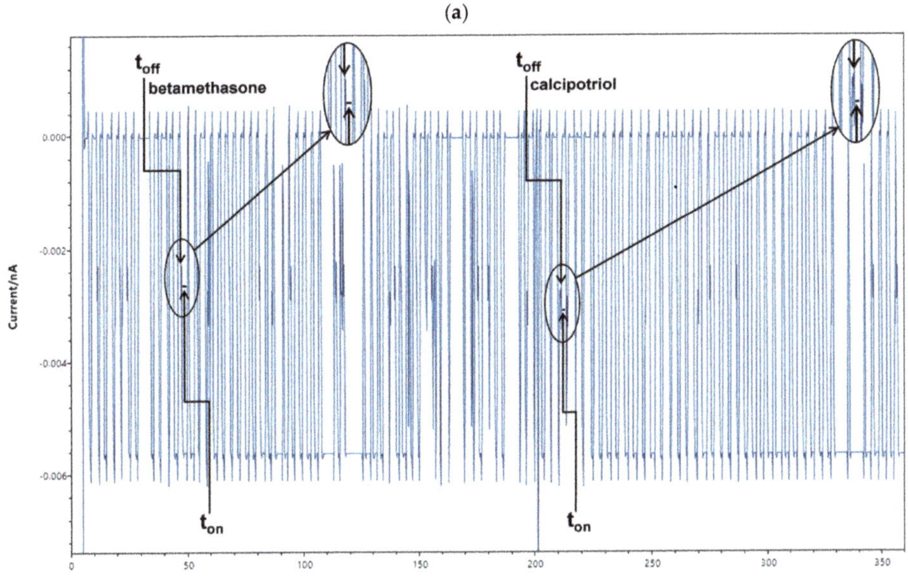

(**b**)

Figure 3. Stochastic diagrams obtained employing the C6A/MWCNT-AuNPs SPE disposable stochastic platform for the simultaneous determination of calcipotriol and betamethasone in: (**a**) pharmaceutical gel sample; (**b**) spiked surface water samples.

Table 4. Simultaneous determination of calcipotriol and betamethasone in a pharmaceutical gel and spiked surface water samples using the proposed C6A/MWCNT-AuNPs SPE disposable stochastic platform (N = 10).

Sample	Calcipotriol and Betamethasone, Added Amount (mol L^{-1})	Recovery (%)	
		Calcipotriol	Betamethasone
Topical pharmaceutical gel	–	99.15 ± 0.03	99.93 ± 0.02
Surface water samples	1.0×10^{-4}	99.21 ± 0.03	99.47 ± 0.04
	1.0×10^{-6}	99.50 ± 0.05	99.30 ± 0.02
	1.0×10^{-8}	99.21 ± 0.03	99.47 ± 0.03
	1.0×10^{-10}	99.10 ± 0.02	99.99 ± 0.02
	1.0×10^{-12}	99.12 ± 0.04	99.97 ± 0.03

The results indicate that the proposed platform exhibits a high degree of reliability for qualitative and quantitative analysis of calcipotriol and betamethasone from pharmaceutical and water samples, as shown by the recorded values for the recoveries and the relative standard deviations. Moreover, the other components present in the sample matrix did not influence the platform's response. Since the platform is connected to a portable device, this could enable on-site analysis. Additionally, cross-contamination of the samples is avoided by using disposable detection platforms. The combination of these features and the fact that the stochastic method requires minimal sample processing makes the proposed approach a convenient alternative for the quality control of pharmaceuticals and for the screening of surface water quality.

4. Conclusions

A disposable stochastic platform based on calix [6]arene and a commercially available screen-printed electrode was proposed for the simultaneous recognition and determination of calcipotriol and betamethasone in two types of samples: a pharmaceutical formulation, and water samples. When employing the proposed platform, it was possible to achieve a very low limit of quantification, an extensive linear concentration range, and very good sensitivity values. All these characteristics facilitated the assay of calcipotriol, and betamethasone in both the pharmaceutical formulation and water samples with high reliability. The response of the platform was not influenced by the complex sample matrices in which the analysis was performed. For these reasons, the proposed disposable stochastic platform represents a promising versatile candidate as a tool for both the pharmaceutical industry (where it can be reliably used for on-site analysis of the pharmaceutical formulations containing calcipotriol and betamethasone as active compounds), as well as the on-site analysis of water, especially that coming out from hospitals and clinics where gel containing calcipotriol and betamethasone is used. The cost-effectiveness (the cost of 100 analyses of simultaneous assay of calcipotriol and betamethasone in either pharmaceutical or water samples is far less than the cost for HPLC analysis of just one of the active compounds) as well as the high reliability of the developed platform make it an excellent candidate for the on-site analysis of pharmaceutical and water samples.

Author Contributions: Conceptualization, R.-I.S.-v.S., J.F.v.S. and H.Y.A.-E.; methodology, B.-M.T., R.-I.S.-v.S., J.F.v.S. and H.Y.A.-E.; validation, B.-M.T., R.-I.S.-v.S., J.F.v.S. and H.Y.A.-E.; formal analysis, R.-I.S.-v.S.; investigation, B.-M.T.; writing—original draft preparation, B.-M.T., R.-I.S.-v.S., J.F.v.S. and H.Y.A.-E.; writing—review and editing, B.-M.T., R.-I.S.-v.S., J.F.v.S. and H.Y.A.-E.; supervision, R.-I.S.-v.S., J.F.v.S. and H.Y.A.-E.; funding acquisition, J.F.v.S. All authors have read and agreed to the published version of the manuscript.

Funding: This work was supported by a grant of the Ministry of Research, Innovation and Digitization, CNCS/CCCDI-UEFISCDI, project number PN-III-P4-ID-PCE-2020-0059, within PNCDI III.

Data Availability Statement: Not applicable.

Conflicts of Interest: The authors declare no conflict of interest.

References

1. Griffiths, C.E.M.; Christophers, E.; Barker, J.N.W.N.; Chalmers, R.J.G.; Chimenti, S.; Krueger, G.G.; Leonardi, C.; Menter, A.; Ortonne, J.P.; Fry, L. A Classification of Psoriasis Vulgaris According to Phenotype. *Br. J. Dermatol.* **2007**, *156*, 258–262. [CrossRef]
2. Fenton, C.; Plosker, G.L. Calcipotriol/Betamethasone Dipropionate A Review of Its Use in the Treatment of Psoriasis Vulgaris. *Am. J. Clin. Dermatol.* **2004**, *5*, 463–478. [CrossRef]
3. Murphy, G.; Reich, K. In Touch with Psoriasis: Topical Treatments and Current Guidelines. *J. Eur. Acad. Dermatol. Venereol.* **2011**, *25*, 3–8. [CrossRef]
4. Mccormack Adis, P.L. Calcipotriol/Betamethasone Dipropionate a Review of Its Use in the Treatment of Psoriasis Vulgaris of the Trunk, Limbs and Scalp. *Drugs* **2011**, *71*, 709–730. [CrossRef] [PubMed]
5. Hamilton, C.M.; Winter, M.J.; Margiotta-Casaluci, L.; Owen, S.F.; Tyler, C.R. Are Synthetic Glucocorticoids in the Aquatic Environment a Risk to Fish? *Environ. Int.* **2022**, *162*, 107163. [CrossRef] [PubMed]
6. Council of Europe. *European Pharmacopoeia Commission; European Directorate for the Quality of Medicines & Healthcare European Pharmacopoeia*, 10th ed.; Council of Europe: Strasbourg, France, 2017; Volume 1.
7. Hassib, S.T.; Mahrouse, M.A.; Elkady, E.F.; Sayed, R.M. Simultaneous HPLC Determination of Betamethasone Esters-Containing Mixtures: Analysis of Their Topical Preparations. *J. Chromatogr. Sci.* **2018**, *56*, 716–723. [CrossRef]
8. Merey, H.A.; El-Mosallamy, S.S.; Hassan, N.Y.; El-Zeany, B.A. Validated Chromatographic Methods for Simultaneous Determination of Calcipotriol Monohydrate and Betamethasone Dipropionate in the Presence of Two Dosage Form Additives. *J. Chromatogr. Sci.* **2019**, *57*, 305–311. [CrossRef]
9. Kannaiah, K.P.; Sugumaran, A. Environmental Benign Analytical Quality by Design and UPLC Method Development for Betamethasone and Calcipotriene in Ointment. *J. Taibah Univ. Sci.* **2021**, *15*, 1156–1167. [CrossRef]
10. Raeis Farshid, B.; Sohrabi, M.R.; Davallo, M.; Raeis Farshid, S. Green Spectrophotometric Method for the Spectral Pattern Recognition Based on Fuzzy Inference System Compared to Continuous Wavelet Transform for the Quantitative Determination of Anti-Psoriasis Drugs in Commercial Skin Ointment Formulation: Comparison with HPLC. *Sustain. Chem. Pharm.* **2023**, *33*, 101107. [CrossRef]
11. Bayley, H.; Cremer, P.S. Stochastic Sensors Inspired by Biology. *Nature* **2001**, *413*, 226–230. [CrossRef]
12. Dinu Gugoasa, L.A.; Stefan-van Staden, R.-I.; van Staden, F.J.; Coros, M.; Pruneanu, S.M. Myoglobin-Silver Reduced Graphene Oxide Nanocomposite Stochastic Biosensor for the Determination of Luteinizing Hormone and Follicle-Stimulating Hormone from Saliva Samples. *Anal. Bioanal. Chem.* **2020**, *412*, 5191–5202. [CrossRef] [PubMed]
13. Stefan-van Staden, R.I.; Musat, O.R.; Gheorghe, D.C.; Ilie-Mihai, R.M.; Cioates Negut, C.; Sfirloaga, P. Ultrasensitive Assay of HER1, HER2, and Heregulin-α in Whole Blood. *Talanta Open* **2022**, *6*, 100151. [CrossRef]
14. Stefan-van Staden, R.I.; Bogea, I.M.; Ilie-Mihai, R.M.; Gheorghe, D.C.; Coroş, M.; Pruneanu, S.M. Stochastic Microsensors Based on Modified Graphene for Pattern Recognition of Maspin in Biological Samples. *Anal. Bioanal. Chem.* **2022**, *414*, 3667–3673. [CrossRef] [PubMed]
15. Stefan-van Staden, R.-I.; Moscalu-Lungu, A.; van Staden, J.F. Nanostructured Materials Used for Pattern Recognition of Bisphenols in Waste Water Samples. *J. Electrochem. Soc.* **2019**, *166*, B903–B907. [CrossRef]
16. Stefan-van Staden, R.I.; Chera-Anghel, I.A.; Gheorghe, D.C.; van Staden, J.F.; Badulescu, M. Smart Portable Device Based on the Utilization of a 2D Disposable Paper Stochastic Sensor for Fast Ultrasensitive Screening of Food Samples for Bisphenols. *Sensors* **2023**, *23*, 314. [CrossRef]
17. Stefan-van Staden, R.I.; Bogea, M.I.; Ilie-Mihai, R.M.; Gheorghe, D.C.; Aboul-Enein, H.Y.; Coros, M.; Pruneanu, S.M. N,S-Decorated Graphenes Modified with 2,3,7,8,12,13,17,18-Octaethyl-21H,23H-Porphine Manganese(III) Chloride-Based 3D Needle Stochastic Sensors for Enantioanalysis of Arginine: A Key Factor in the Metabolomics and Early Detection of Gastric Cancer. *Anal. Bioanal. Chem.* **2022**, *414*, 6521–6530. [CrossRef]
18. Georgescu State, R.; van Staden, J.F.; Stefan-van Staden, R.I.; State, R.N. Electrochemical Platform Based on Molecularly Imprinted Polymer with Zinc Oxide Nanoparticles and Multiwalled Carbon Nanotubes Modified Screen-Printed Carbon Electrode for Amaranth Determination. *Microchim. Acta* **2023**, *190*, 229. [CrossRef]
19. Norizan, M.N.; Moklis, M.H.; Ngah Demon, S.Z.; Halim, N.A.; Samsuri, A.; Mohamad, I.S.; Knight, V.F.; Abdullah, N. Carbon Nanotubes: Functionalisation and Their Application in Chemical Sensors. *RSC Adv.* **2020**, *10*, 43704–43732. [CrossRef]
20. Georgescu State, R.; van Staden, J.F.; State, R.N.; Papa, F. Rapid and Sensitive Electrochemical Determination of Tartrazine in Commercial Food Samples Using IL/AuTiO2/GO Composite Modified Carbon Paste Electrode. *Food. Chem.* **2022**, *385*, 132616. [CrossRef]
21. Stefanov, C.; Cioates Negut, C.; Dinu Gugoasa, L.A.; van Staden, J.F. Gold Nanoparticle-Graphene Quantum Dots Nanozyme for the Wide Range and Sensitive Electrochemical Determination of Quercetin in Plasma Droplets. *Microchim. Acta* **2020**, *187*, 611. [CrossRef]
22. Afkhami, A.; Bahiraei, A.; Madrakian, T. Gold Nanoparticle/Multi-Walled Carbon Nanotube Modified Glassy Carbon Electrode as a Sensitive Voltammetric Sensor for the Determination of Diclofenac Sodium. *Mater. Sci. Eng. C* **2016**, *59*, 168–176. [CrossRef] [PubMed]

23. Shinkai, S.; Araki, K.; Manabe, O. Does the Calixarene Cavity Recognise the Size of Guest Molecules? On the "hole-Size Selectivity" in Water-Soluble Calixarenes. *J. Chem. Soc. Chem. Commun.* **1988**, 187–189. [CrossRef]
24. Goyal, R.N.; Bishnoi, S.; Raj, A.; Rana, S. A Sensitive Voltammetric Sensor for Detecting Betamethasone in Biological Fluids. *Comb. Chem. High Throughput Screen.* **2010**, *13*, 610–618. [CrossRef] [PubMed]
25. Goyal, R.N.; Bishnoi, S. Effect of Single Walled Carbon Nanotube-Cetyltrimethyl Ammonium Bromide Nanocomposite Film Modified Pyrolytic Graphite on the Determination of Betamethasone in Human Urine. *Colloids. Surf. B Biointerfaces* **2010**, *77*, 200–205. [CrossRef]
26. Smajdor, J.; Paczosa-Bator, B.; Baś, B.; Piech, R. High Sensitive Voltammetric Determination of Betamethasone on an Amalgam Film Electrode. *J. Electrochem. Soc.* **2018**, *165*, H646–H651. [CrossRef]
27. Hibbert, D.B.; Korte, E.H.; Örnemark, U. IUPAC Recommendations. Metrological and quality concepts in analytical chemistry (IUPAC Recommendations 2021). *Pure Appl. Chem.* **2021**, *93*, 997–1048. [CrossRef]

chemosensors MDPI

Article

Enantioanalysis of Leucine in Whole Blood Samples Using Enantioselective, Stochastic Sensors

Raluca-Ioana Stefan-van Staden [1,2,*] and Oana-Raluca Musat [1,2]

[1] Faculty of Chemical Engineering and Biotechnologies, Politehnica University of Bucharest, 060042 Bucharest, Romania
[2] Laboratory of Electrochemistry and PATLAB, 202 Splaiul Independentei Str., 060021 Bucharest, Romania
* Correspondence: ralucavanstaden@gmail.com

Abstract: Enantioanalysis of amino acids became a key factor in the metabolomics of cancer. As a screening method, it can provide information about the state of health of patients. The main purpose of the study is to develop a highly reliable enantioanalysis method for the determination of D-, and L-leucine in biological samples in order to establish their role as biomarkers in the diagnosis of breast cancer. Two enantioselective stochastic sensors based on N-methyl-fullero-pyrrolidine in graphite and graphene nanopowder pastes were designed, characterized, and validated for the enantioanalysis of leucine in whole blood. Different signatures were recorded for the biomarkers when the stochastic sensors were used, proving their enantioselectivity. In addition, limits for detection on the order of ag L^{-1} were recorded for each of the enantiomers of leucine when the proposed enantioselective stochastic sensors were used. The wide linear concentration ranges facilitated the assay of the L-leucine in healthy volunteers, and also in patients confirmed with breast cancer. Recoveries of one enantiomer in the presence of the other enantiomer in whole blood samples, higher than 96.50%, proved that the enantioanalysis of enantiomers can be performed reliably from whole blood samples.

Keywords: leucine; enantioanalysis; enantioselective stochastic sensor; whole blood analysis; breast cancer

Citation: Stefan-van Staden, R.-I.; Musat, O.-R. Enantioanalysis of Leucine in Whole Blood Samples Using Enantioselective, Stochastic Sensors. *Chemosensors* **2023**, *11*, 259. https://doi.org/10.3390/chemosensors11050259

Academic Editor: António M. Peres

Received: 30 March 2023
Revised: 17 April 2023
Accepted: 21 April 2023
Published: 22 April 2023

1. Introduction

Amino acids metabolism is of high importance for cell proliferation, as well as for personalized treatment and early diagnosis of cancer [1–4]. Amino acids have an important role in the redox process, energy regulation, biosynthesis, and maintenance of homeostasis [5]. They usually feed cancer cells by providing building blocks for cancer cell growth [5]. The major function of amino acids in mammalian cancer cells is as substrates for new protein synthesis [6].

Racemization of amino acids in the body, while cancer cells are forming, was shown for different amino acids, such as aspartic acid [7]. In addition, distortion of DNA can produce the D—enantiomer of an amino acid. While L-amino acids are the most common in healthy people, the enantioanalysis of different biological samples showed the presence of D-amino acids in patients confirmed with different types of cancer [2–4,8].

The first evidence for the presence of D-amino acids in cancer was given in 1939 by Kögl and Erxleben [9], when they reported that some D-amino acids including D-leucine were found in tumor proteins and that the development of the tumor was facilitated by the presence of these amino acids, including D-leucine. While initially the assay of a chiral compound was done despite its stereochemistry, the evidence that differentiation between enantiomers may conducive for the diagnosis of different diseases. Discovered in 1819 in its impure form, leucine is an amino acid with high implications in carcinogenesis [10]. In 1820, the crystalline form of leucine was extracted from muscles and wool [10]. The name leucine is given by the Greek word leukos, which has the meaning "white"; the motivation of the name is given by the fact that the purification of leucine extracted from natural products provided a white, crystalline state [10]. Leucine is an essential amino acid due to its proven biochemical properties: it contributes to protein synthesis and is involved in metabolomics including cancer metabolomics [10].

Leucine is well known for its involvement in controlling whole blood sugar levels and improving the growth/recovery of muscles and bone tissues. Leucine is also responsible for the synthesis of growth hormones. Leucine also has a role in the prevention of breaking muscle proteins due to injuries like stress [10]. Leucine is part of the branched-chain amino acid valine, isoleucine, and leucine. It is one of the two (the other being lysine) ketogenic amino acids.

Acetyl coenzyme A and acetoacetate are the main metabolic end products produced during leucine metabolism [11]. Leucine and β-hydroxy β-methylbutyric acid (one of the leucine metabolites), have shown a great role in the human body. They induce biosynthesis of proteins by phosphorylation reactions. The metabolism of the dietary leucine takes place in the liver, adipose tissue, and muscle tissue, with the synthesis of sterols and derivatives being a result of the metabolism of leucine in the adipose and muscle tissues [11]. The pathway of L-leucine in the body [11] shows that the branched-chain amino acid aminotransferase enzyme is responsible for the initial transformation of L-leucine into α-ketoisocaproate (known as α-KIC) [11]. Next, mitochondrial enzyme branched-chain α-ketoacid dehydrogenase converts α-KIC into isovaleryl-coenzyme A, which is further transformed using the isovaleryl-coenzyme A dehydrogenase into methylcrotonyl-coenzyme A [11]. L-leucine also contributes to the increase of biotin during deficiency, when the hydroxymethylbutyrate (HMB) can be synthesized from methylcrotonyl-coenzyme A using the enoyl-coenzyme A hydratase as well as an unknown thioesterase enzyme. The chain of reactions involved in the conversion of methylcrotonyl-coenzyme A into hydroxymethylbutyrate-coenzyme A ends with the final conversion into hydroxymethylbutyrate. In the liver an extremely small amount of α-KIC is metabolized using the cytosolic enzyme 4-hydroxyphenylpyruvate dioxygenase (KIC dioxygenase); in addition, in the liver, the main end product is the hydroxymethylbutyrate. The described chain of transformations—starting from L-leucine to α-KIC and then to HMB—is specific to healthy people [11]. Leucine aminomutase transforms a very small amount of L-leucine into β-leucine, which is transformed further into β-ketoisocaproate (β-KIC), β-ketoisocaproyl-coenzyme A, and acetyl-coenzyme A. HMB is transformed into β-hydroxy β-methylbutyryl-coenzyme A which undergoes transformation using the enoyl-coenzyme A hydratase with the main product β-methylcrotonyl-coenzyme A and hydroxymethylglutaryl-coenzyme A [11].

This article proposed the enantioanalysis of leucine in whole blood samples of patients, confirmed with breast cancer versus healthy volunteers. Leucine is an essential amino acid belonging to the branched-chain amino acids. The role of leucine in breast cancer biochemistry and treatment was studied in different occasions [12–17]. Xiao et al. [12] proved that leucine deprivation inhibits proliferation and induces apoptosis of human breast cancer cells. Troup et al. [13] showed that reduced expression of leucine is associated with poor outcomes in node-negative invasive breast cancer. Shennan et al. [14] studied the L-leucine transport in human breast cancer cells. The role of leucine as a zipper [15,16] was shown in studies by Lin et al. [15] and Jeong et al. [16]. The role of D-leucine as a new biomarker for breast cancer was also investigated.

Due to the complexity of the matrix of whole blood, and the low levels of leucine enantiomers in whole blood, the stochastic sensors were selected as new screening tools. Stochastic sensors are known for their reliability recorded in the biomedical analysis of different biological samples (whole blood, saliva, urine, tumoral tissue, cerebrospinal fluid) when amino acids [2–4] or proteins [17–19] were analyzed. They are able to perform qualitative analysis—for identification of the enantiomer [2–4,20–24]—when the signature of the enantiomer (t_{off} value) is used, and also a quantitative analysis by using the t_{on} value measured in between two signatures (t_{off} values) (Scheme 1); they are also able to identify the stereochemistry of a compound, and also, they are able to distinguish the amino acids from a DNA chain.

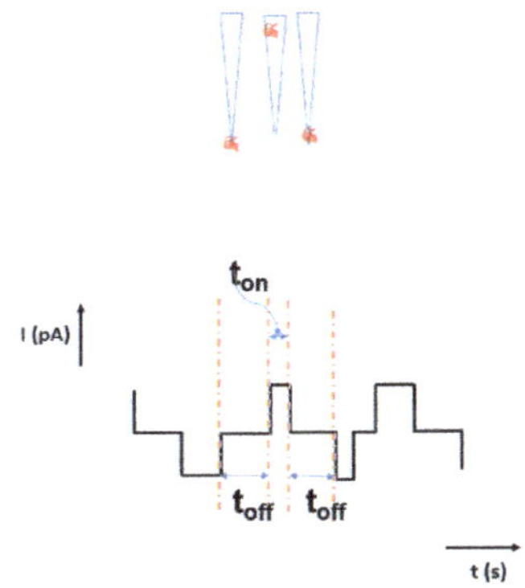

Scheme 1. Current development for stochastic sensors.

The working principle of the stochastic sensors is based on current development. The principle of current development is based on channel conductivity; in the first stage, the enantiomer is getting into the channel and blocking it while the current intensity is dropping to zero (the signature of the enantiomer given by the t_{off} value is characterized in this step); inside the channel, binding and redox processes take place, and equilibrium is achieved (t_{on} value, used for the quantitative assay of the enantiomers, is characterized in this step).

While the first stage is used for identification of the enantiomer, the second stage is used for the determination of the concentration of each enantiomer.

Utilization of the stochastic sensors for biomedical analysis is based on the advantages of the stochastic sensors versus classical electrochemical sensors, which are that the signature of the enantiomers does not depend on the composition of the matrix, but only on the size and geometry of the enantiomer. In addition, the quantitative step takes place inside the channel, and therefore is less influenced by the composition of the sample from where the enantiomer is determined. Graphite and graphene matrices are well known for sensor design due to their conductivity and high reliability when they are used as a paste in the sensor's design [25]. These matrices are able to keep the channel in the best shape needed for obtaining the stochastic signal. The fullerene derivative N-methyl-fulleropyrrolidine was selected due to its capacity to provide the necessary channels for stochastic sensing. Fe_2O_3 was added to the graphite paste in order to improve its conductivity, which facilitated, at the selected potential (125 mV), signatures and values for t_{on} on the order of seconds, and therefore, the reading was done with high reliability.

To date, for the assay of leucine, there was proposed a glassy carbon electrode modified with multiwall carbon nanotubes (presenting a limit of detection of 9 μmol L^{-1}) [26], and high-performance liquid chromatography [27]. The enantioanalysis of leucine was done to date using a low-cost high-performance liquid chromatographic method [28] as well as amperometric biosensors based on diamond paste [29] able to provide limits of detection on the order of pmol L^{-1}.

The applications of electrochemical sensors in biomedical analysis are guaranteed by the following advantages versus other methods of analysis, such as the enzyme-linked immunosorbent assay, high-performance liquid chromatography, electrophoresis, gas chromatography, and spectrometric methods of analysis: non-sampling, or a minimum sampling are required before any measurement is performed; wide working concentration ranges are available; the limits of detection are very low; the selectivity is quite high for the amperometric and potentiometric sensors, and can be improved by using a biological substance such as an enzyme or antibody; the sensitivity is also very high; they are cost-effective as such and also regarding the instrumentation needed to take up the signal and process it; and the time of analysis is far shorter than for any other analytical technique (analysis are usually performed within minutes or even seconds) [30–48].

The design of the sensors used in biomedical analysis influenced the response characteristics of the sensors (e.g., their sensitivity), and also had a role in improving the selectivity of determining the selected biomarkers in highly complex matrices such as biological samples. The utilization of nanomaterials for the design of electrochemical sensors proved to have a high role in the reliable assays of the amino acids in the biological samples [30]. Utilization of molecular imprinted polymers [29] is trending due to the reliable design of sensors, and the possibility of continuous in vivo monitoring of specific biomarkers. Hydrogels were used previously in the design of multifunctional platforms, also offering the possibility of high motion of the biomolecules to the detection side [37]. Utilization of nanorods proved to improve the response characteristics of the electrochemical sensors used for biomedical analysis [47]. Miniaturization of the detection cell to lab-on-a-chip type of design [30,33,37] favored easy transportation and handling of the system, especially for on-site type of measurements. In addition, to avoid cross-contamination, disposable sensors were developed for biomedical analysis [45]. These sensors can be easily be used for point-of-care diagnostics, where fast detection and minimum (preferably no sample preparation) sample preparation is needed [32,35,40]. Smartphones became preferable tools for signal acquisition, processing, and for sending the results to medical doctors and specialized databases and laboratories [35,36]. Commercialization of sensors for biomedical analysis needs smart, portable, and compact detection systems, on which the sensors should easily fit, connected to smartphones [38].

The novelty of the paper is given by the utilization of N-methyl-fullero-pyrrolidine as a modifier of graphite/Fe_2O_3 and of nanographene pastes, for the design of enantioselective stochastic sensors and their utilization for enantioanalysis of leucine in whole blood samples, also by the stochastic method capacity to perform an enantioselective molecular recognition of the enantiomers followed by their quantification, which may facilitate a highly reliable diagnosis of breast cancer. While most of the paper shows the analysis of leucine, and not its enantioanalysis, the main purpose of the study is to develop a highly reliable enantioanalysis method for the determination of D-, and L-leucine in biological samples in order to establish their role as biomarkers in the diagnosis of breast cancer.

2. Materials and Methods

L- and D-leucine, graphite, Fe_2O_3, graphene nanopowder, N-methyl-fullero-pyrrolidine, and paraffin oil were purchased from Sigma Aldrich. All solutions of L- and D-leucine were prepared using phosphate buffer of pH 7.40. The concentration range used for both enantiomers was between 1×10^{-20} and 1×10^{-2} mol L^{-1}.

The design of the enantioselective stochastic sensor was done as follows: 100 mg of graphite/graphene nanopowder were physically mixed with 20 mg of N-methyl-fullero-pyrrolidine. To the graphite powder mixture there were added 10 mg Fe_2O_3. To each of the powders, 30 µL of paraffin oil was added to form a homogeneous paste. Each of the pastes was placed in a nonconducting 3D plastic tube printed in our laboratory using a 3D printer. The inner diameter of each tube was 25 µm, and the length was 1 cm. A Ag wire was used to connect the paste with the external circuit.

The morphology of the designed pastes was investigated using scanning electron microscopy (SEM) (Inspect S, FEI Company, Eindhoven, The Netherlands). In order to obtain a good resolution of the microscopy images, the pastes were analyzed using the LFD detector (low vacuum), at a high voltage (HV) of 30 kV, and magnification of 1600×.

The morphology of the active surface of the stochastic sensors is shown in Figure 1.

In Figure 1, agglomerations of particles and channels in asymmetric formations can be observed.

The stochastic method is based on the utilization of the chronoamperometric technique. A potential of 125 mV versus Ag/AgCl was applied. The AUTOLAB/PGSTAT 12 Potentiostat/Galvanostat (Metrohm, The Netherlands) was used for all measurements. The electrochemical cell comprised the enantioselective stochastic sensor (as a working sensor), an Ag/AgCl reference electrode, and a Pt wire as an auxiliary electrode. The electrochemical measurements were performed at 125 mV vs. Ag/AgCl, and at 25 °C. Diagrams like those shown in Figure 2 were obtained when the whole blood was screened. A standard solution of enantiomers as well as whole blood samples spiked with standard solutions of enantiomers were used for the calibration of the enantioselective stochastic sensors; no differences between the response characteristics of the sensors were recorded when the calibration was done for the standard solution or in whole blood samples. Based on the signatures of the enantiomers of leucine (values of t_{off}), the enantiomers were identified in the diagrams recorded for whole blood samples. In between two signatures, the t_{on} was read and used for the determination of response characteristics of the sensors; linear regression method was used to determine a, and b parameters of the calibration equation in the form of $1/t_{on} = a + b \times C_{enantiomer}$. The unknown concentration of the enantiomer was determined using the calibration equation.

(a) **(b)**

Figure 1. SEM images for (**a**) graphite paste based enantioselective stochastic sensor and (**b**) nanographene paste based enantioselective stochastic sensor.

The whole blood samples obtained from confirmed patients with breast cancer as well as whole blood samples from healthy volunteers were provided by the University Hospital Bucharest under the protocol approved by the University of Medicine and Pharmacy "Carol Davila", from Bucharest (ethics committee approval no 75/2015). No pretreatment of the samples was required before analysis. The samples were placed into an electrochemical cell; the diagram was recorded. First of all, the molecular recognition of L- and D-leucine was performed by identifying their signatures (t_{off} values) in the diagrams. Their concentrations were determined according to the stochastic method described above.

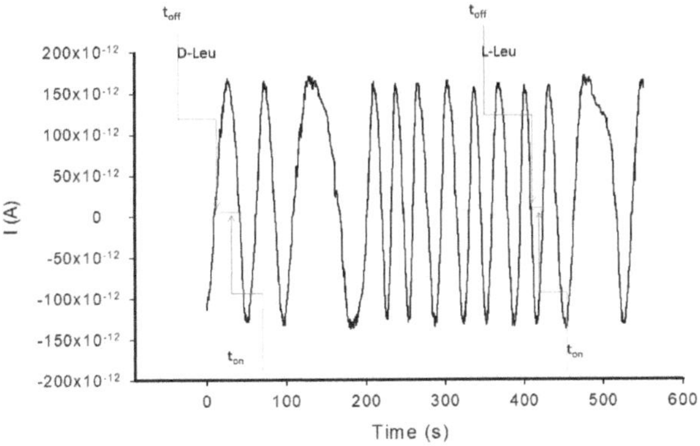

(a)

Figure 2. *Cont.*

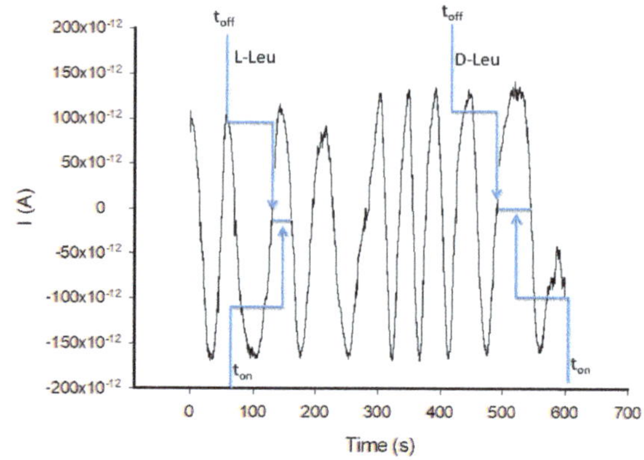

(b)

Figure 2. Examples of diagrams obtained for the screening of whole blood samples using the (**a**) graphite paste based enantioselective stochastic sensor and (**b**) nanographene paste based enantioselective stochastic sensor.

3. Results and Discussions

3.1. Response Characteristics

The stochastic mode was used for the determination of the response characteristics of the enantioselective stochastic sensors. The response characteristics are given in Table 1.

Table 1. Characteristics of response for the enantioselective stochastic sensors used for the enantioanalysis of leucine.

Stochastic Sensor Based on N-Methyl-fullero-pyrrolidine and	Leucine	Signature t_{off} (s)	Equation of Calibration [1], R	Sensitivity, s^{-1} g^{-1} L	Limit of Determination, ag L^{-1}	Linear Concentration Range, g L^{-1}
Graphite/Fe_2O_3	L	0.6	$1/t_{on} = 0.04 + 8.17 \times 10^{13}C$ r = 0.9993	8.17×10^{13}	10.00	1×10^{-17}–1×10^{-5}
	D	0.8	$1/t_{on} = 0.01 + 1.42 \times 10^{13}C$ r = 0.9994	1.42×10^{13}	100.00	1×10^{-16}–1×10^{-9}
Nanographene	L	2.2	$1/t_{on} = 0.02 + 3.44 \times 10^{12}C$ r = 0.9995	3.44×10^{12}	100.00	1×10^{-16}–1×10^{-4}
	D	0.9	$1/t_{on} = 0.02 + 1.35 \times 10^{15}C$ r = 0.9996	1.35×10^{15}	1.00	1×10^{-18}–1×10^{-6}

[1] <C> = mol L^{-1}; <t_{on}> = s.

All response characteristics reported in Table 1 were obtained by applying a potential of 125 mV versus Ag/AgCl (as reference electrode), at a constant temperature of 25 °C.

The signatures of the two enantiomers are different when the same stochastic sensor was used, proving that the sensors are enantioselective, and they can be used for the simultaneous assay of the two enantiomers in biological samples. Linear concentration ranges are wide: twelve decades of concentration for L-leucine, seven decades of concentration for the D-leucine when the sensor based on graphite/Fe_2O_3 was used, and twelve decades of concentration for D-leucine when the sensor based on nanographene was used, allowing the enantioanalysis of leucine in whole blood samples despite people's state of health. Very low limits of detection (determined as the lowest value of concentration from the linear concentration range)—all in the order of ag L^{-1}—were recorded; for the assay of L-leucine the lowest limit of detection was recorded when the enantioselective stochastic sensor based on graphite/Fe_2O_3 was used, while for the assay of D-leucine the lowest limit of detection was recorded for the enantioselective stochastic sensor based on nanographene. L-leucine was determined with the highest sensitivity when the sensor based on graphite/Fe_2O_3 was used, while D-leucine was determined with the highest sensitivity when the sensor based on nanographene was used.

3.2. Selectivity of the Enantioselective Stochastic Sensors

The selectivity of the stochastic sensors is given by the difference between the signatures (t_{off} values) recorded for the enantiomers of leucine and those obtained for CA15-3, CEA, HER2, p53, Ki67, maspin, CA19-9, and L- and D-serine—the usual biomarkers used for the establishment of the diagnostic of breast cancer.

Results shown in Table 2 prove that none of the other biomarkers interfere in the enantioanalysis of leucine despite the matrix used for the stochastic sensor's design.

Table 2. Selectivity of the stochastic sensors.

Stochastic Sensor Based on N-Methyl-fullero-pyrolidine and	CA15-3	CEA	HER2	Maspin	Ki67	CA19-9	p53	L-Leucine	D-Leucine	L-Serine	D-Serine
	Signature (s)										
Graphite/Fe_2O_3	1.1	1.5	2.2	1.9	3.0	2.4	3.5	0.6	0.8	0.2	1.7
Nanographene	0.2	0.6	3.0	2.5	3.2	2.8	1.7	2.2	0.9	1.3	2.0

3.3. Stability and Reproducibility Measurements

Ten enantioselective stochastic sensors from each of the two types designed for the enantioanalysis of leucine were constructed. Response characteristics were measured for one month on a daily basis. The variation in sensitivities of the proposed sensors was lower than 0.15%; accordingly, the design of the proposed stochastic sensors is reproducible. In addition, variations of the sensitivities were lower than 1.20% for measurements made during the one month, proving the one-month stability of the stochastic sensors.

3.4. Enantioanalysis of Leucine in Whole Blood Samples

The whole blood samples collected from the patients were used without any processing. The measurements were performed accordingly with the description of the stochastic method from above. After the diagrams were obtained (Figure 2), the signatures of the L- and D-leucine were identified. For each of the L- and D-enantiomers, readings of t_{on} values (in between two consecutive t_{off} values were done) were performed for the determination of the concentrations of L- and D-leucine. The t_{on} values were introduced in the equations of calibration (Table 1) for the determination of the concentrations of the enantiomers of leucine. Validation tests were performed for the assay of L- and D-leucine in whole blood samples.

To validate the sensors, synthetic mixtures of enantiomers (in different ratios) were spiked into the whole blood samples, and the recovery of enantiomers was performed (Table 3). Determinations of the enantiomers of leucine were performed before and after the addition of the mixtures, in order to calculate the recovery of the known amount added. Different ratios of the enantiomers were selected in order to determine if the ratio of the enantiomers influenced in any way the reliability of the enantioanalysis of leucine (Table 3).

Table 3. Recovery tests of L- and D-leucine in whole blood samples (N = 10).

	Recovery %													
L:D	1:99		1:50		1:25		1:1		25:1		50:1		99:1	
Enantiomer	L	D	L	D	L	D	L	D	L	D	L	D	L	D
Graphite/Fe_2O_3 based sensor	99.10 ± 0.05	96.98 ± 0.02	97.95 ± 0.04	99.18 ± 0.02	99.00 ± 0.02	98.75 ± 0.03	98.82 ± 0.02	98.99 ± 0.01	99.15 ± 0.01	99.10 ± 0.03	99.99 ± 0.01	97.98 ± 0.02	98.16 ± 0.03	99.90 ± 0.02
Nanographene based sensor	99.32 ± 0.02	96.50 ± 0.03	98.00 ± 0.02	99.53 ± 0.03	99.18 ± 0.03	99.65 ± 0.02	98.09 ± 0.01	99.99 ± 0.03	99.13 ± 0.02	99.76 ± 0.02	99.12 ± 0.03	97.00 ± 0.04	97.43 ± 0.02	99.66 ± 0.03

Table 3 shows that high recoveries of the enantiomers were obtained in the whole blood, despite the ratios between L- and D-leucine in the whole blood. This proved the accuracy and high reliability of the measurements performed with the enantioselective stochastic sensors, despite the ratio of which the enantiomers are found in the whole blood sample.

Real whole blood samples collected from patients confirmed with breast cancer and healthy volunteers were analyzed using the enantioselective stochastic sensors. Results obtained are shown in Table 4.

Table 4. Enantioanalysis of leucine in whole blood samples using enantioselective stochastic sensors (N = 10).

Sample No.	State of Health	L-Leucine, pg mL^{-1}		D-Leucine, ng mL^{-1}	
Stochastic Sensor Based on N-Methyl-fullero-pyrrolidine and		Graphite/Fe_2O_3	Nanographene	Graphite/Fe_2O_3	Nanographene
1	Confirmed with breast cancer	8.62 ± 0.02	8.08 ± 0.01	3.52 ± 0.02	3.30 ± 0.03
2		0.48 ± 0.01	0.49 ± 0.03	5.00 ± 0.03	4.75 ± 0.02
3		8.25 ± 0.01	8.71 ± 0.03	0.20 ± 0.01	0.18 ± 0.02
4		15.48 ± 0.02	16.02 ± 0.03	1.87 ± 0.02	1.69 ± 0.03
5		6.11 ± 0.01	5.50 ± 0.03	1.00 ± 0.02	1.00 ± 0.03
6		1.24 ± 0.03	0.98 ± 0.02	2.00 ± 0.01	2.17 ± 0.02
7		3.47 ± 0.01	2.93 ± 0.02	2.60 ± 0.03	2.58 ± 0.01
8		0.08 ± 0.01	0.07 ± 0.02	33.10 ± 0.02	35.01 ± 0.01
9		2.91 ± 0.01	2.26 ± 0.03	5.23 ± 0.03	5.00 ± 0.01
10		8.30 ± 0.03	8.12 ± 0.01	1.17 ± 0.03	1.18 ± 0.01
1	Healthy volunteers	18.21 ± 0.01	18.34 ± 0.02	- *	- *
2		20.58 ± 0.02	20.84 ± 0.01	- *	- *
3		5.48 ± 0.01	5.12 ± 0.03	- *	- *
4		32.25 ± 0.02	32.40 ± 0.01	- *	- *
5		7.12 ± 0.01	7.15 ± 0.03	- *	- *
6		27.16 ± 0.02	27.19 ± 0.01	- *	- *
7		52.01 ± 0.03	51.15 ± 0.02	- *	- *
8		4.89 ± 0.01	4.68 ± 0.03	- *	- *
9		51.97 ± 0.03	52.53 ± 0.01	- *	- *
10		43.47 ± 0.01	43.50 ± 0.02	- *	- *
t-test		2.96		3.01	

* D-leucine was not found.

A paired Student t-test was conducted at a confidence level of 99.90%. At the 99.00% confidence level, the tabulated theoretical value is 4.032. The calculated t-values for each enantiomer of leucine were less than 3.10 (which is less than the tabulated value 4.13), indicating that there is no statistically significant difference between the results obtained using the proposed enantioselective stochastic sensors (Table 4), and that enantioselective stochastic sensors can be relied upon for the molecular identification and quantification of L- and D-leucine in whole blood samples. Furthermore, the D-leucine was only found in the samples collected from confirmed patients with breast cancer, and not in the whole blood from healthy volunteers.

Compared with the sensors proposed earlier for the analysis [24] and enantioanalysis [27] of leucine, the proposed enantioselective stochastic sensors have a far lower limit of detection (on the order of ag L^{-1}). In addition, a wider linear concentration range was recorded for the proposed sensors than for those reported before [24,27]. High selectivity was recorded versus other biomarkers that may be present in the whole blood of patients with breast cancer. High recoveries of one enantiomer in the presence of the other enantiomer as well as the results of paired Student t-tests proved that the enantioselective stochastic sensors can reliably be used for the enantioanalysis of leucine in whole blood samples. Compared with the method proposed earlier by Hormozi Jangi, et al. [49] this screening method provides wider linear concentration ranges, higher sensitivities, and higher accuracy for the enantioanalysis of L- and D-leucine when they are mixed in different ratios.

4. Conclusions

Two enantioselective stochastic sensors were designed, characterized, and validated for the enantioanalysis of leucine in whole blood samples. The utilization of nanographene as a matrix for the sensor's design favored the decrease of the limit of detection of D-leucine to 1 ag L^{-1}. The sensors are not only enantioselective but also selective versus a series of other biomarkers such as CA15-3, CEA, HER2, maspin, Ki67, CA19-9, and p53, which are the main biomarkers used in the diagnosis of breast cancer. The proposed enantioselective stochastic sensors proved to have great features in biomedical analysis. Their utilization for the screening of whole blood samples can bring the screening test very close to a diagnostic test, because the D-leucine was only found in confirmed patients with breast cancer, and not in healthy volunteers. Accordingly, D-leucine may be considered a biomarker for breast cancer.

Author Contributions: Conceptualization, R.-I.S.-v.S. and O.-R.M.; methodology, R.-I.S.-v.S. and O.-R.M.; validation, R.-I.S.-v.S. and O.-R.M.; investigation, O.-R.M.; writing—original draft preparation, R.-I.S.-v.S. and O.-R.M.; writing—review and editing, R.-I.S.-v.S. and O.-R.M.; supervision, R.-I.S.-v.S.; project administration, R.-I.S.-v.S.; funding acquisition, R.-I.S.-v.S. All authors have read and agreed to the published version of the manuscript.

Funding: This work was supported by the Nucleus Program within the framework of the National Plan for Research, Development and Innovation 2022–2027, carried out with the support of the Ministry of Research, Innovation and Digitization, project number PN 23 27 03 01.

Institutional Review Board Statement: The study was conducted in accordance with the Declaration of Helsinki, and approved by the Ethics Committee of the University Hospital Bucharest, approval no 75/2015.

Informed Consent Statement: Written informed consent has been obtained from the patient(s) to perform the research using the collected samples.

Data Availability Statement: No data are available to be shared.

Acknowledgments: The authors want to thank Paula Sfirloaga for performing the SEM analysis of the pastes designed for the construction of the stochastic sensors.

Conflicts of Interest: The authors declare no conflict of interest.

References

1. Ragni, M.; Fornelli, C.; Nisoli, E.; Penna, F. Amino acids in cancer and cachexia: An integrated view. *Cancers* **2022**, *14*, 5691. [CrossRef] [PubMed]
2. Ilie-Mihai, R.M.; Stefan-van Staden, R.I.; Magerusan, L.; Coros, M.; Pruneanu, S. Enantioanalysis of tryptophan in whole blood samples using stochastic sensors—A screening test for gastric cancer. *Chirality* **2020**, *32*, 215–222. [CrossRef] [PubMed]
3. Stefan-van Staden, R.I.; Ilie-Mihai, R.M.; Magerusan, L.; Coros, M.; Pruneanu, S. Enantioanalysis of glutamine-a key factor in establishing the metabolomics process in gastric cancer. *Anal. Bioanal. Chem.* **2020**, *412*, 3199–3207. [CrossRef] [PubMed]
4. Stefan-van Staden, R.I.; Bogea, M.I.; Ilie-Mihai, R.M.; Gheorghe, D.C.; Aboul-Enein, H.Y.; Coros, M.; Pruneanu, S.M. NS Decorated Graphenes Modified with 2,3,7,8,12,13,17,18-Octaethyl-21H,23H-Porphine Manganese (III) Chloride Based 3D Needle Stochastic Sensors for Enantioanalysis of Arginine-a Key Factor in the Metabolomics and Early Detection of Gastric Cancer. *Anal. Bioanal. Chem.* **2022**, *414*, 6521–6530. [CrossRef] [PubMed]
5. Lieu, E.L.; Nguyen, T.; Rhyne, S.; Kim, J. Amino acids in cancer. *Exp. Mol. Med.* **2020**, *52*, 15–30. [CrossRef]
6. Vettore, L.; Westbrook, R.L.; Tennant, D.A. New aspects of amino acid metabolism in cancer. *Br. J. Cancer* **2020**, *122*, 150–156. [CrossRef]
7. Ritz-Timme, S.; Collins, M.J. Racemization of aspartic acid in human proteins. *Ageing Res. Rev.* **2002**, *1*, 43–59. [CrossRef]
8. Bastings, J.J.A.J.; van Eijk, H.M.; Damink, S.W.O.; Rensen, S.S. D-amino acids in health and diseases: A focus on cancer. *Nutrients* **2019**, *11*, 2205. [CrossRef]
9. Kögl, F.; Erxleben, H. Zur ätiologie der malignen tumoren. *Z. Physiol. Chem.* **1939**, *258*, 57–95. [CrossRef]
10. Dawson, R.M.C.; Elliott, D.C.; Elliott, W.H.; Jones, K.M. *Data for Biochemical Research*, 3rd ed.; Clarendon Press: Oxford, UK, 1989.
11. Zanchi, N.E.; Gerlinger-Romero, F.; Guimarães-Ferreira, L.; de Siqueira Filho, M.A.; Felitti, V.; Lira, F.S.; Seelaender, M.; Lancha, A.H. HMB supplementation: Clinical and athletic performance-related effects and mechanisms of action. *Amino Acids* **2011**, *40*, 1015–1025. [CrossRef]

12. Xiao, F.; Wang, C.; Yin, H.; Chen, S.; Fang, J.; Guo, F. Leucine deprivation inhibits proliferation and induced apoptosis of human breast cancer cells via fatty acid synthase. *Oncotarget* **2016**, *7*, 63679–63689. [CrossRef] [PubMed]
13. Troup, S.; Njue, C.; Kliewer, E.V.; Parisien, M.; Roskelley, C.; Chakravarti, S.; Roughley, P.J.; Murphy, L.C.; Watson, P.H. Reduced expression of the small leucine-rich proteoglycans, lumican, and decorin is associated with poor outcome in node-negative iinvasive breast cancer. *Clin. Cancer Res.* **2003**, *9*, 207–214. [PubMed]
14. Shennan, D.B.; Thomson, J.; Gow, I.F.; Travers, M.T.; Barber, M.C. L-leucine transport in human breast cancer cells (MCF-7 and MDA-MB-231): Kinetics, regulation by estrogen and molecular identity of the transporter. *Biochem. Biophys. Acta-Biomater.* **2004**, *1664*, 206–216. [CrossRef] [PubMed]
15. Lin, Y.; Zhou, X.; Peng, W.; Wu, J.; Wu, X.; Chen, Y.; Cui, Z. Expression and clinical implications of basic leucine zipper ATF-like transcription factor 2 in breast cancer. *BMC Cancer* **2021**, *21*, 1062. [CrossRef]
16. Jeong, J.; Park, S.; An, H.T.; Kang, M.; Ko, J. Small leucine zipper protein functions as a negative regulator of estrogen receptor α in breast cancer. *PLoS ONE* **2017**, *12*, e0180197. [CrossRef]
17. Junior, R.P.; Martins Sonehara, N.; Jardim-Perassi, B.V.; Pal, A.; Asad, Y.; Chuffa, L.G.A.; Chammas, R.; Raymand, F.I.; Zuccari, D.A.P.C. Presence of human breast cancer xenograft changes the diurnal profile of amino acids in mice. *Sci. Rep.* **2022**, *12*, 1008. [CrossRef]
18. Stefan-van Staden, R.I.; Gheorghe, D.C.; Pogacean, F.; Pruneanu, S. Simultaneous analysis of MLH1, MSH2, MSH6, PMS2 and KRAS in patients with gastric and colon cancer using stochastic sensors. *Chemosensors* **2022**, *10*, 380.
19. Gheorghe, D.C.; Stefan-van Staden, R.I. Molecular recognition and quantification of HER-3, HER-4 and HRG-α in whole blood and tissue samples using stochastic sensors. *Micromachines* **2022**, *13*, 1749.
20. Jia, W.; Hu, C.; Wang, Y.; Liu, Y.; Wang, L.; Zhang, S.; Zhu, Q.; Gu, Y.; Zhang, P.; Ma, J.; et al. Identification of single-molecule catecholamine enantiomers using a programmable nanopore. *ACS Nano* **2022**, *16*, 6615–6624. [CrossRef]
21. Wang, J.; Prajapati, J.D.; Gao, F.; Ying, Y.L.; Kleinekathofer, U.; Winterhalter, M.; Long, Y.T. Identification of single amino acid chiral and positional isomers using an electrostatically asymmetric nanopore. *J. Am. Chem. Soc.* **2022**, *144*, 15072–15078. [CrossRef]
22. Ensslen, T.; Sarthak, K.; Aksimentiev, A.; Behrends, J.C. Resolving isomeric posttranslational modifications using a biological nanopore as a sensor of molecular shape. *J. Am. Chem. Soc.* **2022**, *144*, 16060–16068. [CrossRef] [PubMed]
23. Liu, Y.; Zhang, S.; Wang, Y.; Wang, L.; Cao, Z.; Sun, W.; Fan, P.; Zhang, P.; Chen, H.Y.; Huang, S. Nanopore identification of alditol epimers and their application in rapid analysis of alditol-containing drinks and healthcare products. *J. Am. Chem. Soc.* **2022**, *144*, 13717–13728. [CrossRef] [PubMed]
24. Kang, X.F.; Cheley, S.; Guan, X.; Bayley, H. Stochastic detection of enantiomers. *J. Am. Chem. Soc.* **2006**, *128*, 10684–10685. [CrossRef]
25. Coros, M.; Pruneanu, S.; Stefan-van Staden, R.I. Recent progress in the graphene-based electrochemical sensors and biosensors. A review. *J. Electrochem. Soc.* **2020**, *167*, 037528. [CrossRef]
26. Rezaei, B.; Mirahmadi Zare, Z. Modified glassy carbon electrode with multiwall carbon nanotubes as a voltametric sensor for determination of leucine in biological and pharmaceutical samples. *Anal. Lett.* **2008**, *41*, 2267–2286. [CrossRef]
27. Horber, F.F.; Khal, J.; Lecavalier, L.; Krom, B.; Haymond, M.W. Determination of leucine and α-ketoisocaproic acid concentrations and specific activity in plasma and leucine specific actitivies in proteins using high-performance liquid chromatography. *J. Chromatogr. B* **1989**, *495*, 81–94. [CrossRef]
28. Tojo, M.; Murakami, M.; Nagata, Y. Simple and low-cos high performance liquid chromatographic method for determination of D- and L-amino acids. *J. Chromatogr. Sci.* **2012**, *50*, 393–395. [CrossRef]
29. Stefan-van Staden, R.I.; Muvhulawa, L.S. Determination of L- and D-enantiomers of leucine using amperometric biosensors based on diamond paste. *Instrum. Sci. Technol.* **2006**, *34*, 475–481. [CrossRef]
30. Behyar, M.B.; Hasanzadeh, M.; Seidi, F.; Shadjou, N. Sensing of amino acids: Critical role of nanomaterials for the efficient biomedical analysis. *Microchem. J.* **2023**, *188*, 108452. [CrossRef]
31. Karasu, T.; Özgür, E.; Uzum, L. MIP-on-a-chip: Artificial receptors on microfluidic platforms for biomedical applications. *J. Pharm. Biomed. Anal.* **2023**, *226*, 115257. [CrossRef]
32. Macovei, D.G.; Irimes, M.B.; Hosu, O.; Cristea, C.; Tertis, M. Point-of-care electrochemical testing of biomarkers involved in inflammatory and inflammatory-associated medical conditions. *Anal. Bioanal. Chem.* **2023**, *415*, 1033–1063. [CrossRef] [PubMed]
33. Özyurt, I.; Uludag, I.; Ince, B.; Sezginturk, M.K. Lab-on-a-chip systems for cancer biomarker diagnosis. *J. Pharm. Biomed. Anal.* **2023**, *226*, 115266. [CrossRef] [PubMed]
34. Aslam, N.; Zhou, H.; Urbach, E.K.; Turner, M.J.; Walsworth, R.L.; Lukin, M.D.; Park, H. Quantum sensors for biomedical applications. *Nat. Rev. Phys.* **2023**, *5*, 157–169. [CrossRef] [PubMed]
35. Udhani, R.; Kothari, C.; Sarvaiya, J. A comprehensive study: Traditional and cutting-edge analytical techniques for the biomarker based detection of the micronutrients & POC sensing directions for next-generation diagnostic. *Crit. Rev. Anal. Chem.* **2023**. [CrossRef]
36. Scott, A.; Pandey, R.; Saxena, S.; Osman, E.; Li, Y.; Soleymani, L. A smartphone operated electrochemical reader and actuator that streamlines the operation of electrochemical biosensors. *ECS Sens. Plus* **2022**, *1*, 014601. [CrossRef]
37. Tong, Z.; Shen, C.; Li, Q.; Yin, H.; Mao, H. Combining sensors and actuators with electrowetting-on-dielectric (EWOD): Advanced digital microfluidic systems for biomedical applications. *Analyst* **2023**, *148*, 1399–1421. [CrossRef]

38. Juanola-Feliu, E.; Colomer-Farrarous, J.; Miribel-Catala, P.; Samitier, J.; Valls-Pasola, J. Market challenges facing academic research in commercializing nano-enabled implantable devices for in vivo biomedical analysis. *Technovation* **2012**, *32*, 193–204. [CrossRef]
39. Hu, Y.; Li, J.; Yu, X.; Yang, J.; Li, Y. A simple jet multifunctional sensing platform inspired by healing-assembly hydrogels serving motion and sweat monitoring. *Sens. Actuators B* **2023**, *378*, 133173. [CrossRef]
40. Fan, R.; Li, Y.; Park, K.W.; Du, J.; Chang, L.H.; Strieter, E.R.; Andrew, T.L. A strategy for accessing nanobody-based electrochemical sensors for analyte detection in complex media. *ECS Sens. Plus* **2022**, *1*, 010601. [CrossRef]
41. Chaudhary, V.; Kaushik, A.; Furukawa, H.; Khosla, A. Review—Towards 5th generation AI and IoT driven sustainable intelligent sensors based on 2D MXenes and borophene. *ECS Sens. Plus* **2022**, *1*, 013601. [CrossRef]
42. Hyodo, T.; Sakata, W.; Ueda, T.; Shimizu, Y. Effects of surface modification of Platinum electrodes with Gold on Hydrogen-sensing properties of diode-type sensors. *ECS Sens. Plus* **2022**, *1*, 013602. [CrossRef]
43. Sornambikai, S.; Amir, H.; Bhuvaneshwari, G.; Ponpandian, N.; Viswanathan, C. Review—Systematic review on electrochemical biosensing of breast cancer miRNAs to develop alternative DCIS diagnostic tool. *ECS Sens. Plus* **2022**, *1*, 021602. [CrossRef]
44. Moonla, C.; Lee, D.H.; Rokaya, D.; Rasitanon, N.; Kathayat, G.; Lee, W.Y.; Kim, J.; Jeerapan, I. Review—Lab-in-a-mouth and advanced point-of-care sensing systems: Detecting bioinformation from the oral cavity and saliva. *ECS Sens. Plus* **2022**, *1*, 021603. [CrossRef]
45. Singh, S.; Wang, J.; Cinti, S. Review—An overview on recent progress in screen-printed electroanalytical (bio)sensors. *ECS Sens. Plus* **2022**, *1*, 023401. [CrossRef]
46. Shekhar, S.; Yadav, A.K.; Khosla, A.; Solanki, P.R. Review—Interleukins profiling for biosensing applications: Possibilities and the future of disease detection. *ECS Sens. Plus* **2022**, *1*, 041601. [CrossRef]
47. Ramírez, S.; Díaz, F.; Silva, C.P.; Pavez, J.; Silva, J.F. Molecular anchoring with 4-mercaptobenzoic acid and 4-aminothiophenol for using active nanorods in the detection of dopamine. *ECS Sens. Plus* **2022**, *1*, 045201. [CrossRef]
48. Manjunatha, C.; Shwetha, K.P.; Athreya, Y.; Kumar, G.; Kamath, M.K.S. Perspective—Supercapacitor-powered flexible wearable strain sensors. *ECS Sens. Plus* **2023**, *2*, 017002. [CrossRef]
49. Hormozi Jangi, S.R.; Akhond, M. Ultrasensitive label-free enantioselective quantification of D-/L-leucine enantiomers with a novel detection mechanism using an ultra-small high-quantum yield N-doped CDs prepared by a novel highly fast solvent-free method. *Sens. Actuators B* **2021**, *339*, 129901. [CrossRef]

chemosensors

Article

Electrochemical Detection of Surfactant-Encapsulated Aqueous Nanodroplets in Organic Solution

Pankyu Kim, Hyeongkwon Moon and Jun Hui Park *

Department of Chemistry, Chungbuk National University, Cheongju 28644, Republic of Korea
* Correspondence: jhp@chungbuk.ac.kr; Tel.: +82-43-261-2287; Fax: +82-43-267-2279

Abstract: We report enhanced electrochemical detection of single water-in-oil emulsion droplets using the nano-impact method. To detect the emulsion droplets, the water molecules in the droplets were directly oxidized (i.e., water splitting) without additional electroactive species when the droplets collided with the ultramicroelectrode. The water molecules in the emulsion droplet cannot be directly electrolyzed in an organic solvent because the emulsifier does not require a hydrophobic electrolyte. To enhance the signal intensity, the electrochemistry of sub-microscale single droplets was investigated considering the charge neutrality and limiting reagent. Therefore, effective electrolysis of the droplets was achieved. Approximately 10% of water molecules in the droplet (55.6 M H_2O) were oxidized based on calculations from the electrochemical peak analysis and DLS measurements.

Keywords: nanodroplet; single entity; collision; nanoelectrochemistry; single particle

Citation: Kim, P.; Moon, H.; Park, J.H. Electrochemical Detection of Surfactant-Encapsulated Aqueous Nanodroplets in Organic Solution. *Chemosensors* **2023**, *11*, 112. https://doi.org/10.3390/chemosensors11020112

Academic Editors: Iulia Gabriela David and Dana Elena Popa

Received: 11 January 2023
Revised: 2 February 2023
Accepted: 2 February 2023
Published: 3 February 2023

1. Introduction

Nanoelectrochemistry has attracted tremendous interest in modern electrochemistry and has been studied using platforms such as nanoparticles [1–4], nanoelectrodes, nanogaps, nanojunctions, nanocavities, nanopores, nanochannels, and nanointerfaces [5–11]. The electrochemical reaction at the nanoscale is significantly different from that in the bulk solution. Recently, nanoelectrochemistry has been applied to single entities, such as emulsion droplets [12], liposomes [13–15], cells [16], micelles [17], nanobubbles [18] and vesicles [19]. Reaction monitoring in a single nano-entity is a challenging topic. A pure water droplet, which acts as a nanoreactor, is temporarily stable when dispersed in an immiscible organic solvent. In a real system, the surfactant is present in one or both liquids. A water droplet can be stabilized in an organic solvent using a surfactant (i.e., an emulsifier), which is called a water-in-oil emulsion (i.e., an inverse emulsion; W/O) [20–24]. In addition, organic media are not always pure and may contain chemical species. For example, gasoline and diesel fuels contain surfactants that act as corrosion inhibitors, dispersants, detergents, and anti-icing additives [25]. Therefore, a water droplet detection system must be developed that considers the surfactants included in the aqueous and/or organic phases.

Recently, single-liquid nanodroplets have been studied electrochemically using the 'single particle collision' (i.e., 'nano impact') method [26–31]. As a proof-of-concept, the suggested collisional detection method focuses on investigating the size distribution of a single droplet. This technique can be used to determine particle properties such as individual droplet size, droplet polydispersity, and droplet concentration in continuous media [32–36]. The reaction occurring inside a single liquid droplet enables single-entity electrochemistry. Studies have utilized reactions in single droplets for the controlled synthesis of single metal/nonmetal nanoparticles [37,38]. The liquid droplets can be dewetted or deformed when they land on the electrode surface, which can be used to control the electrochemical reactions inside single droplets. In single-droplet electrochemistry, the electrical contact area is a crucial factor determining the product of an electrochemical reaction.

In the early stages, single water nanodroplets and single oil nanodroplets were first detected electrochemically by blocking the electrode area via adsorption of the droplet in the presence of redox species in the organic phase [39,40]. This method is a powerful tool for investigating liquid nanodroplets that do not contain redox species. However, these methods, which are based on the blockage of the electrode surface, cannot be used to obtain precise information about single droplets. For droplets of the same size, blockage of the amperometric current can differ depending on the shape of the droplet adsorbed on the electrode surface. In the presence of redox species in the droplets, a comparable electrochemical detection method is also suggested that uses the electrolysis reaction within the droplet [22,33,34]. In general, however, aqueous droplets in organic media seldom contain redox species inside them. Organic solvents may contain either water molecules or a hydrophobic/hydrophilic electrolyte (non-redox active) because the water content in such systems is mostly coming from humidity in the air and electrolytes are pre-dissolved in an organic solvent. As a simplified droplet detection system, water molecules are considered to be redox species that can be directly electrolyzed. In a previous study, we reported the direct electrolysis of water molecules in water nanodroplets in the absence of surfactants and lipophilic electrolytes [21].

However, only 0.81% of the water content in the surfactant-free water droplets was electrolyzed because the electrochemical conditions in the nanodroplets were not properly prepared. In the presence of a surfactant, the electrochemistry can be partially blocked by the surfactant, which can hinder charge (electron and ion) transfer at the water/metal and water/oil interfaces. Furthermore, in the absence of hydrophobic electrolytes in the organic phase, oxidation of water (water splitting) molecules in the emulsion was not observed.

Herein, we report the sensitive electrochemical detection of water-in-oil emulsion droplets using the highly efficient oxidation of water molecules in a droplet. Surfactant-protected emulsion droplet dispersed in DCE solvent is investigated electrochemically based on the water oxidation (water splitting) reaction upon the collisional contact of the droplet on an electrode surface. The importance of a supporting electrolyte outside a single droplet in nanodroplet electrochemistry has been previously reported [20,34].

To consider the reaction efficiency in a single nanodroplet (i.e., water-in-oil emulsion droplet), reaction parameters such as the emulsifier, hydrophobic electrolyte, and hydrophilic electrolyte were carefully considered. By optimizing the reaction environment of a nanodroplet, electrolysis of a single nanodroplet has efficiently proceeded even with a surfactant-protected emulsion droplet. Therefore, an understanding of the reaction environment on nanodroplets is mandatory to detect the single entity electrochemically. In addition, this principle can be applied to elevate the electrolysis efficiency of water, which generates hydrogen gas as a reaction product.

During the water electrolysis in continuous aqueous media, bubbles are produced and cannot be removed rapidly from the electrode surface. The bubble is adsorbed on the electrode surface and covers the active reaction sites of the electrode to attenuate the production of hydrogen gas. This phenomenon would lead to high overpotential and high ohmic potential. Reducing bubbles is one of the key factors for water electrolysis. Injection of reactant water as a droplet could be a promising approach to removing the bubble effect on the electrode surface. Actually, ionic liquid as electrolytes for hydrogen production from water electrolysis was introduced for chemical stability and low reactivity of ionic liquid and effective hydrogen gas production from immiscible water ionic liquid properties [41]. Similarly, a water-in-oil emulsion system can also be utilized in efficient hydrogen gas production to effectively control the bubble adsorption on an electrode surface.

2. Materials and Methods

2.1. Reagents and Materials

Sodium chloride (NaCl, 99%), silver nitrate ($AgNO_3$, 99.9%), tetrabutylammonium hexafluorophosphate ($TBAPF_6$, 98%), and sodium dioctyl sulfosuccinate (AOT, 96%) were obtained from Alfa Aesar. 1,2-Dichloroethane (DCE, 99.8%) and tetrabutylammonium tetrafluoroborate ($TBABF_4$, 99%) were obtained from Sigma-Aldrich (St. Louis, MO, USA). Millipore water (>18 MW·cm) was used in all the experiments. Pt wire (99.99%; diameter,

10 μm) was supplied by Goodfellow (Devon, PA, USA). The Pt wire used as the counter electrode was purchased from Alfa Aesar. The reference electrode was an Ag wire with 10 mM $AgNO_3$ and 100 mM $TBAPF_6$ in acetonitrile (ACN, Thermo Fisher, Cleveland, OH, USA). All the reagents were used as received.

2.2. Electrochemical Instrumentation

Electrochemical experiment setups were composed of CHI 750E bipotentiostat (CH Instruments, Austin, TX, USA) with a three-electrode system. Every electrochemical measurement was performed inside a CHI 200B Faraday cage (CH Instruments, Austin, TX, USA). The three-electrode system is composed of an Ag/Ag^+ nonaqueous reference electrode (10 mM $AgNO_3$ in acetonitrile, MF-2062, BASi, IN, USA), homemade ultramicroelectrode (Pt, Au, C) as a working electrode, and a Pt wire (diameter: 1 mm, length: 100 mm) as a counter electrode. The electrochemical cell was composed of a homemade Teflon cap assembled with a glass vial. A horn sonicator system was used to prepare the emulsion solution using a Q700 ultrasonic processor (Qsonica, Newtown, CT, USA) with a microtip probe at a pulse mode. The average size of the particle is investigated based on the dynamic light scattering (DLS) principle using a NanoBook 90 Plus particle size analyzer (Brookhaven Instruments Corporation, Holtsville, NY, USA).

2.3. Preparation of the Water-in-oil Emulsion

The water-in-oil inverse emulsion solution was prepared according to the following procedures: 10 mL of 100 mM sodium chloride aqueous solution was prepared using distilled water, and 200 μL of 100 mM NaCl aqueous solution was added to 5.0 mL of DCE solution containing 1 mM sodium dioctyl sulfosuccinate (AOT). The aqueous/DCE mixture solution was vortexed for 20 s, then homogenized with a horn ultrasonicator (700 W, 30% of amplitude) in the pulse mode (1 cycle: on for 7 s, and off for 3 s) for 1 min. The molar concentration of the prepared emulsion droplet (C_{em}) was calculated using Equation (1):

$$C_{em} = \frac{n_{em}}{V} = \frac{N_{em}}{N_A V} = \left(\frac{V_W}{V_{em}}\right)\frac{1}{N_A V} \tag{1}$$

where n_{em} is the moles of emulsion droplets, N_{em} is the number of droplets, N_A is Avogadro's constant, V is the volume of the solution, V_W is the volume of the aqueous solution, and V_{em} is the average volume of a single emulsion droplet. As shown in Equation (2), the volume of a single emulsion droplet was estimated from r_{em} which is the average radius of the droplet obtained using DLS. It is assumed that all emulsion droplets are spherical in shape and have the same radius.

$$r_{em} = \sqrt[3]{\frac{3V_{em}}{4\pi}} \tag{2}$$

2.4. Preparation of Pt Ultramicroelectrode (Pt-UME)

The UME was prepared according to a procedure developed in our laboratory. Briefly, a 10 μm Pt wire (Goodfellow) was sealed in a borosilicate glass capillary (1.5 mm Outer Diameter × 0.75 mm Inner Diameter, Sutter Instrument, Novato, CA, USA), which was sonicated in hexane, toluene, IPA, ethanol, and water, respectively. The electrode was then polished with silicon carbide abrasive sandpaper (400, 1000, 1200, 2000, and 2500 grit; Buehler, Lake Bluff, IL, USA) until a mirror-like surface was observed.

The active area was monitored by standard redox electrochemistry in 1 mM ferrocenemethanol (FeMeOH) solution and estimated by the following equation:

$$i_{lim} = 4nD_{Red}R_{UME}C_{Red}$$

where i_{lim} is the limiting current, n is the number of electrons in the redox reaction, D_{Red}, and C_{Red} is the diffusion coefficient and concentration of redox species, respectively, and R_{UME} is the radius of Pt-UME. The obtained electrode area was almost 10 μm in diameter when we assume it has a flat circular shape. Before each electrochemical experiment, all UMEs were polished using 4000-grit SiC sandpaper (R&B Co. Ltd. Daejeon, Korea) and 1 μm, 0.5 μm, and 0.1 μm diamond lapping film (R&B Co. Ltd. Daejeon, Korea) was also used to obtain a mirror-like finish on the UME surface.

3. Results and Discussion

As a detection target, a water-in-oil emulsion droplet is prepared by mixing an aqueous solution of 100 mM NaCl with 1,2-dichloroethane (DCE) through the ultrasonic process. To study the homogeneous water droplets in the organic phase, 1mM sodium dioctyl sulfosuccinate (AOT), which serves as an inverse emulsion stabilizer, was dissolved in the DCE continuous phase. Surfactant-encapsulated water-in-oil droplets were studied, which are a more probable type of water droplet in organic solvents than pure water droplets (Scheme 1).

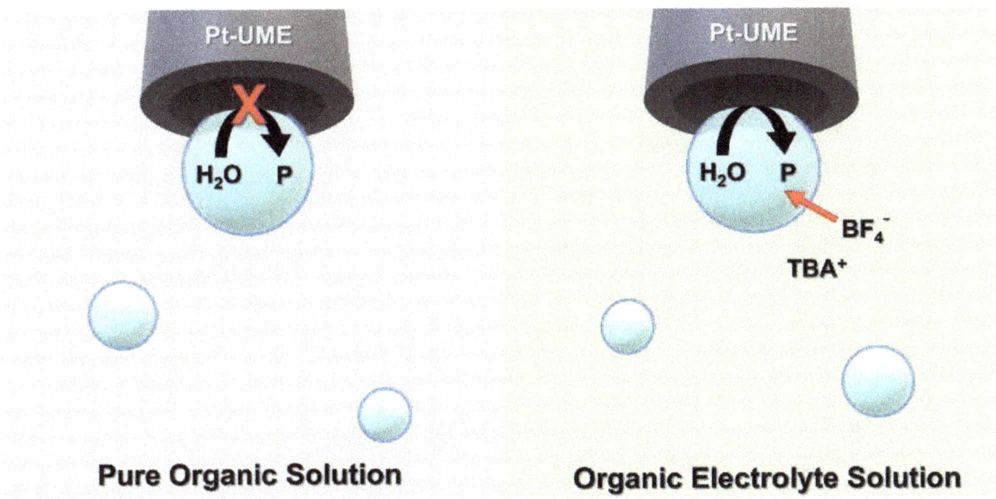

Scheme 1. The schematic illustration of the electrochemical reaction of the water droplet when it contacts the electrode surface. There is no reaction without a supporting electrolyte in organic solution; however, the electrolysis of water molecules in droplets takes place in the presence of a supporting electrolyte in organic solution.

To investigate the electrolysis conditions of the surfactant-encapsulated emulsion droplets, cyclic voltammetry (CV) and chronoamperometry (CA) were conducted in the absence and presence of a hydrophobic electrolyte in the continuous DCE phase. The CVs were scanned from 0 V to 1.2 V (*vs.* Ag/Ag^+), where water oxidation can occur thermodynamically. When the emulsion was injected into pure DCE (Fig. 1a black line), no anodic current response was observed until 1.2 V. However, in the presence of hydrophobic electrolyte in the organic phase, the oxidation of water started from around 0.85 V (Figure 1a, red line), resulting in a drastic current increase. In addition, current spikes were observed above the increasing current trend, owing to the continuous collision of the emulsion droplets. Figure 1b shows the CAs of the emulsions in DCE with and without 10 mM $TBABF_4$ in the organic phase. The electrode potential was maintained under an oxidative condition of 1.0 V *vs.* Ag/Ag^+. When the emulsions were injected into pure DCE (Figure 1b, black line), no spiky peak currents were observed. When the emulsion was injected into a DCE solution containing 10 mM $TBABF_4$, a collisional current spike originating from the collision and adsorption of the emulsion droplet on the UME, followed by simultaneous oxidation of the water molecules in the droplet was observed (Figure 1b, red line). Collisional current spikes were observed only in the presence of hydrophobic electrolytes in the organic phase in both the voltammetric and chronoamperometric measurements. Owing to the addition of the electrolyte to the solution, the resistance of the bulk solution decreased. Therefore, the background current increased slightly owing to the reduced iR overpotential.

$$\frac{1}{2}O_2 + 2H^+ + 2e^- \rightleftarrows H_2O \;\; E^0 = 0.689\ (\mathrm{V}\ vs.\ \mathrm{Ag/Ag^+})\ (\mathrm{R1})$$

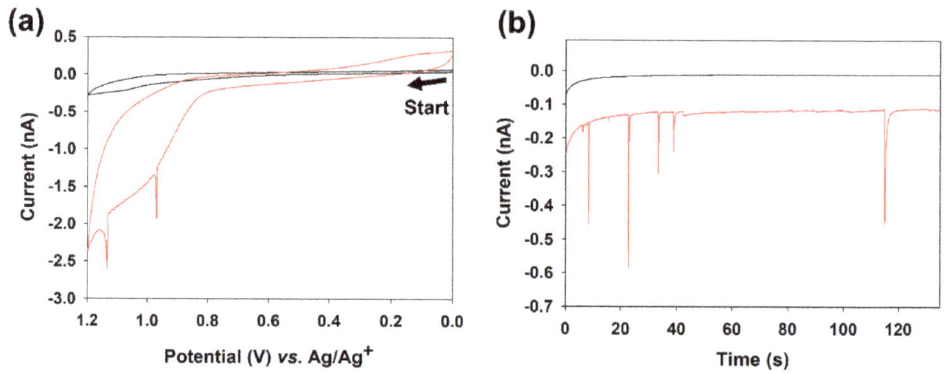

Figure 1. (**a**) CVs of the W/O emulsion droplets (0.1 M NaCl dissolved in aqueous phase) in pure DCE (black line) and in DCE containing 10 mM $TBABF_4$ (red line) on a Pt-UME at a scan rate 50 mV s^{-1}. (**b**) Amperometric i-t curves of the W/O emulsion droplets (0.1 M NaCl dissolved in aqueous phase) in pure DCE (black line) and in DCE containing 10 mM $TBABF_4$ (red line) on a Pt-UME biased at 1.0 V (*vs.* Ag/Ag^+).

To determine a suitable potential for emulsion detection in an organic electrolyte solution, CA was conducted at various potentials in the Pt UME (Figure 2a). The experiments were conducted at potentials from 0.4 V to 1.2 V with 0.2 V intervals. When 0.4 V was applied, anodic peak currents were not observed. At 0.6 V and 0.8 V, the spiky current peaks were observed, resulting from the oxidation of the collided emulsion droplets. However, the current peaks were comparatively small because the overpotential was not sufficiently high for water oxidation. When the potential was increased to 1.0 V and 1.2 V, relatively large spiky peaks were observed. Based on the experimental results shown in Figure 2b, 1.0 V was chosen as the optimal electrode potential for the detection of emulsion droplets using the CA method because it showed high peak currents and maintained a stable background current.

In pure DCE solution, CA was performed at various anodic potentials to observe emulsion droplet collisions in the Pt UME (Figure 3). For the consideration of highly resistive organic solvent without supporting electrolyte, CA experiments were conducted at potentials from 1.0 V to 1.8 V with 0.2 V intervals. However, no collisional current spikes were observed in any of the experiments, including a high anodic potential of 1.8 V. The result indicates that hydrophobic electrolytes are required to detect the AOT-encapsulated emulsion droplets. For the detection of surfactant-free droplets (e.g., pure water droplets only containing hydrophilic electrolytes in an aqueous phase), the water droplets can be detected in the absence of hydrophobic electrolytes in DCE [22]. Although the surfactant species are dissolved in the organic phase, they are also present at the interface of the aqueous and organic solutions. Therefore, the electrochemistry of the aqueous droplet can be affected by the interface material.

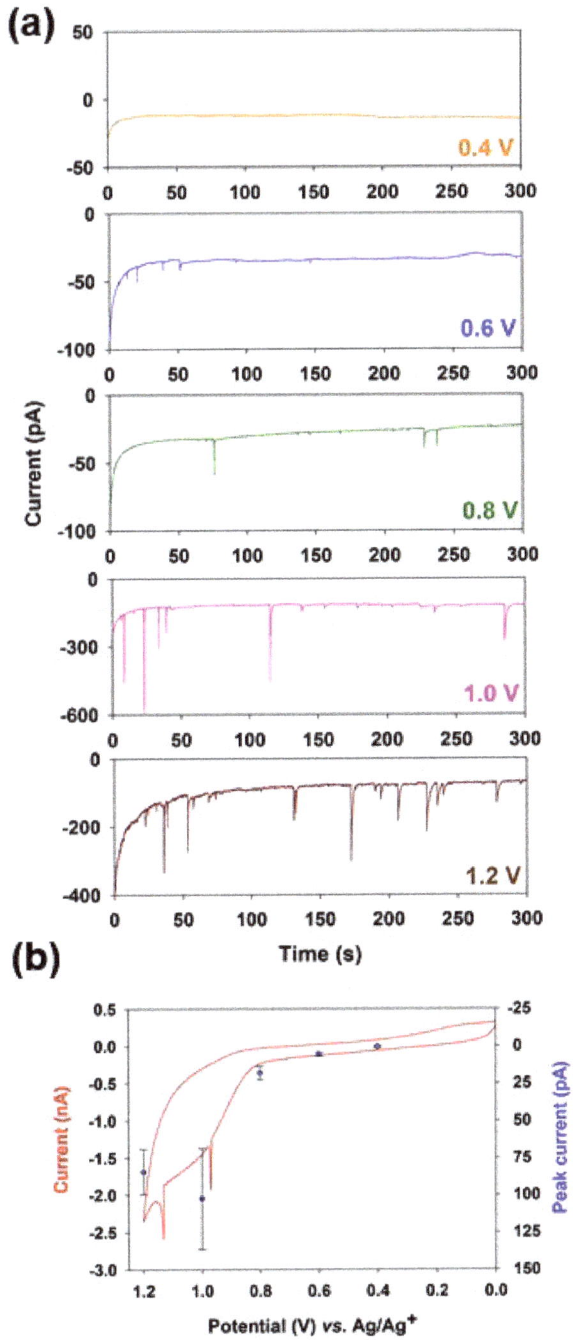

Figure 2. (**a**) Amperometric i–t curves of W/O emulsion droplet on Pt-UME at various applied potentials (0.4, 0.6, 0.8, 1.0, and 1.2 V *vs.* Ag/Ag^+) in DCE electrolyte solution containing 10 mM $TBABF_4$ (**b**) Collisional peak heights obtained from W/O emulsion in the DCE electrolyte solution were plotted together with a cyclic voltammogram of the same solution depending on the applied electrode potentials.

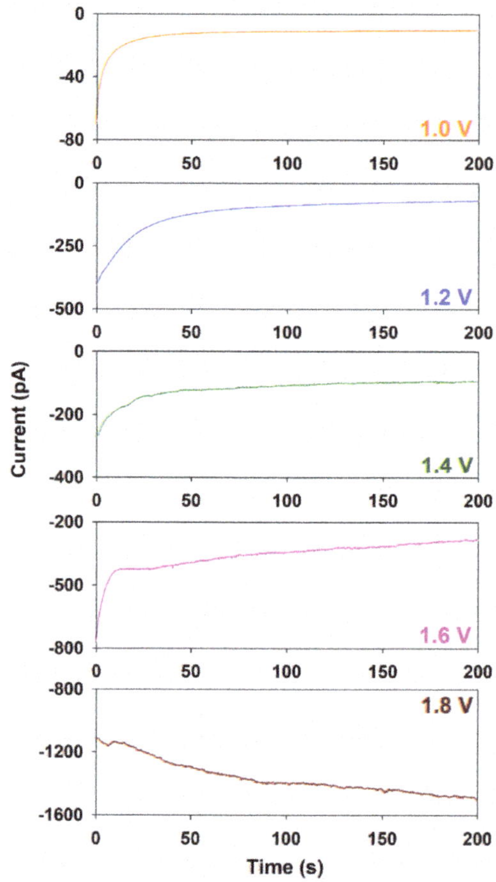

Figure 3. Amperometric i–t curves of W/O emulsion droplets on Pt-UME at various applied potentials (1.2, 1.4, 1.6, 1.8, and 2.0 V *vs.* Ag/Ag^{+}) in pure DCE solution.

Using the CA data obtained from the collision experiments, the collision frequency was measured based on the concentration of the emulsion droplets. A plot of the collision frequency as a function of the emulsion concentration is shown in Figure 4. The upper and lower error bars were calculated as three times the standard deviation. It was observed that the collision frequency was proportional to the injected emulsion droplets when the droplet concentration was less than ca. 300 pM. At lower droplet concentrations, the emulsion droplet behaved as a randomly moving independent particle, and a low probability of interaction was observed between the particles. Therefore, the increased number of emulsion droplets due to the increased emulsion injected suggests that collisions of emulsion droplets on UME are more likely to result from the stochastic diffusion process. The following Equation (3) was derived from a regression analysis of the experimental collision frequencies (Figure 4, blue line).

$$f_{exp} = 0.00044C_{em} \quad (3)$$

where f_{exp} is the collision frequency of the experiment and C_{em} is the concentration of the W/O emulsion.

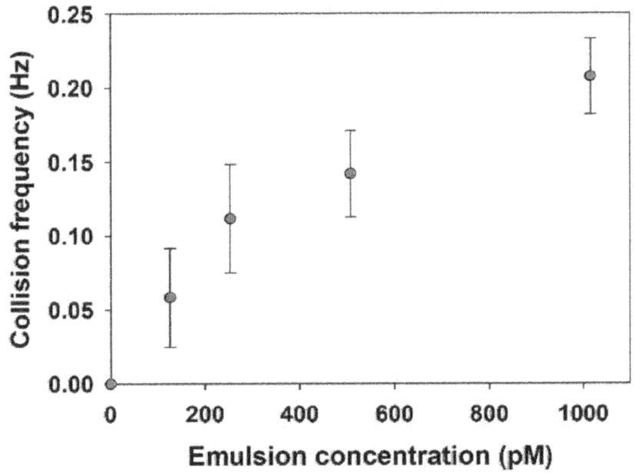

Figure 4. Plot of the collision frequency (Hz) as a function of the W/O emulsion concentration (pM). The error bars indicate the 3 times of the standard deviation of 6 independent measurements. All experiments were conducted on a Pt-UME biased at 1.0 V (*vs.* Ag/Ag^+).

However, when the droplet concentration is increased, a greater chance of interaction is expected between the droplets. Coalescence occurs during the injection of emulsion droplets into an organic medium. Furthermore, in the presence of hydrophobic electrolytes in the organic phase, the droplets can agglomerate to form larger droplets because the electrostatic repulsion between the charged droplets can be reduced at a high ionic strength. In our experiment, at higher emulsion droplet concentrations, the coalescence of the emulsion droplets resulted in a lower collision frequency than that expected from the linear trend.

The theoretical collision frequency can be evaluated by considering the diffusion and migration of W/O emulsion droplets toward the UME. In this system, only the diffusional movement of the droplet was considered because the migration effect can be ignored using supporting electrolytes in an organic solvent. The theoretical collision frequency is given by Equation (4):

$$f_{em} = 4D_{em}C_{em}r_eN_A \tag{4}$$

where f_{em} is the theoretical frequency of the W/O emulsion owing to diffusion toward the working electrode, D_{em} is the diffusion coefficient of the W/O emulsion droplet, C_{em} is the concentration of the W/O emulsion, r_e is the radius of the working electrode, and N_A is Avogadro's number. D_{em} was calculated using the Stokes–Einstein relationship (Equation (5)):

$$D_{em} = \frac{k_BT}{6\pi\eta r_{em}} \tag{5}$$

where k_B is the Boltzmann constant, T is the temperature (298 K), η is the viscosity of water at 25 °C, and r_{em} is the radius of an emulsion droplet. Consequently, the theoretical frequency of collision (f_{em}) from the 127, 254, 508, and 1016 pM emulsion droplet was 2.1, 4.1, 8.3, and 16.5 Hz and the experimental frequency (f_{exp}) was estimated to be 0.058, 0.112, 0.142, and 0.208 Hz, respectively. The f_{em} values of the water droplet collisions were approximately 35–80 times higher than the f_{exp} values. The lower experimental frequency (f_{exp}) might be attributed to the invalid collision of the emulsion droplets on the UME; that is, the collision of droplets did not lead to an electrochemical reaction, possibly owing to incomplete contact on the UME. The adsorption of the injected emulsion droplet on the surface of the electrochemical cell and glass sheath of the electrode also reduced the experimental collision frequency.

Upon adsorption of the emulsion droplet onto the Pt UME, the electrochemical reaction started instantly on the biased Pt UME. The electrolysis of water molecules in the emulsion droplets exhibited a high current magnitude because the tiny droplets contained a highly condensed reactant (i.e., 55.6 M water molecules). The shape of the anodic current peaks showed a sharp increase and exponential decay with time. The current peaks were analyzed to determine the size of each water droplet. Therefore, the number of charges from each collision event was calculated by integrating the peak areas in CA. Assuming a spherical droplet, the diameter of each droplet (d_{drop}) was calculated using Equation (3).

$$d_{drop} = \sqrt[3]{\frac{6Q}{n\pi F C_{redox}}} \tag{6}$$

where Q is the measured charge and n is the number of electrons transferred per molecule. F is the Faraday constant and C_{redox} is the concentration of the redox species in a droplet. In our experiments, the electrochemical redox species is water. Therefore, the water concentration in the droplets was ca. 55.6 M. During oxidation, each water molecule generates two electrons. Therefore, n is 2.

The size distribution of the droplets, calculated using Equation (6), is shown in Figure 5 (blue bars), and that obtained from the DLS measurements is shown in Figure 5 (red line). Previous studies using electroactive redox species have demonstrated similar size distributions using both the electrochemical collision method and DLS measurements [22,34]. However, in our experiments, comparing each average size obtained through the electrochemical collision method (~171 nm) and DLS measurements (~363 nm), the two results indicated considerable differences. The size of the emulsion droplets obtained electrochemically was underestimated because the oxidation of water molecules in the droplets did not proceed completely until the water was depleted. During the oxidation of water in a neutral solution, two water molecules produce one oxygen molecule and four protons. The produced oxygen molecules can form bubbles at the interface of the electrode, which can lead to a loss of electrical connection between the droplet and the UME, hindering the complete electrolysis of the water molecules in the droplet.

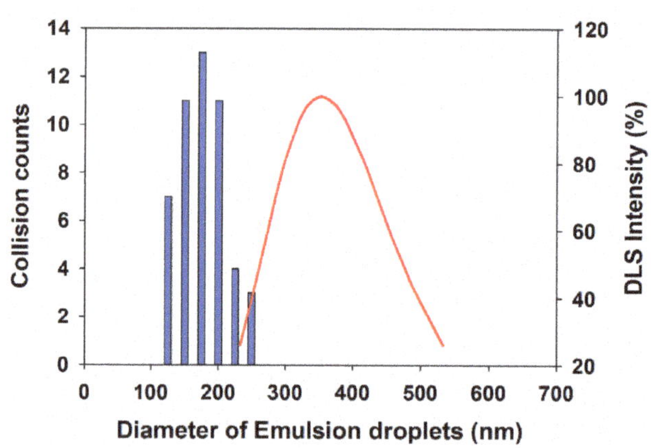

Figure 5. Comparison of the size distribution between the single W/O (water-in-oil) emulsion collision method (blue bars) and DLS measurement (red line).

By comparing the underestimated size and actual size of the droplets, the number of water molecules in the droplets that are oxidized can be calculated. Please note that the average size of the droplets from DLS measurement can contain large errors because the droplets are a polydisperse sample rather than a monodisperse. When a large particle is included in the sample, the large particle can have strong scattering and also can partly disturb the weaker scattering from the small particles. More accurate results could be obtained by using nanoparticle tracking analysis (e.g., Nanosight). If the water molecules in the 363 nm diameter (obtained from DLS) droplets react completely, a 269 pC of charge should be generated. The average charge obtained from the same sample using the electrochemical collision method was 28 pC. Comparing these two charge values, we can conclude that roughly ~10% of water molecules in the droplet were electrochemically oxidized, and the remainder were not oxidized. This result is significantly enhanced from a previous report of the 0.81% electrolysis efficiency in an electrolyte-free organic solvent [21]. Owing to the confined size of the reactor (i.e., the emulsion droplets), it is difficult to pump the reaction product out of the reactor for a continuous reaction in the droplets. If the produced charged product (e.g., protons) is not properly balanced inside the droplet, the electrolysis reaction is halted because charge neutrality inside the droplet is not maintained. The hydrophobic electrolyte anion (BF_4^-), which permeates to the aqueous droplet from the organic electrolyte solution, promotes the electrochemical water oxidation by maintaining the charge balance.

$$2H_2O(aq) + 4BF_4^-(org) \rightleftarrows O_2(aq) + 4H^+(aq) + 4BF_4^-(aq) + 4e^- \quad (R2)$$

Therefore, we established efficient electrolysis conditions for surfactant-stabilized aqueous droplets in organic solvents based on an understanding of the reaction in a single entity. Using this system, we detected aqueous emulsion droplets without the addition of additional redox species. Furthermore, this principle can elevate water electrolysis efficiency under neutral conditions. In the bulk electrolysis of the water continuum, the adsorption of bubbles on the electrode surface results in a high ohmic potential drop and a large reaction overpotential. The efficient electrolysis of water emulsion droplets in an organic medium can be studied by reducing bubble adsorption on the electrode surface and controlling the hydrophobic surfactant on an organic solution.

4. Conclusions

Herein, we describe the electrochemical detection of surfactant-encapsulated aqueous droplets in organic electrolyte solutions. Previous studies on surfactant-free water droplets have utilized the direct oxidation of solvent molecules in a pure organic solution. However, in the presence of a surfactant (e.g., an emulsifier) in an organic solvent, the water molecules are not electrochemically oxidized. In our experiments, the water molecules in the emulsion droplets are electrolyzed using a hydrophobic electrolyte. Although the water oxidation partly proceeds in an electrolyte-free organic solvent (0.81%), more efficient electrolysis occurs in an organic solvent in the presence of hydrophobic electrolytes. By enhancing the electrolysis efficiency, a high signal-to-noise ratio can be obtained for the electrochemical detection of emulsion droplets. Therefore, the electrochemical detection of nano/microdroplets can be utilized in the study of electrochemical sensors. Furthermore, this principle can be applied to elevate water electrolysis efficiency under neutral conditions.

Author Contributions: Conceptualization, J.H.P. and P.K.; methodology, J.H.P., H.M. and P.K.; software, H.M. and P.K.; validation, J.H.P. and H.M.; formal analysis, H.M. and P.K.; investigation, H.M. and P.K.; resources, J.H.P.; data curation, P.K. and H.M.; writing—original draft preparation, P.K. and J.H.P.; writing—review and editing, H.M. and J.H.P.; visualization, H.M. and P.K.; supervision, J.H.P.; project administration, J.H.P.; funding acquisition, J.H.P. All authors have read and agreed to the published version of the manuscript.

Funding: This work was supported by the Basic Science Research Program through the National Research Foundation of Korea (NRF), funded by the Ministry of Science, ICT, and Future Planning (NRF- 2022R1A2C1007062).

Institutional Review Board Statement: Not applicable.

Informed Consent Statement: Not applicable.

Data Availability Statement: Not applicable.

Acknowledgments: We thank Heekyung Park in Chungbuk National University for preparing the graphical abstract.

Conflicts of Interest: The authors declare no conflict of interest.

References

1. Xiong, J.; Chen, Q.; Edwards, M.A.; White, H.S. Ion transport within high electric fields in nanogap electrochemical cells. *ACS Nano* **2015**, *9*, 8520–8529. [CrossRef]
2. Anderson, T.J.; Zhang, B. Single-nanoparticle electrochemistry through immobilization and collision. *Acc. Chem. Res.* **2016**, *49*, 2625–2631. [CrossRef]
3. Vezvaie, M.; Brosseau, C.L.; Lipkowski, J. Electrochemical SERS study of a biomimetic membrane supported at a nanocavity patterned Ag electrode. *Electrochim. Acta* **2013**, *110*, 120–132. [CrossRef]
4. Moon, H.; Park, J.H. In Situ Probing Liquid/Liquid Interfacial Kinetics through Single Nanodroplet Electrochemistry. *Anal. Chem.* **2021**, *93*, 16915–16921. [CrossRef] [PubMed]
5. Jia, R.; Mirkin, M.V. The double life of conductive nanopipette: A nanopore and an electrochemical nanosensor. *Chem. Sci.* **2020**, *11*, 9056–9066. [CrossRef] [PubMed]
6. Ma, K.; Yang, L.; Liu, J.; Liu, J. Electrochemical sensor nanoarchitectonics for sensitive detection of uric acid in human whole blood based on screen-printed carbon electrode equipped with vertically-ordered mesoporous silica-nanochannel film. *Nanomaterials* **2022**, *12*, 1157. [CrossRef] [PubMed]
7. Lin, B.; Hui, J.; Mao, H. Nanopore technology and its applications in gene sequencing. *Biosensors* **2021**, *11*, 214. [CrossRef]
8. Yan, Y.; Chen, G.; She, P.; Zhong, G.; Yan, W.; Guan, B.Y.; Yamauchi, Y. Mesoporous nanoarchitectures for electrochemical energy conversion and storage. *Adv. Mater.* **2020**, *32*, 2004654. [CrossRef]
9. Pollok, N.E.; Rabin, C.; Walgama, C.T.; Smith, L.; Richards, I.; Crooks, R.M. Electrochemical detection of NT-proBNP using a metalloimmunoassay on a paper electrode platform. *ACS Sens.* **2020**, *5*, 853–860. [CrossRef]
10. Zhou, M.; Bao, S.; Bard, A.J. Probing size and substrate effects on the hydrogen evolution reaction by single isolated Pt atoms, atomic clusters, and nanoparticles. *J. Am. Chem. Soc.* **2019**, *141*, 7327–7332. [CrossRef]
11. Liu, S.-C.; Ying, Y.-L.; Li, W.-H.; Wan, Y.-J.; Long, Y.-T. Snapshotting the transient conformations and tracing the multiple pathways of single peptide folding using a solid-state nanopore. *Chem. Sci.* **2021**, *12*, 3282–3289. [CrossRef]
12. Kim, S.D.; Park, J.H.; Ahn, H.; Lee, J.; Shin, C.-H.; Jang, W.-D.; Kim, B.-K.; Ahn, H.S. The discrete single-entity electrochemistry of Pickering emulsions. *Nanoscale* **2022**, *14*, 6981–6989. [CrossRef] [PubMed]
13. Liu, Y.; Xu, C.; Yu, P.; Chen, X.; Wang, J.; Mao, L. Counting and sizing of single vesicles/liposomes by electrochemical events. *ChemElectroChem* **2018**, *5*, 2954–2962. [CrossRef]
14. Luy, J.; Ameline, D.; Thobie-Gautier, C.; Boujtita, M.; Lebègue, E. Detection of Bacterial Rhamnolipid Toxin by Redox Liposome Single Impact Electrochemistry. *Angew. Chem. Int. Ed.* **2022**, *134*, e202111416. [CrossRef]
15. Lebègue, E.; Barrière, F.; Bard, A.J. Lipid membrane permeability of synthetic redox DMPC liposomes investigated by single electrochemical collisions. *Anal. Chem.* **2020**, *92*, 2401–2408. [CrossRef]
16. McCormick, H.K.; Dick, J.E. Nanoelectrochemical quantification of single-cell metabolism. *Anal. Bioanal. Chem.* **2021**, *413*, 17–24. [CrossRef]
17. Cheng, W.; Compton, R.G. Oxygen Reduction Mediated by Single Nanodroplets Containing Attomoles of Vitamin B12: Electrocatalytic Nano-Impacts Method. *Angew. Chem. Int. Ed.* **2015**, *54*, 7082–7085. [CrossRef]
18. Chen, Q.; Luo, L.; Faraji, H.; Feldberg, S.W.; White, H.S. Electrochemical measurements of single H2 nanobubble nucleation and stability at Pt nanoelectrodes. *J. Phys. Chem. Lett.* **2014**, *5*, 3539–3544. [CrossRef]
19. Ho, T.L.; Hoang, N.T.; Lee, J.; Park, J.H.; Kim, B.-K. Determining mean corpuscular volume and red blood cell count using electrochemical collision events. *Biosens. Bioelectron.* **2018**, *110*, 155–159. [CrossRef] [PubMed]
20. Zhang, H.; Sepunaru, L.; Sokolov, S.V.; Laborda, E.; Batchelor-McAuley, C.; Compton, R.G. Electrochemistry of single droplets of inverse (water-in-oil) emulsions. *Phys. Chem. Chem. Phys.* **2017**, *19*, 15662–15666. [CrossRef]
21. Lee, J.; Ho, T.L.; Kim, H.Y.; Park, J.H.; Kim, B.K. Direct electrolysis and detection of single nanosized water emulsion droplets in organic solvent using stochastic collisions. *Electroanalysis* **2019**, *31*, 167–171. [CrossRef]
22. Yang, H.-J.; Kwon, H.; Kim, B.-K.; Park, J.H. Electrochemical detection of single attoliter aqueous droplets in electrolyte-free organic solvent via collision events. *Electrochim. Acta* **2019**, *320*, 134620. [CrossRef]
23. Glasscott, M.W.; Pendergast, A.D.; Dick, J.E. A universal platform for the electrodeposition of ligand-free metal nanoparticles from a water-in-oil emulsion system. *ACS Appl. Nano Mater.* **2018**, *1*, 5702–5711. [CrossRef]
24. Perini, N.; Prado, A.; Sad, C.; Castro, E.; Freitas, M. Electrochemical impedance spectroscopy for in situ petroleum analysis and water-in-oil emulsion characterization. *Fuel* **2012**, *91*, 224–228. [CrossRef]
25. Nelson, R. Application of surfactants in the petroleum industry. *JAOCS* **1982**, *59*, 823A–826A. [CrossRef]
26. Li, X.; Ren, L.; Dunevall, J.; Ye, D.; White, H.S.; Edwards, M.A.; Ewing, A.G. Nanopore opening at flat and nanotip conical electrodes during vesicle impact electrochemical cytometry. *ACS Nano* **2018**, *12*, 3010–3019. [CrossRef]
27. Lee, J.; Kang, Y.; Chang, J.; Song, J.; Kim, B.-K. Determination of Serotonin Concentration in Single Human Platelets through Single-Entity Electrochemistry. *ACS Sens.* **2020**, *5*, 1943–1948. [CrossRef]
28. Choi, Y.; Park, C.; Kang, Y.; Muya, J.T.; Jang, D.P.; Chang, J. Temporally Resolved Electrochemical Interrogation for Stochastic Collision Dynamics of Electrogenerated Single Polybromide Droplets. *Anal. Chem.* **2021**, *93*, 8336–8344. [CrossRef]
29. Deng, H.; Peljo, P.; Huang, X.; Smirnov, E.; Sarkar, S.; Maye, S.; Girault, H.H.; Mandler, D. Ionosomes: Observation of ionic bilayer water clusters. *J. Am. Chem. Soc.* **2021**, *143*, 7671–7680. [CrossRef]

30. Martínez, G.; Merinero, M.; Pérez-Aranda, M.; Pérez-Soriano, E.M.; Ortiz, T.; Villamor, E.; Begines, B.; Alcudia, A. Environmental impact of nanoparticles' application as an emerging technology: A review. *Materials* **2020**, *14*, 166. [CrossRef]
31. Renault, C.; Lemay, S.G. Electrochemical collisions of individual graphene oxide sheets: An analytical and fundamental study. *ChemElectroChem* **2020**, *7*, 69–73. [CrossRef]
32. Jeong, S.; Choi, M.-H.; Jagdale, G.S.; Zhong, Y.; Siepser, N.P.; Wang, Y.; Zhan, X.; Baker, L.A.; Ye, X. Unraveling the Structural Sensitivity of CO2 Electroreduction at Facet-Defined Nanocrystals via Correlative Single-Entity and Macroelectrode Measurements. *J. Am. Chem. Soc.* **2022**, *144*, 12673–12680. [CrossRef] [PubMed]
33. Kim, B.-K.; Kim, J.; Bard, A.J. Electrochemistry of a single attoliter emulsion droplet in collisions. *J. Am. Chem. Soc.* **2015**, *137*, 2343–2349. [CrossRef] [PubMed]
34. Deng, H.; Dick, J.E.; Kummer, S.; Kragl, U.; Strauss, S.H.; Bard, A.J. Probing ion transfer across liquid–liquid interfaces by monitoring collisions of single femtoliter oil droplets on ultramicroelectrodes. *Anal. Chem.* **2016**, *88*, 7754–7761. [CrossRef] [PubMed]
35. Lu, S.-M.; Li, M.-Y.; Long, Y.-T. Dynamic Chemistry Interactions: Controlled Single-Entity Electrochemistry. *J. Phys. Chem. Lett.* **2022**, *13*, 4653–4659. [CrossRef] [PubMed]
36. Sabaragamuwe, S.G.; Madawala, H.; Puri, S.R.; Kim, J. Towards ultralow detection limits of aromatic toxicants in water using pluronic nanoemulsions and single-entity electrochemistry. *Anal. Chim. Acta* **2020**, *1139*, 129–137. [CrossRef] [PubMed]
37. Lee, M.W.; Kwon, D.J.; Park, J.; Pyun, J.C.; Kim, Y.J.; Ahn, H.S. Electropolymerization in a confined nanospace: Synthesis of PEDOT nanoparticles in emulsion droplet reactors. *Chem. Commun.* **2020**, *56*, 9624–9627. [CrossRef]
38. Glasscott, M.W.; Pendergast, A.D.; Goines, S.; Bishop, A.R.; Hoang, A.T.; Renault, C.; Dick, J.E. Electrosynthesis of high-entropy metallic glass nanoparticles for designer, multi-functional electrocatalysis. *Nat. Commun.* **2019**, *10*, 2650. [CrossRef] [PubMed]
39. Kim, B.-K.; Boika, A.; Kim, J.; Dick, J.E.; Bard, A.J. Characterizing Emulsions by Observation of Single Droplet Collisions Attoliter Electrochemical Reactors. *J. Am. Chem. Soc.* **2014**, *136*, 4849–4852. [CrossRef]
40. Hoang, N.T.; Ho, T.L.; Park, J.H.; Kim, B.-K. Detection and study of single water/oil nanoemulsion droplet by electrochemical collisions on an ultramicroelectrode. *Electrochim. Acta* **2017**, *245*, 128–132. [CrossRef]
41. De Souza, R.F.; Padilha, J.C.; Gonçalves, R.S.; Rault-Berthelot, J. Dialkylimidazolium ionic liquids as electrolytes for hydrogen production from water electrolysis. *Electrochem. Commun.* **2006**, *8*, 211–216. [CrossRef]

chemosensors

MDPI

Article

Smart Sensor for Lithium Detection: An Environmental Application

Ilaria Antonia Vitale, Giulia Selvolini and Giovanna Marrazza *

Department of Chemistry "Ugo Schiff", University of Florence, Via della Lastruccia 3, 50019 Sesto Fiorentino (FI), Italy
* Correspondence: giovanna.marrazza@unifi.it; Tel.: +39-055-457-3320

Abstract: In this work, we propose a potentiometric smart sensor for lithium detection in environmental samples based on a screen-printed cell. The graphite screen-printed electrode (GSPE) was first modified by gold and silver nanoparticles to increase the conductivity, and then by an ion-selective membrane, which was directly deposited onto the nanostructured electrode surface. The potentiometric cell, thanks to its small size, was integrated with a portable instrument connected to a smartphone for decentralized analysis. The developed sensor was used in drop- and microflow-configurations and showed a high sensitivity for lithium with a low detection limit (1.6 μM). It was also applied in the analysis of real samples from the industrial recycling of automotive batteries and complex matrices, such as contaminated soils.

Keywords: lithium; microflow; ion-selective membrane; metal nanoparticles; screen-printed electrodes

Citation: Vitale, I.A.; Selvolini, G.; Marrazza, G. Smart Sensor for Lithium Detection: An Environmental Application. *Chemosensors* **2023**, *11*, 77. https://doi.org/10.3390/chemosensors11020077

Academic Editors: Iulia Gabriela David and Dana Elena Popa

Received: 21 December 2022
Revised: 12 January 2023
Accepted: 15 January 2023
Published: 19 January 2023

1. Introduction

Over the past decades, the production and consumption of lithium have increased at an average of approximately 3% per annum. Lithium and its compounds indeed own a strategic importance as they have several industrial applications, among which the most well-known are lithium and lithium-ion batteries (LIBs). LIBs are the primary driving force in the increase of lithium production, as they offer a viable alternative to other rechargeable batteries in terms of costs, number of charge–discharge cycles, and high-performance results [1]. In addition, they have also become a promising option for the fueling of electrical vehicles (EVs) and the reduction of CO_2 emissions, which is increasingly salient due to the societal awareness of climate change [2,3]. Consequently, the exploitation of mineral deposits to provide the raw material required by the LIBs market has been subjected to a rapid increase [4]. Recycling processes (e.g., hydrometallurgical separation) thus represent a sustainable option for the reintroduction of LIB compounds into the economic cycle, reducing both the need for primary raw materials, the generated byproducts, and the impact on the environment, including the health of fauna and flora [5–7]. However, only a few metals with high economic value are recycled from exhausted batteries as such approaches could involve complex processes. This is part of the reason why spent LIBs have been treated as waste ever since the first commercial cells hit the market in the early 1990s: it was often cheap enough to dump old batteries and mine the virgin material [8].

Considering what has just been discussed, the importance of developing a fast and cost-effective analytical technique to help increase both the efficiency of the recycling process and quality control in terms of environmental pollution becomes increasingly urgent.

From this perspective, electrochemical sensors like ion-selective electrodes (ISEs) are a promising alternative to most of the existing analytical techniques for lithium detection (e.g., inductively coupled plasma—ICP, atomic absorption spectroscopy—AAS, ionic chromatography—IC) for an accurate in situ monitoring and analysis [9,10], as they provide rapid results and are small and cost-effective. Lithium detection by potentiometric sensors has been accomplished through various approaches, including those that rely on crystalline membranes. Freitas et al. modified a carbon-paste electrode (CPE) with an aluminum-doped, spinel-type manganese oxide, retrieving a super-Nernstian response towards lithium [11]. Nel-lo et al. used two different solid-state electrolytes with a garnet-type crystal structure for lithium detection in molten Pb-Li alloys as the sensing element [12].

In the case of alkali metals, the coupling of electrochemical techniques to macrocyclic molecules (i.e., ionophores) can yield high sensitivity, reproducibility, low detection limits, and high selectivity due to the complementarity between the electrostatic properties of the macrocyclic cavity and those of the metal cation. Potentiometric ISEs based on an ionophore-containing membrane of polyvinyl chloride (PVC) have been developed for the detection of potassium and sodium ions [13,14]. Regarding lithium detection, the potentiometric approaches based on polymeric ion-selective membranes are reported in Table 1.

Table 1. Lithium detection by potentiometric sensors based on a polymeric ionophore-containing membrane.

Ionophore	Linear Range (M)	Slope (mV/decade)	Real Samples	Ref.
Li ionophore III	10^{-5}–10^{-1}	58	-	[15]
Dendrimer	2.5×10^{-5}–10^{-1}	52	Blood	[16]
Li ionophore VI	2.7×10^{-6}–10^{-1}	48	Serum	[17]
Li ionophore VI	5×10^{-5}–10^{-1}	57	Serum	[18]

Portable devices can also be designed, taking advantage of the aforementioned characteristics, which could allow an accurate in situ detection of lithium, whether in hospitals remote locations, landfill sites, or offshore monitoring [19].

In fact, screen-printed cells (SPCs) opened new opportunities to apply electrochemical techniques for environmental in situ analyses, providing an analytical tool to satisfy the increasing demand of the rapid, sensitive, and selective determination of a wide range of analytes. The most attractive features of these transducers are undoubtedly their low cost and versatility [20]; moreover, to enhance their sensing performance, nano-modified electrode surfaces can be exploited to retrieve a specific electrochemical behavior depending solely on the nature of the modifier [21,22]. Moreover, SPCs integrated into flow injection analysis (FIA), batch injection analysis (BIA), and high-performance liquid chromatography (HPLC) have been adopted in a variety of analytical studies [23]. The combination of SPCs and flow systems increases the throughput of measurements, improves electrodes lifetime, and reduces reagent consumption and waste generation beyond simplifying and streamlining the analysis of the low volumes of samples, which is extremely critical for on-spot environmental analysis [24–28].

Motivated by the aforementioned reasons, the authors in this work propose a smart potentiometric solid-contact sensor for lithium detection based on silver and gold nanoparticles-modified graphite screen-printed electrodes ((Au&Ag)NPs/GSPEs) and a PVC-based ion-selective membrane. It is known that bimetallic nanoparticles possess excellent electrochemical properties due to synergistic and electronic effects: among them, the Ag-Au bimetallic catalysts have drawn much attention due to their electronic, optical, and catalytic properties which differ from those of individual mono metals [29]. To the best of our knowledge, only one example of a potentiometric screen-printed sensor based on platinum and gold nanostructures and a polymeric membrane for lithium detection in human perspiration was reported in the literature; this showed a slightly higher sensitivity but also a higher detection limit compared to our sensor, and it was not integrated with a microflow setup [30,31]. In our case, the use of silver nanoparticles coupled to gold ones enhanced the sensing performance of inexpensive graphite electrodes, lowering the production cost of the sensors with respect to the use of platinum. Moreover, the relative amounts of membrane-cocktail components, which were already known to work properly for lithium detection, were varied in order to improve the features of the sensor; finally, a study on the proper amount of the membrane cocktail to be deposited completes the optimization of the sensor. A good Nernstian response from the sensor (~58 mV/decade) was then retrieved;

the selectivity was also thoroughly investigated. The developed sensors were ready to use and, in contrast to most of the reported ion-selective devices, no electrode conditioning was required before performing the measurements. This ion-selective solid-contact sensor was integrated into a smart setup comprising a pocket instrument connected to a smartphone to perform drop and microflow measurements for on-spot decentralized analysis. To assess the applicability of the developed device, lithium was also detected in waste liquids from an industrial recycling process of exhausted batteries and in contaminated soils, which, to the best of our knowledge, was not performed in any of the reported previous studies.

2. Materials and Methods

2.1. Chemicals

Tetrachloroauric acid ($HAuCl_4$), silver nitrate ($AgNO_3$), sulfuric acid 95–97% (H_2SO_4), sodium nitrate ($NaNO_3$), potassium ferrocyanide $K_4[Fe(CN)_6]$, potassium ferricyanide $K_3[Fe(CN)_6]$, potassium chloride (KCl), 6,6-dibenzyl-1,4,8-11-tetraoxacyclotetradecane (lithium ionophore VI), potassium tetrakis (4-chlorophenyl) borate (K-TCPB), 2-nitrophenyl octyl ether (NPOE), polyvinyl chloride (PVC) high molecular weight, tetrahydrofuran (THF), lithium chloride (LiCl), magnesium chloride ($MgCl_2$), ammonium chloride (NH_4Cl), sodium chloride (NaCl), and calcium chloride ($CaCl_2$) were purchased from Merck (Darmstadt, Germany). All solutions were prepared using MilliQ water (obtained from Milli-Q Water Purification System, Millipore, UK).

2.2. Instrumentation

Screen-printed electrochemical cells (SPCs) based on a graphite-working electrode, a graphite-counter electrode, and a silver pseudo-reference electrode (EcoBioServices srl, Sesto Fiorentino, Italy) were used for the electrochemical experiments. The open circuit potential (OCP) measurements were carried out at room temperature using Sensit Smart portable potentiostat/galvanostat (PalmSens BV, Houten, The Netherlands) connected to an Android© smartphone equipped with PSTouch software for data acquisition and elaboration. Potentiometric measurements were also performed with a commercial lithium ion-selective electrode (DX207-Li ISE half-cell electrode, ref. 51107673) and an Ag/AgCl reference electrode (pH electrode InLab reference, ref. 51343190) purchased from Mettler-Toledo S.p.A. (Milan, Italy). The peristaltic pump needed for flow measurements was purchased from Gilson Inc., USA. Dropsens wall-jet microflow cell (ref. DRP-FLWCL) was purchased from Metrohm (Milan, Italy). Scanning electron microscopy (SEM) analysis was carried out using a Gaia 3 microscope (Tescan a.s., Brno, Czech Republic). SEM images were acquired using an acceleration voltage of 5 kV for the bare-graphite electrodes and 20 kV for the nanostructured electrodes. Energy dispersive X-ray analysis (EDX) was performed to assess the elemental composition of the modified electrode surface.

2.3. Preparation of (Au&Ag)NPs-Modified Graphite Screen-Printed Electrodes

The surface of the graphite screen-printed electrode (GSPE) was modified with silver and gold bimetallic nanoparticles ((Au&Ag)NPs) using cyclic voltammetry (CV) to produce an enhancement of the electrochemical behavior of the GSPE. Firstly, 2.5 mM of $AgNO_3$ in a 0.1 M $NaNO_3$ solution was dropped onto the GSPE surface. The potential was cycled from −1.0 V to +0.5 V vs. Ag/AgCl pseudo-reference electrode for 10 times at a 50 mV/s scan rate. Successively, the AgNPs-modified GSPE was washed for three times with deionized water to remove the excess of free ions from the surface. A solution of 2.5 mM $HAuCl_4$ in 0.5 M H_2SO_4 was dropped onto the AgNPs-modified graphite electrode. The potential was cycled from −0.2 V to +1.6 V vs. Ag/AgCl pseudo-reference electrode for 10 times at a 100 mV/s scan rate. The (Au&Ag)NPs/GSPE was then washed with deionized water for three times.

2.4. Study of the Electrochemical Performance of (Au&Ag)NPs-Modified Graphite Screen-Printed Electrodes

To gain insights into the electrochemical enhancement produced by the nanostructuration of the electrode surface, CV measurements were performed with the (Au&Ag)NPs/GSPE by dropping 50 μL of 5 mM $[Fe(CN)_6]^{-4/-3}$ as a redox-probe (equimolar mixture prepared in 0.1 M KCl) solution.

CV measurements were carried out in a range of potentials between −0.5 V and +1.0 V at different scan rates (25, 50, 100, 125, and 150 mV/s). The current peak height was taken as the electrochemical signal and plotted vs. the square root of the scan rate (V/s), and a linear behavior was retrieved. Then, the electroactive area was calculated from the angular coefficient of the obtained linear regression with the Randles–Sevcik equation [32]:

$$i_p = 0.446nFAC^0\left(\frac{nFvD_o}{RT}\right)^{1/2}, \quad (1)$$

where n is the number of electrons transferred in the redox events, F is the Faraday constant (C/mol), A (cm^2) is the electrode surface area, R is the gas constant (8.314 J/(mol·K)), D_o (cm^2/s) is the diffusion coefficient of the oxidized analyte, and C^0 (mol/cm^3) is the analyte bulk concentration. After each CV measurement, the SPCs were discarded.

SEM analysis was performed to assess the morphological characterization of the nanostructured GSPE surface. In particular, the morphology of the graphite electrode was investigated before and after the electrochemical nanostructuration of the substrate material.

2.5. Preparation of Solid-Contact Lithium Sensor

The lithium-selective membrane was realized using a cocktail consisting of 28% (*w*/*w*) PVC, 1% (*w*/*w*) lithium ionophore VI, 0.2% (*w*/*w*) K-TCPB, and 70.8% (*w*/*w*) NPOE for 100 mg of mixture dissolved in 1 mL of THF. The mixture was left to stir for 1 h with a magnetic stirrer. After this, 5 μL of this solution was drop-casted onto the (Au&Ag)NPs/GSPE. After completing the drop-casting procedure, the sensor was left to air-dry overnight in dark conditions to allow for the complete evaporation of the solvent.

2.6. Lithium Detection

The potentiometric cell was integrated into a smart setup where the portable Sensit Smart potentiostat was connected to a smartphone for OCP measurements, which were performed both in drop and in microflow configurations, without any conditioning of the sensor, for a total running time of 100 s each to assure a stable OCP response. Drop measurements were carried out by depositing 50 μL of standard solution onto the SPC, while flow measurements were performed using the wall-jet microflow cell with a flow rate of 0.2 mL/min. The response time was evaluated as the time required for the OCP to reach 90% of the maximum potential value [33]. The standard solutions were analyzed in the range of 10^{-5}–10^{-1} M using a 0.05 M $MgCl_2$ solution as the supporting electrolyte.

The schematic representation of lithium detection with the described microflow configuration is reported in Figure 1.

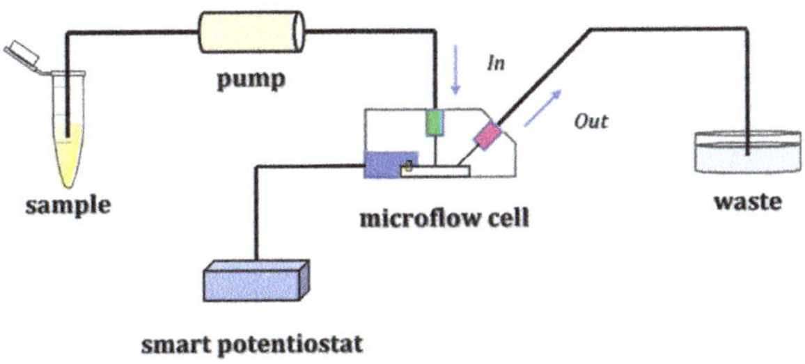

Figure 1. Experimental setup for lithium detection using a microflow configuration.

After each set of measurements, the sensor was washed with a 0.05 M $MgCl_2$ solution to keep the same experimental conditions for the subsequent measurements.

To obtain a calibration curve, OCP values were plotted vs. the logarithm of lithium concentration, and the experimental points were fitted with a linear regression. The detection limit was calculated according to Bakker & Pretsch [34]. Briefly, the potentiometric response is a linear function of the logarithm of the activity of lithium in solution, and its slope is described by the Nernst equation. Below the LOD, it has a constant value, which is ideally defined by the response of the sensor to another ion (i.e., the supporting electrolyte). The potentiometric LOD is then defined as the cross-section of the two linear parts of the response function.

The sensor was stored in dry and dark conditions between day-to-day measurements. In this way, the sensor could be used for at least ten calibrations with good reproducibility, as the relative standard deviation (RSD) was lower than 10%.

2.7. Fixed Interference Method (FIM)

The selectivity of the developed membrane was assessed by evaluating its selectivity coefficient toward potential interfering ions, such as potassium, sodium, ammonium, and calcium ions. OCP measurements were performed in solutions at a constant activity of the interfering ion (i.e., K^+, Na^+, $NH_4{}^+$, Ca^{+2}), a_B, and varying activity of the primary ion (i.e., Li^+), a_A. The measured potential values are plotted vs. the logarithm of the activity of a Li^+ ion. The intersection of the extrapolated linear portions of this plot indicates the value of a_A that is to be used to calculate $K^{pot}{}_{A,B}$ from the following equation:

$$K_{A,B}^{pot} = \frac{a_A}{a_B{}^{\frac{z_A}{z_B}}}, \tag{2}$$

where both z_A and z_B (i.e., the charge of the ions) have the same sign [35].

2.8. Real Samples Analysis

Preliminary experiments to assess the determination of lithium ions in real samples by the developed sensor were also performed. The sensor's response was determined by OCP measurements.

Samples deriving from an industrial process of recycling automotive batteries were diluted at a proper ratio (from 1:5 up to 1:200) in 0.05 M $MgCl_2$ before the analysis with the standard addition method in drop and microflow configurations. Lithium concentration in the samples was then calculated by applying the following equation:

$$C_{sample} = \frac{C_{std}V_{std}}{10^{\frac{\Delta E}{S}}\left(V_{sample} + V_{std}\right) - V_{sample}}, \tag{3}$$

where C_{sample} and V_{sample} are the concentration and the volume of the sample, C_{std} and V_{std} are the concentration (1 M) and the volume of the added lithium standard solution, respectively, S is the slope of the calibration curve, and ΔE is the potential variation due to the addition of the standard.

The application of the developed potentiometric platform for the analysis of lithium in complex matrices such as contaminated soils was also demonstrated. Soil samples (~1 g) were mixed with H_2O_2 and *aqua regia* (HCl/HNO_3 3:1 *v/v*, ~18 g) and then loaded in a microwave oven. Initially, the samples were irradiated for 20 min, to raise their temperature from 20 °C to 200 °C. A maximum power of 1600 W then maintained the temperature constant at 200 °C for at least 3 h. The resulting suspension was then filtered and diluted as required in the supporting electrolyte, and the obtained solution was used for lithium-standards preparation. The results were then compared with those obtained by a commercial lithium ISE.

3. Results and Discussion

3.1. (Au&Ag)NPs Electrodeposition and CV Characterization

The electrochemical performance of graphite screen-printed electrodes (GSPEs) in terms of conductivity and electroactive surface area was enhanced by the electrodeposition of silver and gold nanoparticles ((Au&Ag)NPs) using cyclic voltammetry. Figure 2 shows the typical cyclic voltammograms recorded during the electrodeposition of AgNPs (a) and AuNPs (b), respectively. The deposition scans show that the current of all peaks appearing in the voltammograms increases with an increase in the number of cycles up to the 10th cycle. No significant difference was observed in the peak current if the number of cycles was increased from 10 to 20 in both cases. Therefore, 10 cycles were used in the following experiments for AgNPs and AuNPs electrodeposition.

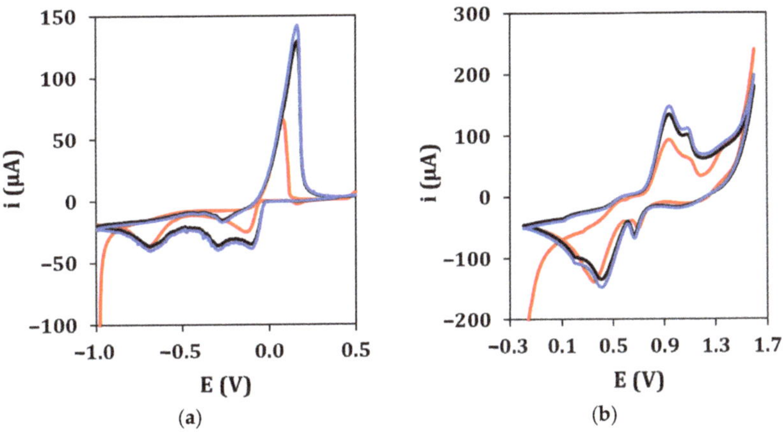

Figure 2. (**a**) AgNPs electrodeposition from a 2.5 mM $AgNO_3$ solution in 0.1 M $NaNO_3$ (•: 1st cycle, •: 10th cycle, •: 20th cycle). Experimental conditions: potential range from −1.0 V to +0.5 V, 50 mV/s scan rate. (**b**) AuNPs electrodeposition from a 2.5 mM $HAuCl_4$ solution in 0.5 M H_2SO_4 (•: 1st cycle, •: 10th cycle, •: 20th cycle). Experimental conditions: potential range from −0.2 V to +1.6 V, 100 mV/s scan rate.

After that, the electrochemical properties of the unmodified GSPEs and nanostructured GSPEs were characterized using the CV technique in presence of a 5 mM ($[Fe(CN)_6]^{-4/-3}$ redox probe (an equimolar solution in 0.1 M KCl) to assess the aforementioned enhancement of the electrochemical performance of the nanostructured electrodes. The electroactive area of the developed platform was calculated by applying the Randles–Sevcik equation to the angular coefficient of the linear regression obtained by plotting the anodic (i_{pa}) and cathodic (i_{pc}) current peak heights vs. the square root of the scan rate (Figure 3) for the bare (a) and nanoparticle-modified (b) electrodes.

The electroactive area values obtained for both platforms are reported in Table 2.

From the obtained results, the electrodeposition of, first, silver and then gold nanoparticles causes an increase in the conductivity of the electrode surface, as noble metals were deposited onto the screen-printed graphite, and in its electroactive area, as the anodic and cathodic peak currents increase when passing from the bare GSPE to the nanostructured (Au&Ag)NPs/GSPE. This is also confirmed by the fact that the electroactive area values of the nanoparticle-modified electrodes are higher when compared to that of the bare graphite, even after taking into consideration their geometric area (7.1 mm^2). From the point of view of potentiometric detection, the electrodeposited metal layer works as the solid contact between the membrane and the substrate electrode, assuring stable potential readings due to reversible charge transfer between the ionically conducting membrane and the electronically conducting support [36,37]. Moreover, an increase in the electroactive area could be helpful in the first place in anchoring the ion-selective membrane to the sensing surface, as the nanostructuration brought on by the electrodeposition of the metal nanoparticles could produce defects while anchoring points.

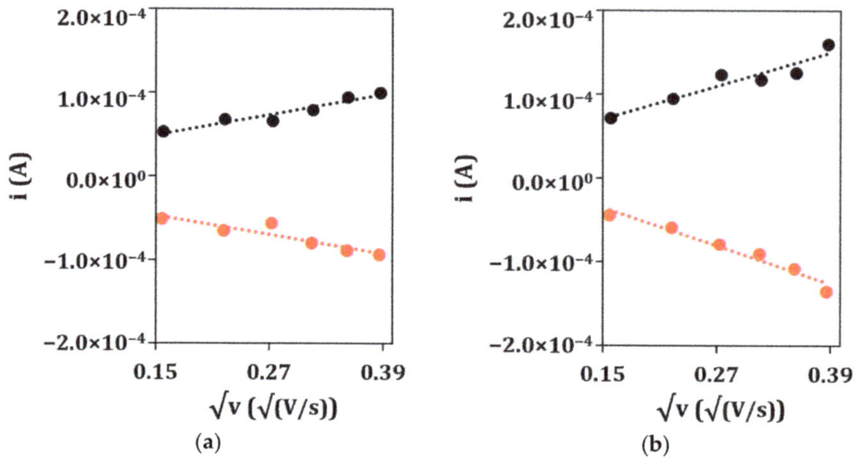

Figure 3. Plots of anodic (•) and cathodic (•) peak currents vs. the square root of the scan rate for (**a**) bare GSPE and (**b**) (Au&Ag)NPs/GSPE.

Table 2. Electroactive-area values of unmodified GSPEs and (Au&Ag)NPs/GSPEs, as calculated from the cyclic voltammograms performed in $[Fe(CN)_6]^{-4/-3}$ redox probe.

	A_{anodic} (mm^2)	$A_{cathodic}$ (mm^2)	$\bar{A}$ (mm^2)
GSPE	5.5	5.1	5.3 ± 0.3
(Au&Ag)NPs/GSPE	9.0	10.3	9.6 ± 0.9

3.2. Morphological Characterization of the Nanostructured Surfaces

As shown in Figure 4, the morphology of GSPEs was investigated before (a) and after (b) the electrochemical nanostructuration of the substrate material with (Au&Ag)NPs.

The presence of (Au&Ag)NPs with dimensions smaller than 100 nm, which are randomly distributed on the entire surface of the graphite screen-printed electrode, is visible. To assess the effective modification of the graphite surface, an EDX analysis of the nanostructured GSPEs was also carried out (Figure 4c). Three characteristic gold bands centered at 2.2, 9.7, and 11.5 keV, and two silver bands around 3.0 keV were found when the (Au&Ag)NPs-modified screen-printed electrodes were investigated, confirming the presence of gold and silver nanoparticles on the sensor surface.

3.3. Optimization of the Amount of Lithium Selective Membrane

The homogeneous coverage of the electrode surface with the ion-selective membrane is an important prerequisite to achieve an optimal sensor response. To identify the correct volume of the membrane cocktail to cast on the electrode surface, the responses of a batch of sensors towards lithium detection, each one obtained after having dropped a different volume of membrane solution, were evaluated. The linear regressions obtained for different amounts of the membrane cocktail are reported in Table 3.

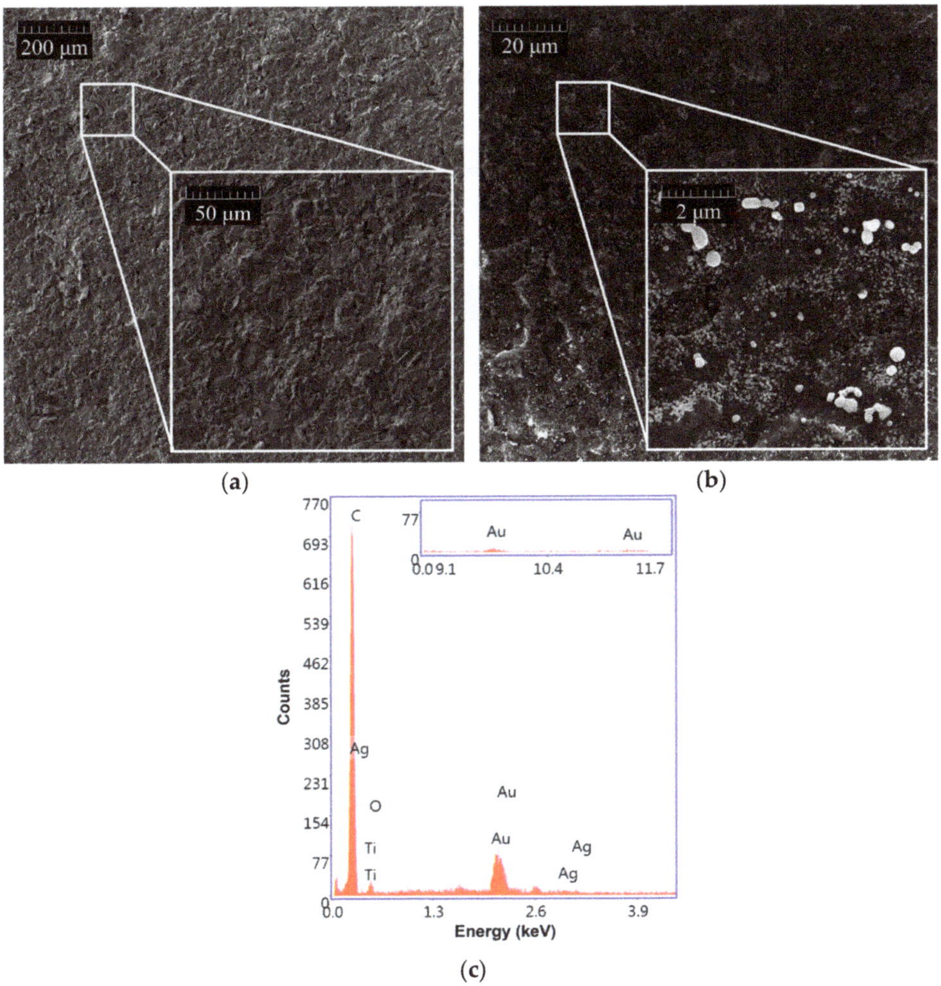

Figure 4. SEM morphologies of (**a**) bare GSPE and (**b**) (Au&Ag)NPs-modified GSPE. (**c**) EDX spectrum of (Au&Ag)NPs/GSPE.

Table 3. Linear regressions obtained for lithium detection using sensors covered with different volumes of the ion-selective membrane cocktail (n = 5).

Volume (µL)	Slope (mV/Decade)
3	47 ± 4
5	58 ± 2
7	46 ± 1

When the electrode surface was covered with 3 µL or 7 µL of the membrane solution, the sensor showed a non-Nernstian behavior. In addition, when 3 µL were used, the repeatability of the measurements among different electrodes was also poor, because the volume was not enough for a homogenous coverage of the electrode surface and for providing sufficient binding sites for lithium detection. When 7 µL were used, the response was slower due to the higher thickness, and lithium ions bound to the outer binding sites are likely too far from the sensing surface and cannot have any more influence on the signal itself. On the other hand, when the (Au&Ag)NPs/GSPE surface was covered with 5 µL of membrane solution, the sensor showed a Nernstian behavior and a good repeatability of the measurements. Therefore, the volume of 5 μL of membrane cocktail solution was selected for further experiments.

3.4. Optimization of the Supporting Electrolyte Concentration

To obtain a stable response of the sensor for lithium detection, a $MgCl_2$ solution was used as the supporting electrolyte. The supporting electrolyte concentration was optimized by analyzing lithium standard solutions containing different concentrations of $MgCl_2$ (0.01, 0.05, 0.1, 0.5 M).

The obtained calibration curves are shown in Figure 5.

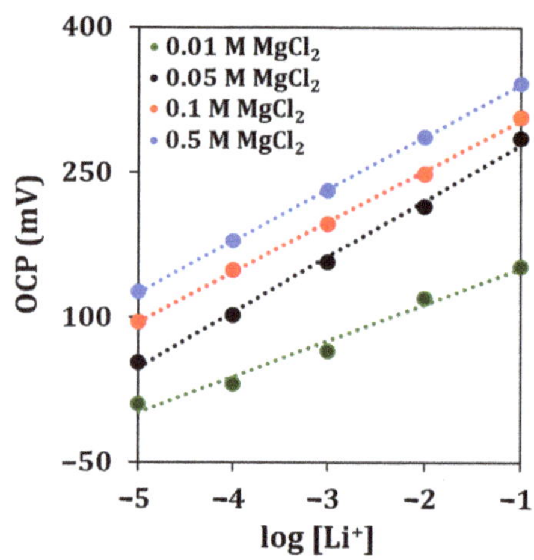

Figure 5. Lithium calibration curves obtained with different concentrations of the supporting electrolyte solution by OCP measurements (n = 5).

The linear regressions obtained for different concentrations of the supporting electrolyte ($MgCl_2$) are reported in Table 4.

Table 4. Linear regressions obtained for lithium detection in standard solutions containing different concentrations of the supporting electrolyte (n = 5).

[$MgCl_2$] (M)	Slope (mV/Decade)
0.01	37 ± 13
0.05	56 ± 2
0.1	52 ± 1
0.5	54 ± 1

The highest sensitivity for lithium detection, almost approaching a Nernstian behavior, was retrieved when a concentration of 0.05 M of $MgCl_2$ was used. Therefore, this concentration of supporting electrolyte solution was selected for all the subsequent experiments.

3.5. Lithium Detection in-Drop Configuration

Under optimized experimental conditions, lithium in-drop detection was performed by OCP measurements. The response time and stability of the signals were evaluated by recording the potential vs. the time during the analysis of solutions at different concentrations of lithium for a total running time of 100 s (Figure 6a). The OCP values measured in correspondence of a fixed time of 90 s for each of the analyzed lithium concentrations were plotted vs. the logarithm of lithium concentration, and a linear regression was used to fit the experimental points (Figure 6b).

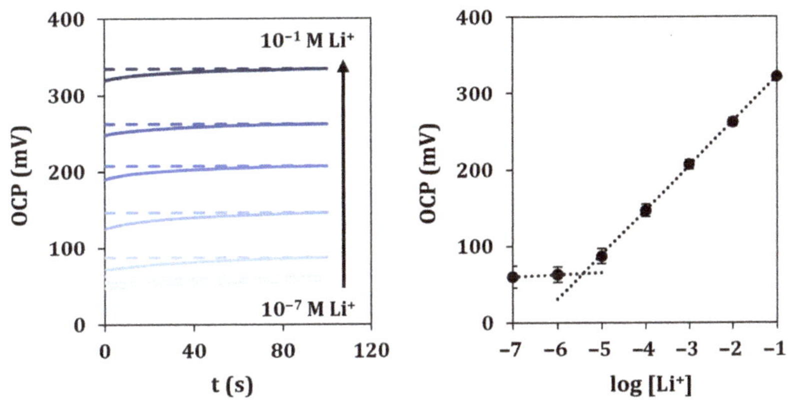

(a) (b)

Figure 6. In-drop lithium detection in standard solutions. (**a**) Recorded OCP signals vs. time for a total duration of 100 s, and (**b**) corresponding calibration curve. Each measurement was repeated at least 5 times using the same sensor.

It can be observed that the average response time for all the standard solutions is 45 s, calculated as reported in Section 2.6. A linear response was retrieved in the range 10^{-5}–10^{-1} M, described by the equation y = (58 ± 3)x + (380 ± 7), with a good correlation coefficient (R^2 = 0.9998). The limit of detection (LOD) was calculated as described in Section 2.6 and found equal to 1.6 μM. The repeatability of the sensor response was evaluated by multiple measurements of the calibration curve (n = 5), obtaining %RSD values lower than 4%. The same ion-selective cell was tested for a period of 48 h, and it showed a long lifetime. A loss of performance could be observed only after 36 h due to different events, such as the detachment of the membrane. Moreover, the sensor showed high working stability, as observed by the low potential drift (3 mV/h) obtained after 5 h of continuous OCP response when measuring a 0.1 M lithium solution. Lastly, the storage stability of the sensor was outstanding, as its analytical performance in terms of sensitivity remained almost constant for at least 1 month after preparation in dry and dark storage conditions at room temperature.

3.6. Interference Study

The most important characteristic of an ion-selective membrane is its response toward the ion of interest in the presence of other potentially interfering ions, which is measured in terms of the potentiometric selectivity coefficient ($K^{pot}{}_{A,B}$). Lithium calibration curves were carried out in presence of 50 mM of the interfering ion. The calculated values of log($K^{pot}{}_{A,B}$), one for each interfering cation, are shown in Table 5.

Table 5. The selectivity coefficients (log $K^{pot}{}_{A,B}$) of lithium sensors for different interfering ions obtained by fixed-interference method.

Interferent Ion	Log($K^{pot}{}_{A,B}$)
K^+	−1.89 ± 0.72
Na^+	−2.45 ± 0.51
$NH_4{}^+$	−2.55 ± 1.02
Ca^{+2}	−4.82 ± 0.94

As it could be expected, the sensor showed the highest selectivity towards Ca^{+2}, while the lowest coefficient values were those obtained for the smallest ions, namely, K^+ and Na^+, which have similar dimensions to Li^+ ions, thus, resulting in a stronger interference.

3.7. Lithium Detection in Microflow Configuration

With the aim of developing a smart setup to be applied in environmental analysis for the continuous monitoring of lithium, the presented sensor was integrated into a microflow configuration comprising a peristaltic pump and a wall-jet flow cell. Standard solutions were pumped at different flow rates to evaluate the sensor response toward lithium detection.

The response time changed significantly depending on the flow rate. To assess the most suitable one, different flow rates were tested, and the response time was calculated as reported in Section 2.6. The obtained results are reported in Table 6.

Table 6. Response time in respect of the flow rate for lithium detection ($[Li^+]$ = 0.01 M) with the microflow setup.

Flow Rate (mL/min)	Response Time (s)
0.10	600
0.15	500
0.20	50
0.40	300

The optimal flow rate was 0.2 mL/min, as in this condition the sensor showed the fastest response time. The use of different flow rates resulted in higher response time values of about 10 min when using a flow rate of 0.10 mL/min and 5 min when using a flow rate of 0.40 mL/min. This was probably due to the fact that a higher flow rate does not allow the establishment of a proper binding equilibrium between lithium ions and the ion-selective membrane, thus, the sensor requires a longer time to detect lithium. On the other hand, a lower flow rate is likely insufficient for providing an efficient binding of lithium ions considering that, in the microflow analysis, the volume of lithium standard solutions which instantly comes into contact with the sensing surface is much smaller with respect to that used in the drop configuration.

A typical recording of the OCP with the microflow setup in the concentration range of 10^{-5}–10^{-1} M of lithium ions, together with the relative calibration curve, is shown in Figure 7.

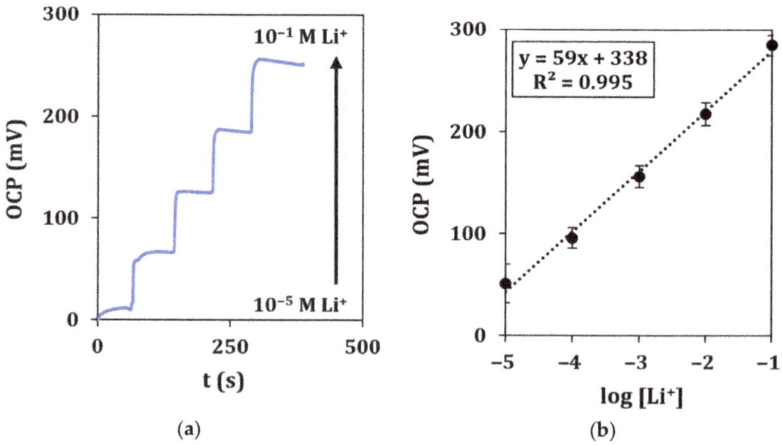

Figure 7. Microflow lithium detection in standard solutions. (**a**) Recorded OCP signals vs. time for a total duration of 500 s at a flow rate of 0.2 mL/min and (**b**) corresponding calibration curve. Each measurement was repeated at least 5 times using the same sensor.

Calibration curves show a similar behavior with respect to drop measurement, as described by the equation y = (59 ± 5)x + (338 ± 10), with a LOD of 2.0 μM.

3.8. Sensor Application in Environmental Samples Analysis

Having verified the suitability of the sensor for detecting lithium in standard solutions, preliminary experiments on real industrial samples were then carried out.

Lithium was detected in several samples derived from an industrial process of recycling automotive batteries. The determination of the analyte was carried out by the standard addition method on the samples diluted to a proper ratio in the supporting electrolyte and by performing OCP measurements in drop and microflow configurations under the aforementioned optimized conditions. The results were compared with those obtained from the commercial ISE using batch measurements (Table 7).

Table 7. Lithium concentrations determined in industrial samples from recycling automotive batteries. Each measurement was repeated at least 3 times using the same sensor. The samples were tested with the developed sensor through both drop measurement and the microflow setup; the results were compared with those obtained in batch with the commercial ISE.

Samples	Li^+ (mg/L)		
	Developed Sensor		Commercial ISE
	Drop	Microflow	Batch
1	128 ± 6	134 ± 12	120 ± 2
2	45 ± 9	60 ± 10	55 ± 1
3	88 ± 4	91 ± 8	87 ± 2
4	290 ± 6	302 ± 13	283 ± 9

As observed, the results obtained with the proposed method, both in drop and microflow configurations, partially overlap with those deriving from the commercial ISE analysis, demonstrating that the two methods have the same precision.

To reinforce the established applicability of the proposed method, lithium was also detected in a complex matrix obtained from soils, which could mimic those derived from grounds accidentally polluted by the byproducts of recycling processes. Given the complexity of the solution obtained from soil digestion, the dilution of the matrix was preliminarily optimized by performing potentiometric measurements with the commercial ISE (Table 8).

Table 8. Lithium detection in contaminated soils at different dilutions. Each measurement was repeated at least 5 times with the commercial ISE.

Dilution Ratio	Slope (mV/Decade)
1:10	45 ± 6
1:20	48 ± 5
1:50	52 ± 3
1:100	55 ± 3

As observed, the optimal dilution ratio was 1:100 since the slope of the corresponding calibration curve has the highest value. Thus, this dilution ratio was used for all the following experiments.

As soil usually contains an elevated amount of interfering ions such as sodium, potassium, and calcium [38], together with other potentially interfering species, the concentration of the supporting electrolyte was re-optimized. An imbalance in the ionic strength (with respect to that of standard solutions) could affect the stability, the reproducibility, and the sensitivity of the measurement. The resulting calibration curve is shown in Figure 8.

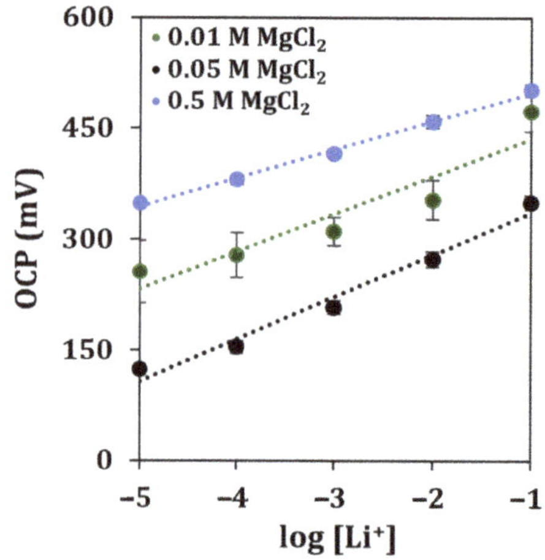

Figure 8. Lithium detection in contaminated soils diluted at a 1:100 ratio in different concentrations of the supporting electrolyte. Each measurement was repeated at least 5 times using the same sensor.

It can be observed that the highest sensitivity was obtained using a solution of $MgCl_2$ at a concentration of 0.05 M, while the highest R^2 score was retrieved when the supporting electrolyte was 0.5 M. To evaluate the operability of the developed sensor once more, the soil sample was contaminated by spiking it with one of the byproducts of the industrial process of recycling batteries, obtaining a recovery of 81%.

Taking into account the good agreement between the results obtained with the proposed method and those deriving from the commercial ISE, it can be asserted that the developed sensor could be applied for lithium monitoring in industrial waste samples and contaminated soils.

4. Conclusions

In this paper, we developed a potentiometric smart sensor for lithium detection based on bimetallic nanoparticle-modified graphite screen-printed electrodes. Gold and silver nanoparticles were quickly deposited using cyclic voltammetry; the obtained platform was electrochemically and morphologically characterized, and the results were compared with the bare-graphite electrode. Increased conductivity and improved surface area were retrieved after the deposition of (Au&Ag)NPs. An appropriate amount of an optimized ion-selective membrane cocktail was then deposited directly onto the surface of the nanostructured graphite electrodes. Lithium was detected in standard solutions over a wide range (10^{-5}–10^{-1} M) with low detection limits of 1.6 μM and 2.0 μM, both in-drop and with a microflow configuration, respectively. The selectivity towards lithium detection in the presence of interfering ions was also evaluated. The absence of the need for conditioning time for the developed sensor has many advantages for its application as a ready-to-use and disposable device for real-sample analysis. Real samples from an industrial process of recycling automotive batteries and a complex matrix deriving from contaminated soils were analyzed, and the results were in accordance with those provided by a commercial lithium selective electrode. The applicability of the proposed method was then reinforced, as lithium was also detected in a complex matrix obtained from contaminated soils. This kind of analysis highlights the importance of the use of screen-printed modified electrodes to monitor all the steps of the process and to assure a complete recovery of the battery materials.

Author Contributions: Conceptualization, I.A.V., G.S. and G.M.; methodology, I.A.V. and G.M.; investigation, I.A.V.; data curation, I.A.V. and G.S.; writing—original draft preparation, I.A.V.; writing—review and editing, I.A.V., G.S. and G.M.; supervision, G.M.; project administration, G.M.; funding acquisition, G.M. All authors have read and agreed to the published version of the manuscript.

Funding: This research was funded by Regione Toscana (grant number 3553.04032020.158000224) within the project RI.B.AT (Riciclo Integrato Batterie AutoTrazione) in the framework of POR FESR 2014/2020.

Institutional Review Board Statement: Not applicable.

Informed Consent Statement: Not applicable.

Data Availability Statement: No new data were created or analyzed in this study. Data sharing is not applicable to this article.

Conflicts of Interest: The authors declare no conflict of interest.

References

1. Kennedy, B.; Patterson, D.; Camilleri, S. Use of lithium-ion batteries in electric vehicles. *J. Power Sources* **2000**, *90*, 156–162. [CrossRef]
2. Velázquez-Martínez, O.; Valio, J.; Santasalo-Aarnio, A.; Reuter, M.; Serna-Guerrero, R. A Critical Review of Lithium-Ion Battery Recycling Processes from a Circular Economy Perspective. *Batteries* **2019**, *5*, 68. [CrossRef]
3. Islam, M.T.; Iyer-Raniga, U. Lithium-Ion Battery Recycling in the Circular Economy: A Review. *Recycling* **2022**, *7*, 33. [CrossRef]
4. Ambrose, H.; Kendall, A. Understanding the future of lithium: Part 1, resource model. *J. Ind. Ecol.* **2020**, *24*, 80–89. [CrossRef]
5. Aral, H.; Vecchio-Sadus, A. Lithium: Environmental Pollution and Health Effects. In *Encyclopedia of Environmental Health*; Nriagu, J.O., Ed.; Elsevier B.V.: Amsterdam, The Netherlands, 2011; pp. 499–508. ISBN 978-0-444-52272-6.
6. Wanger, T.C. The Lithium future-resources, recycling, and the environment. *Conserv. Lett.* **2011**, *4*, 202–206. [CrossRef]
7. Mossali, E.; Picone, N.; Gentilini, L.; Rodrìguez, O.; Pérez, J.M.; Colledani, M. Lithium-ion batteries towards circular economy: A literature review of opportunities and issues of recycling treatments. *J. Environ. Manag.* **2020**, *264*, 110500–110511. [CrossRef]
8. Ferrara, C.; Ruffo, R.; Quartarone, E.; Mustarelli, P. Circular Economy and the Fate of Lithium Batteries: Second Life and Recycling. *Adv. Energy Sustain. Res.* **2021**, *2*, 2100047–2100059. [CrossRef]
9. Tavakolian-Ardakani, Z.; Hosu, O.; Cristea, C.; Mazloum-Ardakani, M.; Marrazza, G. Latest Trends in Electrochemical Sensors for Neurotransmitters: A Review. *Sensors* **2019**, *19*, 2037. [CrossRef]
10. Suherman, A.L.; Rasche, B.; Godlewska, B.; Nicholas, P.; Herlihy, S.; Caiger, N.; Cowen, P.J.; Compton, R.G. Electrochemical Detection and Quantification of Lithium Ions in Authentic Human Saliva Using $LiMn_2O_4$-Modified Electrodes. *ACS Sens.* **2019**, *4*, 2497–2506. [CrossRef]
11. Freitas, B.H.; Amaral, F.A.; Bocchi, N.; Teixeira, M.F.S. Study of the potentiometric response of the doped spinel $Li_{1.05}Al_{0.02}Mn_{1.98}O_4$ for the optimization of a selective lithium ion sensor. *Electrochim. Acta* **2010**, *55*, 5659–5664. [CrossRef]
12. Nel-lo, M.; Ferrer, Ò.; Colominas, S.; Abellà, J. Lithium sensors based on $Li_6La_3Ta_{1.5}Y_{0.5}O_{12}$ and $Li_6BaLa_2Ta_2O_{12}$ garnet electrolytes for molten lead alloys. *Sens. Actuators B Chem.* **2021**, *339*, 129831–129837. [CrossRef]
13. Paczosa-Bator, B. All-solid-state selective electrodes using carbon black. *Talanta* **2012**, *93*, 424–427. [CrossRef]
14. Mazzaracchio, V.; Serani, A.; Fiore, L.; Moscone, D.; Arduini, F. All-solid state ion-selective carbon black-modified printed electrode for sodium detection in sweat. *Electrochim. Acta* **2021**, *394*, 139050–139059. [CrossRef]
15. Ermolenko, Y.; Yoshinobu, T.; Mourzina, Y.; Furuichi, K.; Levichev, S.; Vlasov, Y.; Schöning, M.J.; Iwasaki, H. Lithium sensor based on the laser scanning semiconductor transducer. *Anal. Chim. Acta* **2002**, *459*, 1–9. [CrossRef]
16. Gupta, V.K.; Chandra, S.; Agarwal, S.; Lang, H. Lithium-selective potentiometric sensor based on a second generation carbosiloxane dendrimer. *Sens. Actuators B Chem.* **2005**, *107*, 762–767. [CrossRef]
17. Coldur, F.; Andac, M. All-Solid-State Polyvinyl Chloride Membrane Lithium-Selective Electrode with Improved Selectivity and Its Application in Serum Lithium Assay. *Sens. Lett.* **2011**, *9*, 1738–1744. [CrossRef]
18. Coldur, F.; Andac, M. A Flow-Injection Potentiometric System for Selective and Sensitive Determination of Serum Lithium Level. *Electroanalysis* **2013**, *25*, 732–740. [CrossRef]
19. Kamenica, M.; Kothur, R.; Willows, A.; Patel, B.; Cragg, P. Lithium Ion Sensors. *Sensors* **2017**, *17*, 2430. [CrossRef]
20. Li, M.; Li, Y.-T.; Li, D.-W.; Long, Y.-T. Recent developments and applications of screen-printed electrodes in environmental assays—A review. *Anal. Chim. Acta* **2012**, *734*, 31–44. [CrossRef]
21. Cinti, S.; Arduini, F.; Carbone, M.; Sansone, L.; Cacciotti, I.; Moscone, D.; Palleschi, G. Screen-Printed Electrodes Modified with Carbon Nanomaterials: A Comparison among Carbon Black, Carbon Nanotubes and Graphene. *Electroanalysis* **2015**, *27*, 2230–2238. [CrossRef]

22. Waheed, A.; Mansha, M.; Ullah, N. Nanomaterials-based electrochemical detection of heavy metals in water: Current status, challenges and future direction. *TrAC Trends Anal. Chem.* **2018**, *105*, 37–51. [CrossRef]
23. Squissato, A.L.; Munoz, R.A.A.; Banks, C.E.; Richter, E.M. An Overview of Recent Electroanalytical Applications Utilizing Screen-Printed Electrodes Within Flow Systems. *ChemElectroChem* **2020**, *7*, 2211–2221. [CrossRef]
24. Kanso, H.; González García, M.B.; Llano, L.F.; Ma, S.; Ludwig, R.; Fanjul Bolado, P.; Santos, D.H. Novel thin layer flow-cell screen-printed graphene electrode for enzymatic sensors. *Biosens. Bioelectron.* **2017**, *93*, 298–304. [CrossRef] [PubMed]
25. Brownlee, B.J.; Bahari, M.; Harb, J.N.; Claussen, J.C.; Iverson, B.D. Electrochemical Glucose Sensors Enhanced by Methyl Viologen and Vertically Aligned Carbon Nanotube Channels. *ACS Appl. Mater. Interfaces* **2018**, *10*, 28351–28360. [CrossRef]
26. Henríquez, C.; Cerdà, V. Construction of a New Flow-through Cell for Screen Printed Electrodes. *Electroanalysis* **2020**, *32*, 1323–1328. [CrossRef]
27. Sarakhman, O.; Švorc, Ľ. A Review on Recent Advances in the Applications of Boron-Doped Diamond Electrochemical Sensors in Food Analysis. *Crit. Rev. Anal. Chem.* **2022**, *52*, 791–813. [CrossRef]
28. Marzouk, S.A.M.; Alyammahi, A.R.; Fanjul-Bolado, P. Development and Characterization of Novel Flow Injection, Thin-Layer, and Batch Cells for Electroanalytical Applications Using Screen-Printed Electrodes. *Anal. Chem.* **2021**, *93*, 16690–16699. [CrossRef]
29. Arvinte, A.; Crudu, I.-A.; Doroftei, F.; Timpu, D.; Pinteala, M. Electrochemical codeposition of silver-gold nanoparticles on CNT-based electrode and their performance in electrocatalysis of dopamine. *J. Electroanal. Chem.* **2018**, *829*, 184–193. [CrossRef]
30. Criscuolo, F.; Taurino, I.; Stradolini, F.; Carrara, S.; De Micheli, G. Highly-stable Li+ ion-selective electrodes based on noble metal nanostructured layers as solid-contacts. *Anal. Chim. Acta* **2018**, *1027*, 22–32. [CrossRef]
31. Criscuolo, F.; Taurino, I.; Carrara, S.; De Micheli, G. A novel electrochemical sensor for non-invasive monitoring of lithium levels in mood disorders. In Proceedings of the 2018 40th Annual International Conference of the IEEE Engineering in Medicine and Biology Society (EMBC), Honolulu, HI, USA, 18–21 July 2018; pp. 3825–3828.
32. Elgrishi, N.; Rountree, K.J.; McCarthy, B.D.; Rountree, E.S.; Eisenhart, T.T.; Dempsey, J.L. A Practical Beginner's Guide to Cyclic Voltammetry. *J. Chem. Educ.* **2018**, *95*, 197–206. [CrossRef]
33. Maccà, C. Response time of ion-selective electrodes. *Anal. Chim. Acta* **2004**, *512*, 183–190. [CrossRef]
34. Bakker, E.; Pretsch, E. Potentiometric sensors for trace-level analysis. *TrAC Trends Anal. Chem.* **2005**, *24*, 199–207. [CrossRef]
35. Umezawa, Y.; Bühlmann, P.; Umezawa, K.; Tohda, K.; Amemiya, S. Potentiometric Selectivity Coefficients of Ion-Selective Electrodes. Part I. Inorganic Cations (Technical Report). *Pure Appl. Chem.* **2000**, *72*, 1851–2082. [CrossRef]
36. Kałuża, D.; Michalska, A.; Maksymiuk, K. Solid-Contact Ion-Selective Electrodes Paving the Way for Improved Non-Zero Current Sensors: A Minireview. *ChemElectroChem* **2022**, *9*, e202100892. [CrossRef]
37. Lyu, Y.; Gan, S.; Bao, Y.; Zhong, L.; Xu, J.; Wang, W.; Liu, Z.; Ma, Y.; Yang, G.; Niu, L. Solid-Contact Ion-Selective Electrodes: Response Mechanisms, Transducer Materials and Wearable Sensors. *Membranes* **2020**, *10*, 128. [CrossRef]
38. Robinson, B.H.; Yalamanchali, R.; Reiser, R.; Dickinson, N.M. Lithium as an emerging environmental contaminant: Mobility in the soil-plant system. *Chemosphere* **2018**, *197*, 1–6. [CrossRef]

chemosensors

Article

Bulk Modification of Sonogel–Carbon with Polyaniline: A Suitable Redox Mediator for Chlorophenols Detection

Paloma Calatayud-Macías, David López-Iglesias, Alfonso Sierra-Padilla, Laura Cubillana-Aguilera *, José María Palacios-Santander * and Juan José García-Guzmán

Department of Analytical Chemistry, Institute of Research on Electron Microscopy and Materials (IMEYMAT), Faculty of Sciences, Campus de Excelencia Internacional del Mar (CEIMAR), University of Cadiz, Campus Universitario de Puerto Real, Polígono del Río San Pedro S/N, 11510 Puerto Real, Spain
* Correspondence: laura.cubillana@uca.es (L.C.-A.); josem.palacios@uca.es (J.M.P.-S.)

Abstract: In this work, the development and characterization of a new ceramic material modified with polyaniline powder obtained by a high-power ultrasound sol-gel route is presented. A preliminary screening of the conducting polymer electroactivity was performed by means of cyclic voltammetry in free analyte solution. Remarkable figures of merit for 4-chloro-3-methylphenol (PCMC) determination, selected as the model organic analyte, was obtained with the developed material: the sensitivity and the limit of detection were 2.40 $\mu A/\mu M \cdot cm^2$ and 0.69 μM, respectively. The developed device was also successfully applied in the electrochemical determination of PCMC in water samples collected from different sources, obtaining recovery values ranging from 92% to 105%. The electrochemical performance of the device for the detection of other chlorophenols of interest was better in comparison with the bare electrode in all cases, due to the presence of the bulk modifier in the material. Therefore, the electrode material can be promoted for electrochemical assays of different chlorophenols in buffer and real water media for environmental monitoring.

Keywords: Sonogel–Carbon; polyaniline; bulk modifier; 4-chloro-3-methylphenol; spiked water

Citation: Calatayud-Macías, P.; López-Iglesias, D.; Sierra-Padilla, A.; Cubillana-Aguilera, L.; Palacios-Santander, J.M.; García-Guzmán, J.J. Bulk Modification of Sonogel–Carbon with Polyaniline: A Suitable Redox Mediator for Chlorophenols Detection. *Chemosensors* **2023**, *11*, 63. https://doi.org/10.3390/chemosensors11010063

Academic Editors: Iulia Gabriela David and Dana Elena Popa

Received: 7 December 2022
Revised: 10 January 2023
Accepted: 12 January 2023
Published: 13 January 2023

1. Introduction

Presently, conducting polymers are widely investigated due to their high electrical conductivity and tailored nanostructures. Particularly, polyaniline (PANI) emerged as a suitable option for several applications, such as supercapacitors, batteries and electrochemical devices [1,2]. The electrodeposition of polyaniline films onto gold, carbon or indium tin oxide (ITO) electrode surfaces is widely employed for the determination of diverse analytes [3–5]. Regarding the analytical application, the conducting form of polyaniline (emeraldine salt) possesses an insulating form (emeraldine base) when suffering deprotonation processes at high pH values [6]. In some pieces of research, self-doped polyaniline composites, obtained by copolymerization of polyaniline derivatives, can retard the deprotonation at high pH values, obtaining remarkable results in neutral buffer solutions [7–10].

Despite their excellent electrochemical behavior, surface fouling and the degradation of the deposited film upon use are common drawbacks for analyses [11]. With the aim of minimizing fouling phenomena, carbon paste materials modified with bulk polyaniline were developed. In this sense, silicon dioxide capped with gold nanoparticles (SiO_2@AuNPs) [12], carbon nanofibers coated with polyaniline and platinum [13], carbon nanofibers coated with gold nanoparticles and polyaniline and modified with butyl-3-methylimidazolium hexafluorophosphate [14], graphene nanosheets with multi-walled carbon nanotubes (MWCNT) and polyaniline [15], copper oxide/graphene oxide/polyaniline [16], polyaniline/starch/MWCNT [17] and Prussian blue/polyaniline [18] nanocomposites were employed for electrochemical purposes.

Our group is devoted to the development of electroactive bulk materials with interesting mechanical and electrical properties using sol-gel technology assisted by high-power ultrasound. In this regard, their employment for the electrochemical sensing of different metals and compounds is reported in a review published in the previous decade [19]. The modification of Sonogel–Carbon (SNG–C) surfaces by drop casting was reported in this work, leading to significant improvements in the sensitivity and selectivity of electrochemical oxidation processes of target analytes. The entrapment of tyrosinase within the conducting polymer layer electrodeposited on the SNG–C surface using a sinusoidal current procedure was also reported in this work, allowing the quantitation of the polyphenol indices of several alcoholic beverages and minimizing the influence of sugars in the electrochemical response.

The inclusion of bulk modifiers into the SNG–C matrix constitutes another feasible modification route. The modifier can play a conducting role in the material, leading to higher sensitivities for the determination of diverse analytes. Furthermore, the electroactivity of the surface can be restored mechanically after performing each electrochemical assay, minimizing the surface fouling phenomena and, hence, increasing the lifetime of the device. In this sense, β-cyclodextrin, furoylthioureas, cysteine and carbon nanopowder-modified materials were developed, obtaining remarkable electrochemical features in the electrochemical determination of benchmark analytes in buffer and real matrices [20–22].

The development of bare sol-gel materials based on conducting polymers, poly-(3,4-ethylenedioxythiophene) (PEDOT) and polyaniline, were also synthesized and subsequently applied in electrochemical analyses, providing excellent analytical results [23,24]. Among these works, the manufacturing of a ceramic polyaniline sol-gel composite by using a two-step procedure is noteworthy to mention. The first step involves the sonication of aniline with ammonium persulfate in acid medium, leading to polyaniline dispersion. Subsequently, the formation of the silicon oxide network was promoted through the sonication of the silane/graphite mixture. After synthesis, the composite provided excellent results for 4-chloro-3-methylphenol (PCMC) determination. Thus, the analytical utility of bare and bulk-modified sonogel-based materials for the electrochemical determination of diverse analytes is fully demonstrated in several pieces of research.

In the present work, the bulk modification of Sonogel–Carbon with polyaniline powder is reported. The hybrid composite preserves the electroactivity of the conducting polyaniline embedded in the silicon oxide network, leading to remarkable electrochemical results for model organic pollutant detection, namely of 4-chloro-3-methylphenol. The developed sensor was employed in the quantitation of PCMC in spiked water samples, obtaining suitable recovery values. Furthermore, the antifouling capability of the modified composite could be pointed out based on repeatability studies. With the aim of evaluating further applicability of the modified composite in the detection of different hazardous pollutants, the electrochemical sensing towards other chlorophenols is also investigated.

2. Materials and Methods

2.1. Reagents

All the reagents used were received without further purification. 4-Chloro-3-methylphenol, 2,4-dimethylphenol, pentachlorophenol and 2-chlorophenol were obtained from Merck (Germany, Darmstadt). 2,4,6-Trichlorophenol was purchased from Fluka (Spain, Barcelona). To synthesize Sonogel–Carbon-PANI electrodes, graphite powder from Alfa Aesar (Germany, Johnson Matthey GmbH), methyltrimethoxysilane (MTMOS) from Merck (Germany, Darmstadt), HCl from Panreac (Spain, Barcelona), aniline from Riedel-De-Haën (Germany, Seelze) and ammonium persulfate from Sigma-Aldrich (USA) were used. For the preparation of the buffer solutions, potassium chloride, boric acid and sodium chloride were purchased from Merck (Germany, Darmstadt) and glacial acetic acid, potassium hydroxide, ethanol and disodium tetraborate decahydrate were from Panreac (Spain, Barcelona). Lastly, potassium hexacyanoferrate (II) from Sigma-Aldrich (USA, St Louis, MO) was used to characterize the electrode material.

2.2. Instrumentation

The high-power ultrasound-assisted synthesis of the electrode materials was performed using a MISONIX S-4000 ultrasonic generator (USA, ME, Farmingdale) coupled with a 13 mm diameter titanium tip. The infrared spectra of the developed materials in powder form were recorded with an IRAffinity-1S infrared spectrophotometer from Shimadzu (Japan, Shimadzu) coupled with an attenuated total reflectance (ATR) unit. The electrochemical measurements were made using a PGTSTAT12 potentiostat/galvanostat autolab coupled to a 663 VA stand from Metrohm (The Netherlands, Utrecht). A three-electrode cell with the following composition was employed: a platinum wire as a counter electrode, Ag/AgCl (3 M KCl) as the reference electrode and the Sonogel–Carbon-based composites as the working electrodes.

2.3. Synthesis of Polyaniline Powder and Graphite/Polyaniline Composites

To synthesize the polyaniline powder, the following procedure was applied: Firstly, polyaniline dispersion was obtained using a procedure described in a previous work [24]. In brief, a precursor mixture containing 982 µL of a 0.25 M of ammonium persulfate in 1 M HCl solution and 18 µL of commercial aniline was sonicated under high-power ultrasound for 60 s. The as-synthesized black–green dispersion was centrifuged for 5 min at 4000 rpm (centrifugal acceleration = 1790 g). Subsequently, the supernatant was removed and the powder was dried in an oven for 16 h.

The graphite/polyaniline composites, ranging from 1 to 20% of PANI, were prepared by adding the corresponding amount of graphite and polyaniline powder. Afterwards, the mixture was stirred in the vortex for 30 s.

2.4. Synthesis and Elaboration of Sonogel–Carbon–Polyaniline Electrodes

The synthesis of the Sonogel–Carbon modified with polyaniline was carried out using the high-power ultrasound-assisted sol-gel technology procedure. Briefly, 500 µL of MTMOS and 100 µL of 0.2 M HCl solution were sonicated under high-power ultrasound for 10 s. Afterwards, 500 mg of the graphite powder/polyaniline mixture was added to the sonosol. After preparation of the material, the elaboration of the working electrodes was carried out following the procedure reported for Sonogel–Carbon electrodes [25]. The glass capillary tubes, i.d. 1.15 mm (geometric area: 1.04×10^{-2} cm^2), were filled with the carbon/polyaniline material and, subsequently, were left to dry for one day. After that, the capillary tubes were polished mechanically with P1200 silicon carbide paper (Struers, Germany).

2.5. Measuring Procedure

The electrochemical measurements were carried out as follows: Firstly, 25 mL of buffer solution was added to an electrochemical cell. The electrode surface was previously conditioned by means of cyclic voltammetry in free analyte buffer solution by recording eight cyclic sweeps from −0.2 V to 0.9 V and vice versa at 50 mV/s. After performing the conditioning step, the background was recorded using the differential pulse voltammetry (DPV) technique in free analyte solution. Afterwards, an aliquot of the stock solution was added to the electrochemical cell and the system was stirred for 30 s and, subsequently, the corresponding voltammogram was recorded. The bare electrode, Sonogel–Carbon without PANI (SNG–C), was also employed as the working electrode for some comparative electrochemical measurements.

2.6. Real Sample Analysis

The analysis of several water samples collected from different sources was performed following a similar procedure reported in other works [26]. In brief, the water samples were spiked with 25 µM of PCMC and diluted 10 times with the buffer solution to reach 2.5 µM. The analytical quantitation was performed using the standard addition method. The PCMC concentration added to the electrochemical cell ranged from 2.0 to 6.0 µM.

3. Results

3.1. Effect of the Modifier on the Mechanical/Electrochemical Features of the Conducting Composite

First of all, the effect of the graphite powder/polyaniline proportion on the final properties of the material was studied. For this purpose, Sonogel–Carbon electrodes modified with different proportions of graphite powder and polyaniline (PANI) mixtures, ranging from 1% to 20% of PANI, were prepared following the synthesis procedure reported in Section 2. After preparation, formulations higher than 4% of PANI offered liquid consistencies, leading to difficulties in the manufacturing of the electrodes, such as problems in the filling and compactness steps. Furthermore, no adherence between the glass capillary tube and the material was noticed after the drying step, suggesting high volume contraction. On the contrary, formulations with low PANI content showed a consistency similar to that of the original Sonogel–Carbon material, with better mechanical features for the manufacturing of electrochemical sensors. Thus, ceramic formulations with low polyaniline proportions were selected for further studies.

The electrochemical evaluation of Sonogel–Carbon-modified composites was made by means of DPV using a model chlorophenol, namely 4-chloro-3-methylphenol (PCMC). Figure 1A shows the differential pulse voltammograms recorded with carbon–ceramic composites at different proportions of PANI ranging from 1% to 4%. The corresponding differential pulse voltammograms displayed remarkable symmetric oxidation peaks, ascribed to their suitable electrochemical features. Regarding the electrochemical response, the best peak height was achieved at 2% PANI (see Figure 1B) and, hence, this material presents the greatest electrochemical performance for PCMC detection. The low variability of the electrochemical responses provided with this formulation are also noteworthy, and can be directly related to exceptional reproducibility. Therefore, this ceramic–polyaniline formulation, namely SNG–C–2% PANI, was selected for further studies.

3.2. Electrochemical Characterization

3.2.1. Characterization of the Redox Process of Polyaniline

The redox process of polyaniline in the material was studied by means of cyclic voltammetry in 1 M HCl free analyte solution to assess the proper modification of the Sonogel–Carbon electrode. The cyclic voltammograms recorded with the modified electrode displayed the main waves ascribed to polyaniline, all of them absent in the voltammogram recorded with the unmodified Sonogel–Carbon electrode (see Figure S1 and the inset): two anodic peaks around 0.2 V and 0.9 V, corresponding to leucoemeraldine–emeraldine and emeraldine–pernigraniline oxidation, respectively, and two cathodic peaks located at 0.05 V and 0.7 V, corresponding to the reduction counterparts [27]. Therefore, the presence of conducting-form polyaniline within the Sonogel–Carbon matrix was confirmed.

3.2.2. Electrochemical Behavior Using Ferrocyanide System

The electrochemical behavior of the ferrocyanide ion process was subsequently studied with the SNG–C–2% PANI electrodes. Figure S2 shows the cyclic voltammograms recorded at different scan rates in the presence of 5 mM potassium hexacyanoferrate (II) in 0.5 M KCl, showing high electrochemical reversibility. The relationship between the peak current and the square root of the scan rate is linear, with a regression coefficient of 0.999 (see Figure S2A), indicating a preferably diffusion-controlled process. The unmodified material was also characterized using a ferrocyanide probe for comparison.

The quasi-reversible behavior of the ferrocyanide probe could be monitored through the Randles–Sevcik equation (Equation (1)), as shown in the research work cited below [28].

$$I_p = 0.436\, nFA_{electroactive}C\sqrt{\frac{nFDv}{RT}} \tag{1}$$

where I_p is the current peak, n is the number of electrons involved in the electrochemical reaction, A is the electroactive area, C is the concentration of the redox probe, D is the diffusion coefficient, v is the scan rate, F is the Faraday constant and T is the temperature.

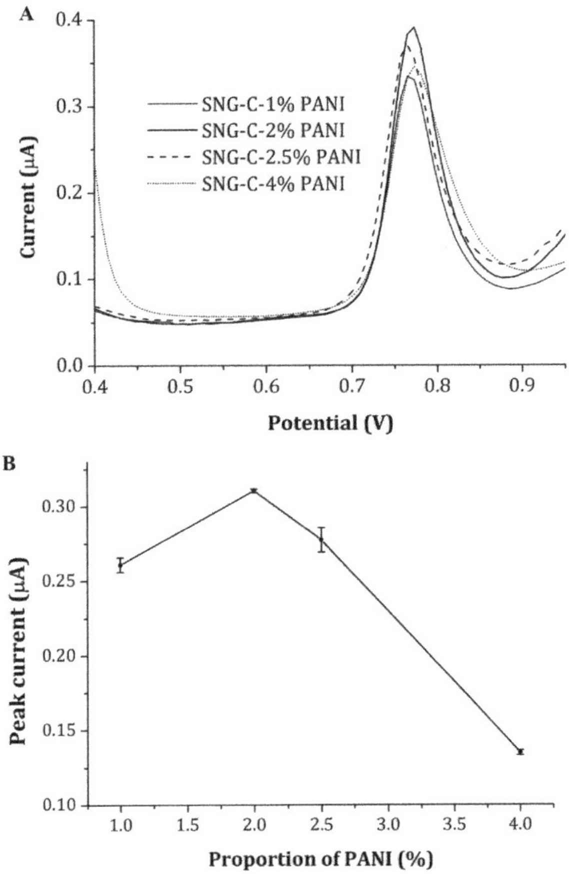

Figure 1. (**A**) Differential pulse voltammograms recorded with formulations with different proportions of polyaniline ranging from 1% to 4%, in presence of a 10 µM 4-chloro-3-methylphenol solution. (**B**) Peak heights recorded using different proportions of polyaniline in a 10 µM 4-chloro-3-methylphenol solution. A buffer solution containing 0.1 M acetic/acetate at pH 4 with 0.5 M KCl was used as electrolytic medium.

According to the previous equation, the electroactive area can be calculated. As can be explained in a previous article, only the voltammetric peaks corresponding to the forward scan were considered.

$$A_{electroactive} = \frac{I_p / v^{1/2}}{0.436\, nFC\sqrt{\frac{nFD}{RT}}} \quad (2)$$

The slopes between the current peak and the square root of the scan rate for both materials ($I_p/v^{1/2}$) were determined at low scan rates, from 25 mV/s to 125 mV/s, using the corresponding linear regression equations (see Figure S2B). Regarding the diffusion coefficient, 6.5×10^{-6} cm^2/s was reported in the literature for a similar ferrocyanide/potassium chloride electrochemical system [29,30]. Based on this information, the electroactive area estimated for the modified material is 7.2×10^{-3} cm^2, while the one corresponding for the unmodified device is 6.9×10^{-3} cm^2. No substantial increase in the electroactive surface area was observed after the modification of the material, which is in consonance with results previously reported with a similar bulk-modified composite [22].

3.3. Structural Characterization by Fourier Transformed Infrared Spectroscopy (FTIR)

The developed modified material was structurally characterized by means of Fourier transform infrared spectroscopy. The polyaniline powder and the unmodified Sonogel composite were also characterized for comparison purposes.

The IR spectra recorded with both Sonogel–Carbon-based materials show the presence of bands corresponding to Si–CH_3 and Si–O vibrations, ascribed to the silicon oxide network and non-hydrolysable methyl groups from the silane precursor (see Figure 2): 770 cm^{-1}, attributed to Si–CH_3 stretching, 1025 cm^{-1} and 1125 cm^{-1}, attributed to Si–O–Si stretching and 1275 cm^{-1}, ascribed to Si–CH_3 deformation. These assignments are in consonance with those recorded for carbon–ceramic oxide materials reported in the literature [25,31]. Although no polyaniline bands were appreciated in the SNG–C-modified spectrum, the electroactivity of this conducting polymer is confirmed by cyclic voltammetry.

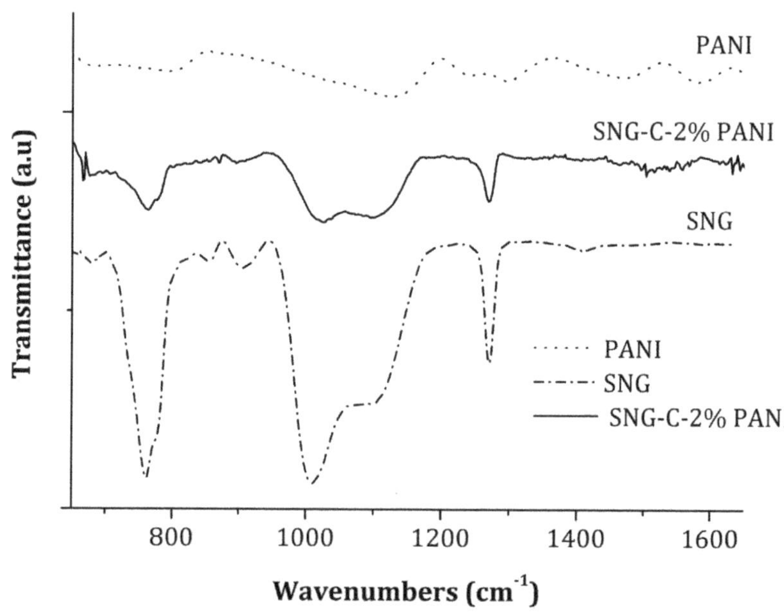

Figure 2. FTIR spectra recorded with different conducting materials: (dotted line) PANI, (dashed line) SNG–C–2% PANI and (solid line) SNG.

Thus, the absence of polyaniline bands in this formulation can be ascribed to its low proportion in the material.

3.4. Electrochemical Assessment of PCMC Using SNG–C-Modified Material

3.4.1. Preliminary Electrochemical Detection

The detection of PCMC, used as a benchmark chlorophenol, was investigated with SNG–C–2% PANI by means of DPV. In order to study the role of the conducting polymer in electrochemical detection, the unmodified electrode (SNG–C) was also used. Figure 3 shows the DPV recorded with both materials in the presence of 10 μM of PCMC in 0.1 M acetic/acetate buffer (ABS) at pH 4 with 0.5 M KCl. The peak height provided with the SNG–C–2% PANI is higher than the one recorded with the bare material, around 25% of the electrochemical response, addressing the superior sensitivity of the modified composite for PCMC detection. This enhancement on the electrochemical response provided by the modified composite may be ascribed to a favorable polymer–analyte interaction.

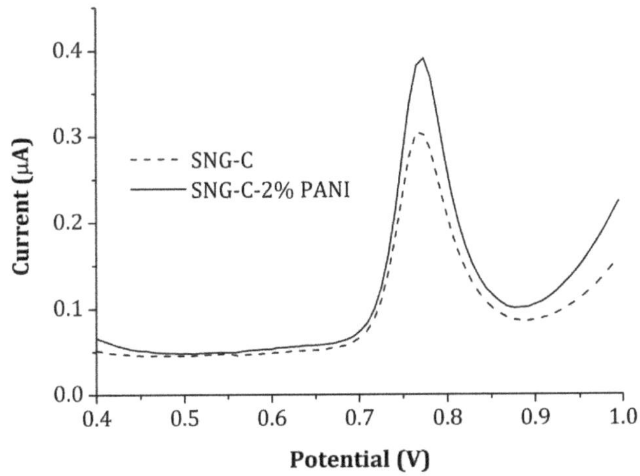

Figure 3. Differential pulse voltammograms recorded with SNG–C–2% PANI and SNG–C in presence of 10 μM of PCMC in 0.1 M acetic/acetate buffer (ABS) at pH 4 with 0.5 M KCl.

According to these results, we can assume that electroactivity of polyaniline is retained within the conducting polymer composite material.

3.4.2. Effect of pH on PCMC Electro-Oxidation

A preliminary screening of the buffer solution employed for electroanalytical purposes was performed. Figure 4 shows differential pulse voltammograms recorded with SNG–C–2% PANI in the presence of the same concentration of 4-chloro-3-methylphenol diluted in three buffer solutions at different pH values. In all cases, 0.5 M KCl solution was used as the supporting electrolyte.

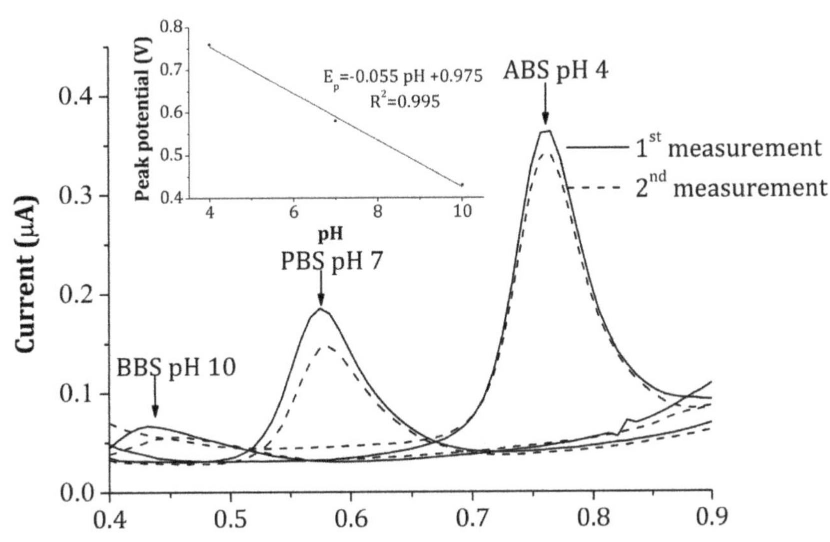

Figure 4. Differential pulse voltammograms recorded with SNG–C–2% PANI in presence of the same concentration of 4-chloro-3-methylphenol using three different buffer solutions at different pH: acetic/acetate buffer (ABS) at pH 4, phosphate buffer (PBS) at pH 7 and borate buffer (BBS) at pH 10. The inset reports the linear relationship between the peak potential and pH.

The peak potential shifts with the pH, attributed to the participation of the protons in the redox reaction. The slope between the peak potential and the pH is 55 mV, close to 59 mV (see inset, Figure 4) ascribed to one-proton–one-electron processes [32]. Concerning the electrochemical response, the buffer solution at pH 4 led to the best peak height for PCMC detection. Additionally, no major change in the peak current was observed after performing the second measurement in acid solution. Thus, this medium is selected for electrochemical assays of the target chlorophenol.

3.4.3. Effect of PCMC Concentration

The developed material was employed in the voltammetric determination of PCMC ranging from 0.7 to 7.0 μM using DPV under the instrumental conditions detailed in Section 2. The current densities at different PCMC concentrations (calculated using the geometrical area of the electrode) recorded with the SNG–C–2% PANI composite are plotted as a function of the PCMC concentration (see Figure 5A). The calibration curve considering three measurements at each PCMC concentration value is shown in the corresponding inset. The relationship between the peak current density and the concentration of PCMC can be described by the following equation: j = 2.40 [PCMC] − 0.78 (R^2 = 0.991), where *j* is the current density (expressed in μA/cm^2) and *[PCMC]* is the concentration of the studied analyte (expressed in μM). The sensitivity was determined as the slope of the calibration curve, obtaining 2.40 ± 0.18 μA/cm^2. The detection limit (LOD) was calculated by LOD = 3 × S_q/m, where S_q is the standard deviation of the intercept and *m* is the slope of the linear equation. The value obtained was 0.69 μM.

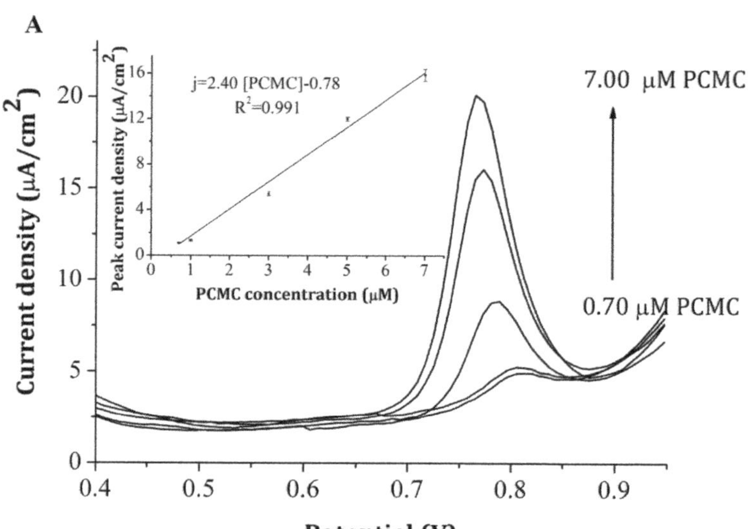

Figure 5. *Cont.*

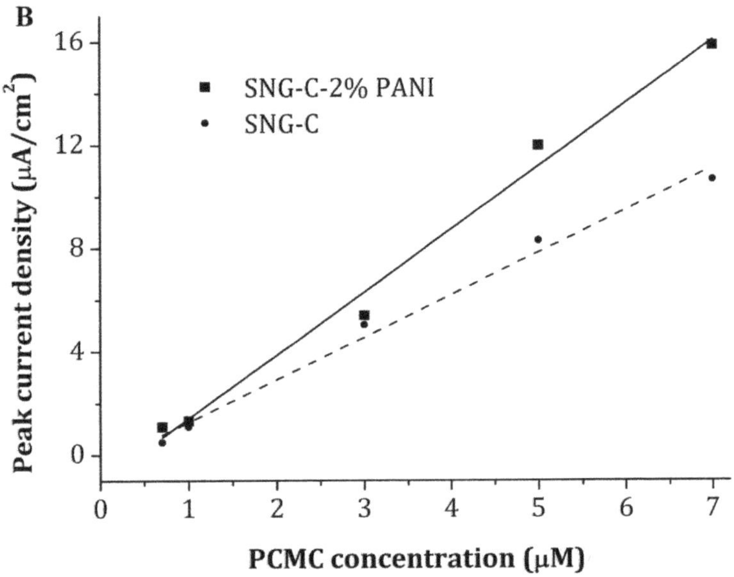

Figure 5. (**A**) Differential pulse voltammograms recorded with SNG–C–2% PANI in presence of different concentrations of PCMC. The inset reports the relevant calibration plots. (**B**) Calibration curve recorded with SNG–C–2% PANI and SNG–C in presence of different concentrations of 4-chloro-3-methylphenol in 0.1 M acetic/acetate with 0.5 M KCl at pH 4.

The electrochemical performance for PCMC determination provided with the modified material was compared with the unmodified one to demonstrate the usefulness of the modification process. Figure 5B exposes the average calibration curve obtained with both electrode materials (considering three measurements for each point) under the same instrumental conditions. As can be seen from the figure, the sensitivity provided with SNG–C–2% PANI is greater than the one obtained with SNG–C, which can be ascribed to the presence of the polymer. This finding is in consonance with what was observed in the preliminary test with PCMC (Section 3.4.1). The detection limit of SNG–C was also calculated, obtaining 0.71 μM, which is slightly higher than the one associated with the modified device.

A comparison of the analytical parameters for PCMC determination with other electrode materials was carried out. The sensitivity obtained with the developed device is comparable to those reported with other sensors (see Table 1), which can be attributed to high electron transfer capability.

Table 1. Figures of merits for the determination of 4-chloro-3-methylphenol obtained using several electrochemical sensors.

	Sensitivity [μA/μM·cm²]	**LOD [μM]**	**Reference**
SNG–C–2% PANI	2.40	0.69	This work
MWCNT–GCE	1.48	8.8	[33]
GP–CCE	0.28	2.71	[26]
MWCNT–CCE	1.41	0.71	[26]
BDD	0.08	0.46	[34]
UiO-66-NH_2@PEDOT/GA/GCE	1.87×10^{-3}	0.20	[32]

GP–CCE: Graphite–carbon ceramic electrode; BDD: Boron-doped diamond electrode; MWCNT–GCE: Multiwalled carbon nanotubes–glassy carbon electrode; UiO-66-NH_2@PEDOT/GA/GCE: PEDOT-modified electrode with UiO-66-NH_2 and graphene aerogel; MWCNT–CCE: Multiwalled carbon nanotubes–carbon ceramic electrode.

On the other hand, LOD is lower than the concentration value permitted in aquatic environments established by the United States Environmental Protection Agency (US-EPA), around 3.5 μM [35]. In this way, it can be deduced that the material may be used for PCMC environmental monitoring in aquatic matrices. Additionally, no organic solvent consumption in the manufacturing process should be highlighted. For all these reasons, the bulk modification of Sonogel–Carbon with polyaniline seems to be promising for the electrochemical sensing of PCMC.

3.4.4. Reproducibility and Repeatability Studies

The reproducibility of the electrochemical measurements in the same PCMC solution using seven different SNG–C–2% PANI electrodes was studied. The coefficient of variation was 4%, indicating that the fabrication process provides electrode materials with similar electrochemical performance.

The repeatability of the electrochemical measurements was also evaluated by performing nine successive measurements in the same PCMC solution. The coefficient of variation between the peak heights was 3.9%. Based on these results, an excellent repeatability can be concluded, which can be ascribed to the resistance of fouling by oxidation products of PCMC. Hence, no surface cleaning beyond the initial polishing is required.

3.4.5. Effect of Salts and Phenols on PCMC Detection

The influence of some typical ion species, i.e., calcium, sodium, magnesium and cadmium salts, on the electrochemical response was examined. Measurements in the presence of 7 μM of PCMC and higher concentration of these salts (100-fold of calcium and sodium salts and 10-fold of cadmium and magnesium ones) were performed by triplicate. The electrochemical response of the device is remarkably similar in presence of the analyte and different salts (see Figure 6A), suggesting a low influence of these compounds in the determination of 4-chloro-3-methylphenol. Moreover, the variation coefficient values are lower than 6% in all electrochemical assays, confirming the repeatability of the electrochemical measurements discussed in the previous subsection and the absence of fouling on the electrode surface. Therefore, no influence of diverse salts on PCMC detection was achieved.

The electrochemical analysis of the target chlorophenol (PCMC) was performed in the presence of another substituted phenol, 2,4-dimethylphenol (DMP). Voltammetric assays were carried out in the presence of different PCMC concentrations and a fixed concentration of 2,4-DMP. As derived from the voltammograms shown in Figure 6B, peaks corresponding to PCMC and DMP oxidation are located at different potentials. The oxidation peak of PCMC increased with the PCMC concentration, while DMP oxidation peak remained invariable in all the experiments. These results suggest a negligible adsorption of DMP at the micromolar concentration level, similar to the findings derived from the electrochemical PCMC assays shown in Section 3.4.4.

Hence, the minimization of fouling phenomena provided with the developed device for PCMC determination could be extended to other substituted phenols. Furthermore, no interference on the electrochemical detection of the target chlorophenol in the presence of DMP could be concluded. This finding opens the door to studying the possible simultaneous determination of both phenols.

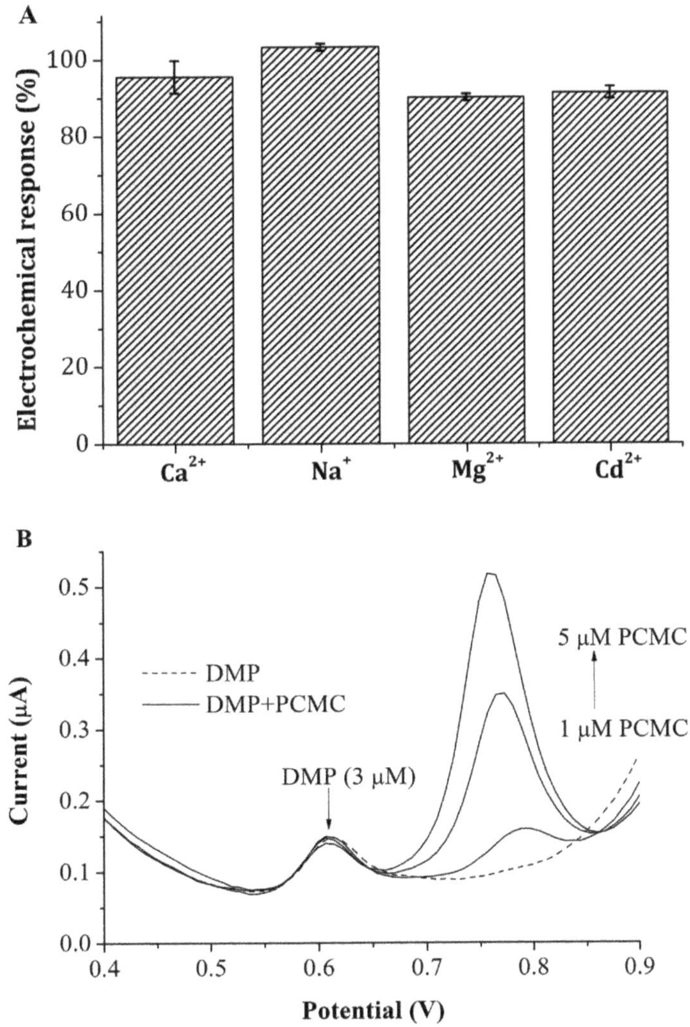

Figure 6. (**A**) Electrochemical responses obtained with SNG–C–2% PANI in presence of 7 μM of 4-chloro-3-methylphenol and 700 μM of calcium and sodium salts and 70 μM of magnesium and cadmium salts. (**B**) Differential Pulse voltammograms recorded with the developed material in presence of different concentrations of PCMC and 3 μM of DMP. A buffer solution containing 0.1 M acetic/acetate at pH 4 with 0.5 M KCl was used as electrolytic medium.

3.4.6. Spiked Water Analysis

The electrochemical determination of PCMC in water collected from different sources in Spain was conducted using DPV. The experimental results derived from this study in terms of recovery percentages are shown in Table 2.

Table 2. Electrochemical determination of PCMC in spiked water samples collected from different areas of Andalusia (Spain).

Water from Diverse Sources	C_{Added} (μM)	$C_{Determined}$ (μM)	Recovery (%)
Fountain water (Ubeda, Jaén)	2.50	2.49	99.6
Well water (Chiclana de la Frontera, Cádiz)	2.50	2.32	92.3
Tap water (Puerto Real, Cádiz)	2.50	2.47	98.8
Mineral water	2.50	2.62	105

The recovery values for all water samples analyzed in this study range from 92% to 105%, which are suitable according to analytical standards. Hence, the electrode material developed in this work is promising for the quantitation of PCMC in different spiked water samples and, hence, it is suitable for environmental monitoring.

3.5. Electrochemical Detection of Other Chlorophenols of Interest

The applicability of detection other chlorophenols of interest with the developed ceramic composite was examined. The electrochemical detection of 2-chlorophenol (2-CP), 2,4,6-trichlorophenol (TCP) and pentachlorophenol (PCP), was assessed in this work. The bare material, Sonogel–Carbon, was also employed under the same instrumental conditions for comparison.

Figure 7 shows the differential pulse voltammograms recorded with the developed material and the unmodified material in the presence of 7 μM of target chlorophenols in 0.1 M ABS with 0.5 M KCl at pH 4. Higher oxidation peak heights in comparison with those obtained with the bare material are noticed, which can be ascribed to the presence of the conducting form of the polymer within the matrix. Based on this study, the developed material is promising for the detection of diverse chlorophenols in the electrochemical frame.

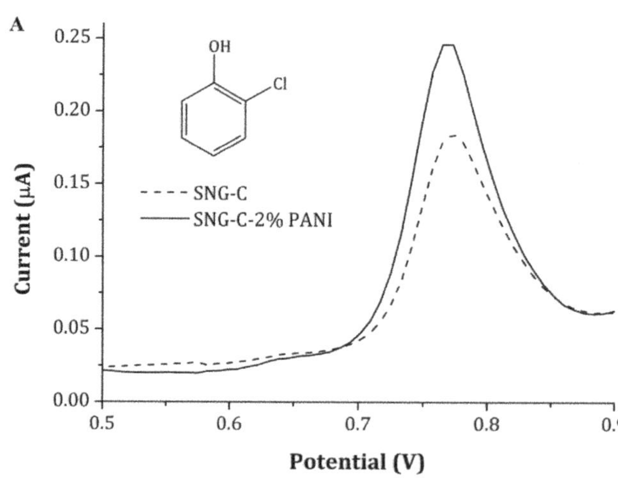

Figure 7. *Cont.*

B

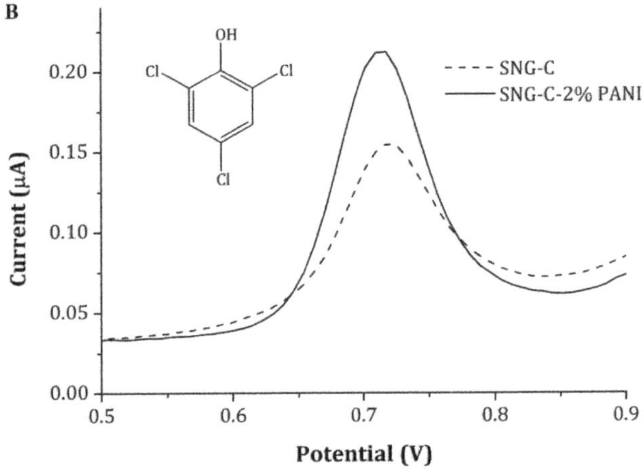

C

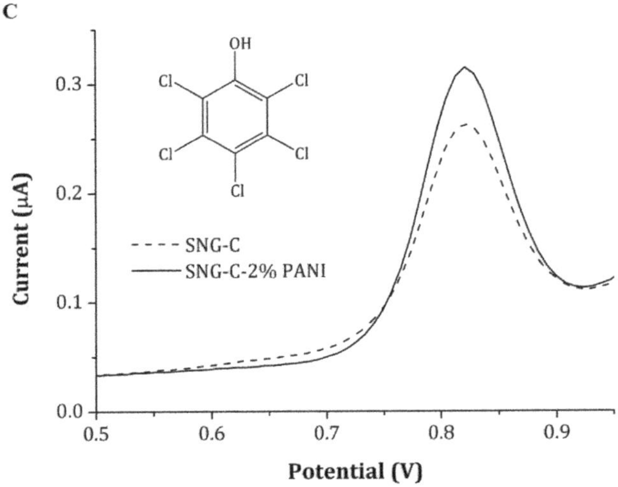

Figure 7. Differential pulse voltammograms recorded with SNG–C–2% PANI in presence of 7 μM of different chlorophenols: (**A**) 2-chlorophenol (2-CP), (**B**) 2,4,6-trichlorophenol (TCP) and (**C**) pentachlorophenol (PCP). A buffer solution containing 0.1 M acetic/acetate with 0.5 M KCl at pH 4 was used as electrolytic medium.

4. Conclusions

The development of a Sonogel–Carbon material modified with a conducting polymer (PANI) was carried out using an environmentally friendly route based on high-power ultrasound. The polyaniline powder was embedded in the silicon oxide matrix, maintaining its electroactivity within the material. Better figures of merit for PCMC determination were obtained with the modified composite in comparison with those obtained with the bare material, which can be attributed to a favorable polymer–analyte interaction. Analytical results were also very satisfactory in comparison with other sensors reported in the literature. Furthermore, no surface fouling in PCMC analysis can be evidenced, making this sensor suitable for successive electrochemical analyses. The electrochemical detection of PCMC in the presence of different salts and one substituted phenolic compound can be also stated with the developed device. Moreover, the Sonogel–Carbon and polyaniline modification approach could be employed in the manufacturing of other modified devices, such as screen-printed electrodes. On the basis of these results, polyaniline could be proposed as an adequate bulk modifier of the sol-gel carbon matrix for the environmental monitoring of chlorophenols.

Supplementary Materials: The following supporting information can be downloaded at https://www.mdpi.com/article/10.3390/chemosensors11010063/s1. Figure S1. Cyclic voltammograms recorded with SNG–C and SNG–C–2% PANI in free analyte 1 M HCl solution; Figure S2: (A) Cyclic voltammograms recorded with SNG–C–2% PANI in presence of 5 mM of potassium hexacyanoferrate (II) in 0.5 M KCl at different scan rates. The inset reports the relationship between the peak intensity and the square root of the scan rate; (B) relationship between the anodic peak intensity and the square root of the scan rate in presence of ferrocyanide recorded from 25 mV/s to 125 mV/s with SNG–C and SNG–C–2% PANI composites.

Author Contributions: Conceptualization, P.C.-M., D.L.-I. and L.C.-A.; methodology, P.C.-M., D.L.-I. and J.J.G.-G.; validation, P.C.-M., A.S.-P. and J.M.P.-S.; formal analysis, P.C.-M., D.L.-I. and J.J.G.-G.; investigation, P.C.-M., A.S.-P. and D.L.-I.; resources, L.C.-A. and J.M.P.-S.; data curation, P.C.-M. and D.L.-I.; writing—original draft preparation, P.C.-M., D.L.-I. and J.J.G.-G.; writing—review and editing, All.; visualization, All.; supervision, L.C.-A. and J.M.P.-S.; project administration, D.L.-I. and J.M.P.-S.; funding acquisition, L.C.-A. and J.M.P.-S. All authors have read and agreed to the published version of the manuscript.

Funding: The authors thank Junta de Andalucia (PAIDI), Institute of Research on Electron Microscopy and Materials (IMEYMAT, Polybiosens and Nano4(bio)sens projects), and Programa de Fomento e Impulso de la Investigación y de la Transferencia de la Universidad de Cádiz 2020–2021, for the project PR2020-013 (Proyectos de Investigación-Puente 2020). We also thank Agencia Estatal de Investigación (AEI), Ministerio de Ciencia e Innovación of Spain, and FEDER funds (EU) for the "Multibioanalysis" research project (Proyecto de Generación del Conocimiento, PID2021-122578NB-I00) financed by MCIN/AEI/10.13039/501100011033 and by FEDER, UE "ERDF A way of making Europe".

Institutional Review Board Statement: Not applicable.

Informed Consent Statement: Not applicable.

Data Availability Statement: Not applicable.

Acknowledgments: The authors thank the TEP-243 research group for the help provided in registering the FTIR spectra.

Conflicts of Interest: The authors declare no conflict of interest.

References

1. Xiang, H.; Deng, N.; Zhao, H.; Wang, X.; Wei, L.; Wang, M.; Cheng, B.; Kang, W. A Review on Electronically Conducting Polymers for Lithium-Sulfur Battery and Lithium-Selenium Battery: Progress and Prospects. *J. Energy Chem.* **2021**, *58*, 523–556. [CrossRef]
2. Simotwo, S.K.; Kalra, V. Polyaniline-Based Electrodes: Recent Application in Supercapacitors and Next Generation Rechargeable Batteries. *Curr. Opin. Chem. Eng.* **2016**, *13*, 150–160. [CrossRef]
3. Chen, X.; Guo, B.; Hu, P.; Wang, Y. A Non-Enzymatic Hydrogen Peroxide Sensor Based on Gold Nanoparticles/Carbon Nanotube/Self-Doped Polyaniline Hollow Spheres. *Electroanalysis* **2014**, *26*, 1513–1521. [CrossRef]
4. Patil, D.S.; Pawar, S.A.; Patil, S.K.; Salavi, P.P.; Kolekar, S.S.; Devan, R.S.; Ma, Y.R.; Kim, J.H.; Shin, J.C.; Patil, P.S. Electrochemical Performance of Potentiodynamically Deposited Polyaniline Electrodes in Ionic Liquid. *J. Alloys Compd.* **2015**, *646*, 1089–1095. [CrossRef]
5. Shen, K.; Ran, F.; Zhang, X.; Liu, C.; Wang, N.; Niu, X.; Liu, Y.; Zhang, D.; Kong, L.; Kang, L.; et al. Supercapacitor Electrodes Based on Nano-Polyaniline Deposited on Hollow Carbon Spheres Derived from Cross-Linked Co-Polymers. *Synth. Met.* **2015**, *209*, 369–376. [CrossRef]
6. Andreev, V.N. Effect of Solution Acidity on the Electrochemical Behavior of Nafion–Polyaniline Films. *Russ. J. Electrochem.* **2005**, *41*, 200–202. [CrossRef]
7. Sanchis, C.; Ghanem, M.A.; Salavagione, H.J.; Morallón, E.; Bartlett, P.N. The Oxidation of Ascorbate at Copolymeric Sulfonated Poly(aniline) Coated on Glassy Carbon Electrodes. *Bioelectrochemistry* **2011**, *80*, 105–113. [CrossRef]
8. Gicevicius, M.; Kucinski, J.; Ramanaviciene, A.; Ramanavicius, A. Tuning the Optical pH Sensing Properties of Polyaniline-Based Layer by Electrochemical Copolymerization of Aniline with O-Phenylenediamine. *Dye. Pigment.* **2019**, *170*, 107457. [CrossRef]
9. Zhao, X.; Bai, W.; Yan, Y.; Wang, Y.; Zhang, J. Core-Shell Self-Doped Polyaniline Coated Metal-Organic-Framework (SPAN@UIO-66-NH2) Screen Printed Electrochemical Sensor for Cd^{2+} Ions. *J. Electrochem. Soc.* **2019**, *166*, B873. [CrossRef]

10. Dkhili, S.; López-Bernabeu, S.; Huerta, F.; Montilla, F.; Besbes-Hentati, S.; Morallón, E. A Self-Doped Polyaniline Derivative Obtained by Electrochemical Copolymerization of Aminoterephthalic Acid and Aniline. *Synth. Met.* **2018**, *245*, 61–66. [CrossRef]
11. Tsakova, V.; Seeber, R. Conducting Polymers in Electrochemical Sensing: Factors Influencing the Electroanalytical Signal. *Anal. Bioanal. Chem.* **2016**, *408*, 7231–7241. [CrossRef] [PubMed]
12. Weng, C.J.; Chen, Y.L.; Chien, C.M.; Hsu, S.C.; Jhuo, Y.S.; Yeh, J.M.; Dai, C.F. Preparation of Gold Decorated SiO2@polyaniline Core-Shell Microspheres and Application as a Sensor for Ascorbic Acid. *Electrochim. Acta* **2013**, *95*, 162–169. [CrossRef]
13. Rajpurohit, A.S.; Punde, N.S.; Rawool, C.R.; Srivastava, A.K. Application of Carbon Paste Electrode Modified with Carbon Nanofibres/Polyaniline/Platinum Nanoparticles as an Electrochemical Sensor for the Determination of Bezafibrate. *Electroanalysis* **2018**, *30*, 571–582. [CrossRef]
14. Afzali, M.; Jahromi, Z.; Nekooie, R. Sensitive Voltammetric Method for the Determination of Naproxen at the Surface of Carbon Nanofiber/gold/polyaniline Nanocomposite Modified Carbon Ionic Liquid Electrode. *Microchem. J.* **2019**, *145*, 373–379. [CrossRef]
15. Kalambate, P.K.; Rawool, C.R.; Srivastava, A.K. Fabrication of Graphene Nanosheet–multiwalled Carbon Nanotube–polyaniline Modified Carbon Paste Electrode for the Simultaneous Electrochemical Determination of Terbutaline Sulphate and Guaifenesin. *New J. Chem.* **2017**, *41*, 7061–7072. [CrossRef]
16. Afzali, M.; Mostafavi, A.; Shamspur, T. Square Wave Voltammetric Determination of Anticancer Drug Flutamide Using Carbon Paste Electrode Modified by CuO/GO/PANI Nanocomposite. *Arab. J. Chem.* **2020**, *13*, 3255–3265. [CrossRef]
17. Gautam, V.; Singh, K.P.; Yadav, V.L. Polyaniline/MWCNTs/starch Modified Carbon Paste Electrode for Non-Enzymatic Detection of Cholesterol: Application to Real Sample (cow Milk). *Anal. Bioanal. Chem.* **2018**, *410*, 2173–2181. [CrossRef]
18. Patel, B.R.; Noroozifar, M.; Kerman, K. Prussian Blue-Doped Nanosized Polyaniline for Electrochemical Detection of Benzenediol Isomers. *Anal. Bioanal. Chem.* **2020**, *412*, 1769–1784. [CrossRef]
19. Bellido-Milla, D.; Cubillana-Aguilera, L.M.; El Kaoutit, M.; Hernandez-Artiga, M.P.; Hidalgo-Hidalgo De Cisneros, J.L.; Naranjo-Rodríguez, I.; Palacios-Santander, J.M. Recent Advances in Graphite Powder-Based Electrodes. *Anal. Bioanal. Chem.* **2013**, *405*, 3525–3539. [CrossRef]
20. Izaoumen, N.; Cubillana-Aguilera, L.M.; Naranjo-Rodríguez, I.; Hidalgo-Hidalgo de Cisneros, J.L.; Bouchta, D.; Temsamani, K.R.; Palacios-Santander, J.M. β-Sonogel-Carbon Electrodes: A New Alternative for the Electrochemical Determination of Catecholamines. *Talanta* **2009**, *78*, 370–376. [CrossRef]
21. Cubillana-Aguilera, L.M.; Palacios-Santander, J.M.; Estevez-Hernandez, O.L.; Naranjo-Rodriguez, I.; Hidalgo-Hidalgo De Cisneros, J.L. 1-Furoylthiourea-Sonogel-Carbon Electrodes: Structural and Electrochemical Characterization. *Talanta* **2010**, *82*, 129–136. [CrossRef]
22. Pigani, L.; Rioli, C.; López-Iglesias, D.; Zanardi, C.; Zanfrognini, B.; Cubillana-Aguilera, L.M.; Palacios-Santander, J.M. Preparation and Characterization of Reusable Sonogel-Carbon Electrodes Containing Carbon Black: Application as Amperometric Sensors for Determination of Cathecol. *J. Electroanal. Chem.* **2020**, *877*, 114653. [CrossRef]
23. López-Iglesias, D.; García-Guzmán, J.J.; Zanardi, C.; Palacios-Santander, J.M.; Cubillana-Aguilera, L.; Pigani, L. Fast Electroanalytical Determination of Cannabidiol and Cannabinol in Aqueous Solution Using Sonogel-Carbon-PEDOT Devices. *J. Electroanal. Chem.* **2020**, *878*, 114591. [CrossRef]
24. López-Iglesias, D.; Fanelli, F.; Marchi, L.; Alcántara, R.; Cocchi, M.; Cubillana-Aguilera, L.; Palacios-Santander, J.M.; García-Guzmán, J.J. Ceramic Polyaniline-Carbon Composite Obtained by Ultrasound-Assisted Sol–gel Route: Electrochemical Performance towards Environmental Pollutants. *J. Electroanal. Chem.* **2022**, *905*, 115971. [CrossRef]
25. Cubillana-Aguilera, L.M.; Palacios-Santander, J.M.; Naranjo-Rodríguez, I.; Hidalgo-Hidalgo-De-Cisneros, J.L. Study of the Influence of the Graphite Powder Particle Size on the Structure of the Sonogel-Carbon Materials. *J. Sol-Gel Sci. Technol.* **2006**, *40*, 55–64. [CrossRef]
26. Robak, J.; Burnat, B.; Leniart, A.; Kisielewska, A.; Brycht, M.; Skrzypek, S. The Effect of Carbon Material on the Electroanalytical Determination of 4-Chloro-3-Methylphenol Using the Sol-Gel Derived Carbon Ceramic Electrodes. *Sens. Actuators B Chem.* **2016**, *236*, 318–325. [CrossRef]
27. Song, E.; Choi, J.-W. Conducting Polyaniline Nanowire and Its Applications in Chemiresistive Sensing. *Nanomaterials* **2013**, *3*, 498–523. [CrossRef] [PubMed]
28. Ferrari, A.G.M.; Foster, C.W.; Kelly, P.J.; Brownson, D.A.C.; Banks, C.E. Determination of the Electrochemical Area of Screen-Printed Electrochemical Sensing Platforms. *Biosensors* **2018**, *8*, 53. [CrossRef] [PubMed]
29. Moldenhauer, J.; Meier, M.; Paul, D.W. Rapid and Direct Determination of Diffusion Coefficients Using Microelectrode Arrays. *J. Electrochem. Soc.* **2016**, *163*, H672–H678. [CrossRef]
30. Klymenko, O.V.; Evans, R.G.; Hardacre, C.; Svir, I.B.; Compton, R.G. Double Potential Step Chronoamperometry at Microdisk Electrodes: Simulating the Case of Unequal Diffusion Coefficients. *J. Electroanal. Chem.* **2004**, *571*, 211–221. [CrossRef]
31. Tsai, Y.C.; Vo, T.G.; Chiang, C.Y. Engineering the Surface Wettability of a Ceramic Carbon Electrode for Improved Hydrogen Evolution Performance of a Molybdenum Sulfide Electrocatalyst. *Sustain. Energy Fuels* **2020**, *4*, 4018–4029. [CrossRef]
32. Tian, Q.; Xu, J.; Zuo, Y.; Li, Y.; Zhang, J.; Zhou, Y.; Duan, X.; Lu, L.; Jia, H.; Xu, Q.; et al. Three-Dimensional PEDOT Composite Based Electrochemical Sensor for Sensitive Detection of Chlorophenol. *J. Electroanal. Chem.* **2019**, *837*, 1–9. [CrossRef]
33. Baranowska, I.; Bijak, K. Voltammetric Determination of Disinfectants at Multiwalled Carbon Nanotube Modified Glassy Carbon Electrode. *J. Anal. Chem.* **2013**, *68*, 891–895. [CrossRef]

34. Brycht, M.; Lochyński, P.; Barek, J.; Skrzypek, S.; Kuczewski, K.; Schwarzova-Peckova, K. Electrochemical Study of 4-Chloro-3-Methylphenol on Anodically Pretreated Boron-Doped Diamond Electrode in the Absence and Presence of a Cationic Surfactant. *J. Electroanal. Chem.* **2016**, *771*, 1–9. [CrossRef]
35. OoW Office of Science and Technology. *Update of Human Health Ambient Water Quality Criteria: 3-Methyl-4-Chlorophenol 59-50-57*; EPA (Unites States Environmental Protection Agency): Washington, DC, USA, 2015.

chemosensors

MDPI

Article

Potentiometric Sensor System with Self-Calibration for Long-Term, In Situ Measurements

Zhehao Zhang [1], Elena Boselli [1] and Ian Papautsky [1,2,*]

[1] Department of Biomedical Engineering, University of Illinois Chicago, Chicago, IL 60607, USA
[2] NSF Center for Advanced Design and Manufacturing of Integrated Microfluidics (CADMIM), Chicago, IL 60607, USA
* Correspondence: papauts@uic.edu; Tel.: +1-312-413-3800

Abstract: We built an integrated solid-contact ion-selective electrode (SCISE) system with the functionality of self-calibration. A multiplexed SCISE sensor (K^+ and NO_3^- vs. Ag/AgCl) was fabricated on printed-circuit board (PCB) substrates and was subsequently embedded into a microfluidic flow cell for self-calibration and flow-through analysis. A PCB circuit that includes modules for both sensor readout and fluid control was developed. The sensors showed a fast and near-Nernstian response (56.6 for the K^+ electrode and −57.4 mV/dec for the NO_3^- electrode) and maintained their performance for at least three weeks. The sensors also showed a highly reproducible response in an automated two-point calibration, demonstrating the potential for in situ monitoring. Lastly, the sensor system was successfully applied to measure mineral nutrients in plant sap samples.

Keywords: ion-selective electrode; printed circuit board; in situ measurement; potentiometry; self-calibration

Citation: Zhang, Z.; Boselli, E.; Papautsky, I. Potentiometric Sensor System with Self-Calibration for Long-Term, In Situ Measurements. *Chemosensors* **2023**, *11*, 48. https://doi.org/10.3390/chemosensors11010048

Academic Editors: Iulia Gabriela David and Dana Elena Popa

Received: 13 December 2022
Revised: 30 December 2022
Accepted: 1 January 2023
Published: 5 January 2023

1. Introduction

Ion-selective electrodes (ISE) are commonly used to develop fast, portable, and cost-effective analytical devices. Such devices have found wide applications in point-of-care testing and medicine [1–4], wearables [5,6], agriculture [7], and environmental monitoring [8,9]. In recent years, the development of solid contact (SC) materials as ion-to-electron transducers has driven the field towards next-generation sensors that are miniaturized, rugged, and calibration- and maintenance-free [10]. Despite the progress, the majority of these solid contact ISEs (SCISEs) still need to be calibrated as their conventional counterparts with an inner filling solution.

Sensors are calibrated to correct for sensitivity loss, baseline drift, and inter-sensor variability. Usually, calibration is performed by the user before and/or after each measurement. Although manual calibration may be sufficient for single-use sensors, it is cumbersome for recurring and continuous measurements; moreover, it is not feasible for in situ measurements in agricultural, environmental, or geochemical analysis where sensors need to be deployed in the field and work autonomously. In these scenarios, automatic calibration at the point of need not only saves time and effort but ensures the accuracy and precision of measurement.

Only a few potentiometric sensor systems with a self-calibration function have been reported to date [2,8,9,11]. They typically employ a flow cell that introduces calibrating reagents to the sensor using fluidic components (e.g., pumps and valves) and corresponding fluid-control modules. In addition, such flow cell arrangement allows for electrode cleaning and sample pretreatment and often leads to lower detection limits than the stationary approach [12,13]. For example, the commercial Abbott i-STAT blood analyzer performs on-site calibration of the sensor through a series of pumping and valving mechanisms [2]; however, since the sensor was intended for single use, its calibration was limited to once per test. More recently, Cuatero et al. [8,9] developed a submersive probe based on a

SCISE-embedded flow cell to monitor seawater in hourly measurements followed by a one-point calibration. However, the sensor readout and fluid control relied on dedicated commercial instruments, which significantly increased the cost and complexity of the system; in addition, the proprietary nature of these commercial systems does not allow them to be easily transferred into other applications. Therefore, a more integrated, open-source sensor system is highly desired, especially for long-term, in situ measurements.

In this work, we developed a potentiometric sensor system with self-calibration capability by integrating a multiplexed SCISE sensor with electronics and fluidic components. Importantly, both the sensor readout and fluid control modules are combined into a single PCB circuit, which makes the system cost-effective, portable, and adaptable. Both the circuit and the multiplexed sensor (K^+ and NO_3^- sensors and Ag/AgCl reference) were fabricated using printed-circuit board (PCB) technology [14]. The sensor was further embedded into a microfluidic flow cell and its potentiometric response was characterized under different flow conditions. To prepare the sensor for future in situ measurements, we then demonstrated the long-term operation and automated two-point calibration. Finally, the sensor was validated with plant sap samples.

2. Materials and Methods

2.1. Materials and Reagents

Mesoporous carbon black (MCB, average pore diameter 6.4 nm), potassium ionophore I (valinomycin), potassium tetrakis(4-chlorophenyl)-borate (KTPB), tridodecylmethylammonium nitrate (TDDMA-NO_3), polyvinyl chloride (PVC, high molecular weight), tetrahydrofuran (THF), and 2-nitrophenyl octyl ether (NPOE) were purchased from Sigma Aldrich (MA, US). Salts of NaCl, KNO_3, KCl were purchased from Fisher Scientific. Silver plating solution (Technic Silver Cyless RTU) was purchased from Technic Inc. All chemicals were of analytical or industrial grade. Deionized (DI) water was used to prepare all aqueous solutions. Double-sided (#9474LE) and single-sided (#9964) adhesive tapes were obtained from 3M. Tygon tubing (0.02″ ID × 0.06″ OD) was purchased from Cole-Parmer.

Suspension of mesoporous carbon black (MCB) was prepared by adding 45 mg MCB and 5 mg PVC as binder into 2 mL THF. Prior to each use, the suspension was vortexed for 30 s, sonicated for 1 h, and then vortexed again for 30 s to disperse the nanoparticles. The K^+ ISE cocktail was prepared by mixing 22.2 mg K^+ ionophore I, 7.0 mg KTPB, 320 mg PVC, and 660 μL NPOE into 10 mL THF and stirring until all components were fully dissolved. For the NO_3^- cocktail, 10 mg TDDMA-NO_3, 330 mg PVC, and 660 μL NPOE were subsequently added to 10 mL THF and stirred until fully dissolved.

2.2. Sensor System

The potentiometric sensor system consists of the multiplexed SCISE sensor and a PCB circuit to control fluidics and perform both sensor readout and calibration. This is schematically illustrated in Figure 1A. A buffer amplifier (LTC6079, Analog Devices, Wilmington, MA, USA) and a 3-channel, 16-bit ADC (analog-to-digital converter, AD7792, Analog Devices) were used to read sensor signals. For fluid control, a DC motor driver based on DRV8830 (Texas Instrument, Dallas, TX, USA) and two DC-DC boost converters based on LTC3122 (Analog Devices) were used to activate pump and valves, respectively. The PCB circuit and the sensor were produced following standard PCB workflow. Briefly, double-sided PCB substrates (FR-4) with connecting traces were designed in CAD software (EasyEDA) and then sent out to a PCB manufacturer for fabrication (Shenzhen JDB Technology, Shenzhen, China). An Arduino circuit (Nano33BLE or Nano33IoT) was used for programming and data transmission (Bluetooth or WiFi) via its microcontroller. The entire circuit (custom PCB + Arduino) can be powered by battery (4.5-21VDC) or USB (3.3VDC).

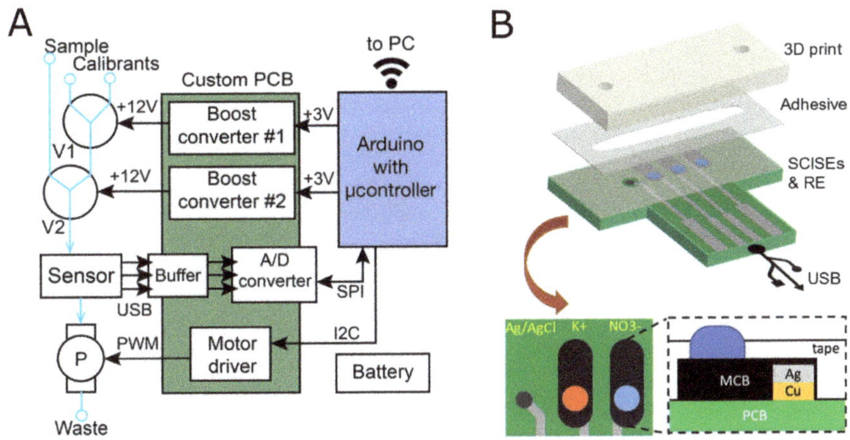

Figure 1. Design of the integrated SCISE sensor system. The system consists of a multiplexed SCISE sensor and a custom PCB for sensor readout and calibration: (**A**) Block diagram of the system illustrates the custom PCB circuit (center) and the supporting fluidic components. Arduino (right side) was used for device control and data transmission. Solenoid valves (V1, V2) and peristaltic pump (P) formed the fluidic circuit (left side). (**B**) Schematic illustration of the multiplexed sensor showing microfluidic integration, sensor layout, and the cross-sectional view of the sensing layers. Note the lateral shift of the ion-selective membrane on top of the mesoporous carbon black (MCB) layer to prevent water layer formation.

Miniaturized peristaltic pump (RP-Q1, H12 × W14 × L30 mm) and solenoid valves (EXAKN-3, φ14.0 × H42.3 mm) were obtained from Takasago Fluidic Systems (Nagoya, Japan). During operation, the pump was placed downstream of the sensor to pull fluid across the sensor, with an adjusted output of 0.2–0.45 mL/min. The 3-way solenoid valves were connected in series and allowed selection of up to 3 reagents.

2.3. Sensor Fabrication

Multiplexed SCISE sensors were prepared based on a process developed previously [14]. The PCB substrates were fabricated using standard PCB technology as described above. In this case, immersion silver was applied as the surface finish. The K^+ and NO_3^- SCISEs were prepared by subsequently drop-casting the MCB suspension (4 µL) and the membrane cocktails (2.8 µL) onto the sensing areas defined on the PCB substrate, using adhesive tapes (3M #9964) for patterning. The sensor structure and layer stacking are illustrated in Figure 1B. The resulting dimensions of the ISEs were 0.06 mm × 1.5 mm (h × w). The ISEs were left to dry in ambient air at room temperature for at least one day before use.

Silver/silver chloride (Ag/AgCl) reference electrodes were prepared by electroplating on the original Ag layer of the PCB, using an industrial plating solution (Technic). To avoid contamination, electroplating was performed before preparing SCISEs. To plate silver, a DC current (−25 µA) was applied for 40 min with a potentiostat (Gamry Reference 600+) and a gold disc (BASi) as counter electrode. After plating, the electrode was cleaned with abundant isopropyl alcohol and DI water to remove organic residues. Then, the electrode was chloridized in 0.1 M KCl by applying a DC current (+ 25 µA) for 20 min, using the same cell setup as silver plating.

2.4. Microchannel Fabrication

The sensor was enclosed into a straight microchannel (17 mm × 2.5 mm × 0.17 mm) constructed by adhesive bonding of a 3D-printed enclosure directly onto the sensor PCB (Figure 1B). The enclosure contained two through-holes as input/output ports of 1.675 mm in diameter to match the OD of Tygon tubing. The enclosure was designed in SolidWorks and printed using a stereolithography printer (Form 2, Formlabs, Somerville, MA, USA), with a vertical printing orientation (slightly tilted). The printed part was cleaned with isopropyl alcohol and cured in ambient air at room temperature overnight. For better visualization, one side of the printed enclosure was polished using sandpaper (up to 15,000 grit) to produce a transparent surface finish. Bonding of the enclosure to PCB was accomplished using double-sided tape (3M #9474LE), which was patterned in the shape of the channel with a laser cutter (HP2436, Boss Laser, FL). The channel was assembled by hand pressing all the layers with the sensor as the substrate, the tape as the spacer (sidewalls), and the 3D print as the top. After assembly, leakage test was conducted by running continuous flow of dye solution through the channel using a syringe pump (New Era Pump Systems, Inc., Farmingdale, NY, USA), with stepwise increase in flow rate (0.05 to 10 mL/min). At each flow rate, channel integrity was examined visually.

2.5. Sensor Characterization

Potentiometric response of the ISEs was recorded (vs. the electroplated Ag/AgCl) in the microchannel, using either the custom PCB circuit developed here or a dual-channel pH meter (Model 225, Denver Instruments). A series of KNO_3 (10^{-7} to 10^{-1} M) was prepared as the electrolyte standard, with NaCl added to provide constant Cl^- background ($[Cl^-]$ = 10 mM) for the Ag/AgCl reference. For testing hydrodynamic conditions, a constant flow (0.45 mL/min) was driven across the sensor for about 5 min and then the flow was stopped for another 5 min before switching to a higher concentration. At each step, the average sensor response in the last 30 s was used for calibration. For long-term characterization as well as self-calibration, a programmed sequence consisting of 25 s flow period and 60 s static period was implemented, while switching between 10^{-3} and 10^{-2} M KNO_3. At each concentration, the sensor's response was taken by averaging data from the last 10 s of the static period.

Potentiometric response of the electroplated Ag/AgCl electrode was measured against a commercial Ag/AgCl electrode (MW-2030, BASi) in bulk solution (1 mM KNO_3 + 10 mM NaCl), using either the pH meter (Denver Instruments) or a bipotentiostat (WaveDriver 20, Pine Instrument, PA). Data were processed and visualized in LabVIEW, MATLAB, or Arduino IoT Cloud in case of WiFi transmission. At the end of each test, the microchannel was flushed with air and the sensors were stored in dry conditions and against light.

2.6. Plant Sap Test

Xylem sap from maize (*Zea mays*) was collected following the root pressure method described previously [15]. Briefly, plants were either well-watered or not watered 2–3 h prior to collection. Then, an incision was made by surgical blade right above the first stem node to remove the top part. Silicone tubes were inserted into the cut site to collect root exudate. The cut site was rinsed with DI water to remove tissue residues and the first ~100 µL sap was discarded. Collection last for ~4 h, which typically resulted in a sap volume of 1–5 mL per plant. The collected sap was stored at freezer temperature (−20 °C) until further testing.

During testing, sap samples were pre-diluted 10× with a solution of 1 mM KNO_3 + 10 mM NaCl. After an initial conditioning phase (5–10 min), a program of pump and valve actions was implemented such that the sensor was calibrated before and after each sap measurement with standard solutions (1/10 mM KNO_3 + 10 mM NaCl). The EMF value at each step was calculated by averaging the last 10 data points at stopped-flow. Ion concentrations (K^+ and NO_3^-) from the original sample were calculated by $C_{unkown} = (C_{dilute} - 0.9) \times 10$ (mM), where C_{dilute} was the concentration of diluted sample. For testing with commercial K^+ and NO_3^- ISEs (9319BN, 9307BN, ThermoOrion, Waltham, MA, USA), the original sap samples were diluted 10–20× and measured following both 'Direct Calibration' and 'Known Addition' protocols according to the manufacturer [16]. The accuracy of our measurements was calculated by dividing the concentrations measured with our and the commercial sensors.

3. Results and Discussion

3.1. System Design

The key feature of the sensor system is the ability to perform automatic sensor calibrations at low cost and in a small form factor. The fully assembled sensor system is shown in Figure 2A. The 3D-printed enclosure is approx. 8 cm square, and about 5 cm deep; however, the size of the enclosure is primarily driven by the size of the used fluid reservoirs and a much more compact arrangement should be possible, especially with lower volume or external fluid compartments. Instead of relying on external fluid control instruments that are usually bulky and expensive, we integrated flow-control modules into the custom PCB circuit (Figure 2B). The motor driver controls the pump speed via pulse-width modulation (PWM). The boost converters control the switching of valves by stepping up voltage input (+3VDC) to the rated voltage of solenoids (+12VDC) while featuring output disconnect during shutdown to save power. Miniaturized peristaltic pump and solenoid valves were selected due to their compact sizes and lower costs than common syringe pumps and rotary valves. While microfluidic pumps and valves have also been widely developed, they increase the total cost of sensor fabrication and tend to suffer leakage issues [17,18].

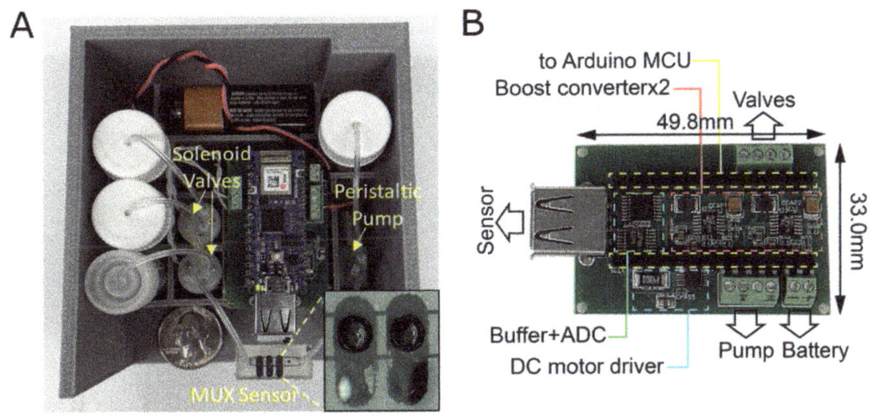

Figure 2. Integrated SCISE sensor system for long-term, in situ measurements: (**A**) Photograph of the fully assembled system. The Arduino board was connected to the custom PCB by stacking. (**B**) Photograph of the custom PCB circuit showing the layout of functional modules.

Power consumption is important for battery-powered devices. The system here draws a current of ~7/30/250 mA, respectively, during default (measuring alone), pumping, and solution switching (when both pump and valve are activated) modes. With a single 9 V battery, the cut-off voltage of the device is found to be ~7.5 V (below which the boost converters could not generate enough current to activate the solenoid valves). Based on these results, we estimate that with a single battery charge the device can continuously run for up to 40 h or about 50 automatic calibrations (see Methods and also below). The operating life can be further extended by using a high-capacity battery such as a power bank.

3.2. Sensor Fabrication and Microfluidic Enclosure

The multiplexed sensor containing SCISEs for K^+ and NO_3^- ions and a Ag/AgCl reference electrode was fabricated on PCB substrates due to the cost-effectiveness of the process and its potential for high-throughput manufacturing. Mesoporous carbon black (MCB) was used as the transducing layer to stabilize the potentiometric response of SCISEs due to its high surface area and good conductivity [14,19]. To suppress the formation of a water layer under the membrane, the ion-selective membranes were shifted laterally from the underlying conducting layer, as shown in Figure 1B, and were discussed in detail in our earlier work [14].

To enable self-calibration capability (and flow-through analysis), the sensor was integrated with a microfluidic flow cell constructed with adhesive tapes and 3D printing. While a variety of microfabrication techniques, such as soft lithography, hot embossing, and injection molding, can be used to make microfluidic channels [20–25], the simplicity and versatility of pressure-sensitive adhesives and 3D printing make them ideal for rapid prototyping of larger microfluidic structures [26,27]. As the spacer layer, the double-coated tape (#9474LE) has proven to be leak-free and chemically inert to the sensor during months of measurement. The 3D-printed top part provided a direct tubing connection via through-holes. Due to the limited resolution of our low-cost 3D printer (~50 μm), it was difficult to consistently print perfectly matched inlets for the tubing and occasional bubbles were introduced at the tubing interface. Nevertheless, when an air-tight interface was formed, the microchannel can sustain flow rates of at least 10 mL/min, which is sufficient for many microfluidic applications and the sensor system herein (where flow rate $\leq$ 0.45 mL/min was used).

3.3. Sensor Performance

A reliable reference electrode is a prerequisite for a stable potentiometric sensor; thus, we first tested the potentiometric response of the electroplated Ag/AgCl electrodes, as shown in Figure 3. The electrodes displayed an average drift of -2.7 ± 0.8 mV (-0.044 mV/min) within the first hour and a baseline reproducibility of 112.8 ± 4.0 mV. Such moderate drift and reproducibility are deemed sufficient for fast analysis, although long-term measurements would require regular calibration. Better performance may be achieved on PCB substrates by electroless silver plating [28] or ink-jet printing [29].

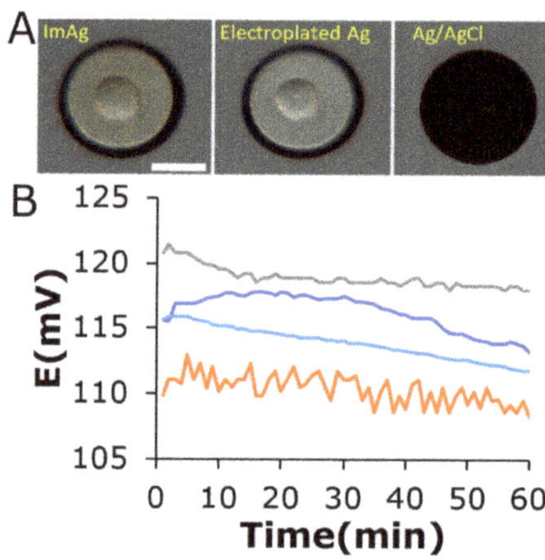

Figure 3. (**A**) Electroplated Ag/AgCl reference electrode. Photographs (left to right) of the original immersion silver layer of PCB, electroplated silver, and finished Ag/AgCl electrode after chloridization. Scale bar = 0.5 mm. (**B**) Potentiometric response of Ag/AgCl reference electrodes (n = 4) against commercial silver chloride electrode (Ag/AgCl 3 M NaCl). Electrolyte: 1 mM KNO_3 + 10 mM NaCl.

The dynamic response of the sensor (SCISE vs. electroplated Ag/AgCl) when subjected to alternating flow and static conditions showed that hydrodynamics did not exert much influence on the response. Results in Figure 4A show that at concentrations > 10^{-5} M, the potential differences (ΔE) between 0.45 mL/min flow and static phases were less than 3 mV. The influence was larger at low concentrations (<10^{-6} M), especially for the K^+ sensor which showed ΔE >20 mV and a much slower response in the static phase. These results can be explained by the leaching of primary ions (K^+ or NO_3^-) from the membrane that raised their local concentrations in the solution. On the other hand, the flow keeps the local concentration to the actual value by constantly refreshing the solution. Despite some flow sensitivity, the sensors showed a generally rapid response (~20 s) to concentration changes (Figure 4A, inset).

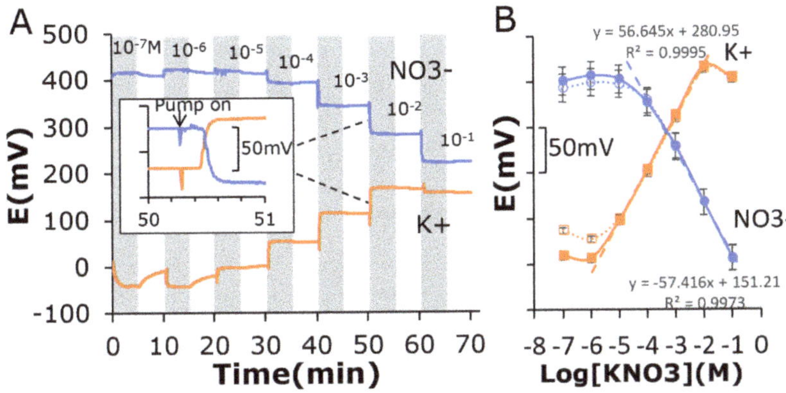

Figure 4. Potentiometric response of the multiplexed sensor (K^+ and NO_3^- electrodes and Ag/AgCl reference) in microchannel under alternating flow (shaded) and static conditions: (**A**) Real-time sensor response. Flow rate: 0.45 mL/min. Electrolyte: KNO_3 + 10 mM NaCl. Inset: enlarged view showing sensor's response to changing concentration. (**B**) Calibration curves of the NO_3^- and K^+ ISEs (3 electrodes for each ion). Solid lines indicate flow condition and dashed lines indicate static condition.

Both the K^+ and NO_3^- electrodes displayed near-Nernstian response over the three-decade concentration range, which covers typical ranges in biological and environmental samples. Figure 4B shows that the K^+ electrodes exhibited a slope of 56.6 mV/dec from 10^{-5} to 10^{-2} M, and the NO_3^- electrodes displayed −57.4 mV/dec from 10^{-4} to 10^{-1} M. The limits of detection (LOD) were calculated to be 1.9×10^{-6} M and 3.6×10^{-5} M for K^+ and NO_3^- electrodes, respectively, with the flow condition showing a slightly lower LOD than the static condition. With NaCl (10 mM) as the background electrolyte, selectivity (LogP) can also be derived and was determined to be −3.72 for the K^+ electrode (vs. Na^+) and −2.44 for the NO_3^- electrode (vs. Cl^-). These values agree with the results from capillary-based electrodes in the bulk solution [19], suggesting that neither the PCB substrate nor the adhesive tape affected sensor selectivity. In the upper range, the signal of K^+ electrodes eventually flat out beyond 10 mM due to the so-called co-ion interference of NO_3^- [30]. Further extension of the upper limit may be achieved by increasing the concentration of ionic sites (TPB) in the membrane.

Evaluation of the long-term performance showed that the sensors retained most of their sensitivity (95% for NO_3^- and 91% for K^+ ISEs) after 3 weeks (Figure 5A). The gradual decay may be linked to the leaching or breakdown of membrane ingredients (e.g., plasticizer or ionic sites), which was thought to have also resulted in decrease in sensor conductivity and capacitance [14]. In addition, the sensors have maintained their selectivity over the testing period, with the NO_3^- ISEs displaying selectivity coefficients of −2.4 to −2.5 (vs. Cl^-) and the K^+ ISEs of −3.7 to −4.2 (vs. Na^+) (Figure 5B). The maintenance of selectivity ensured that the sensors would function properly in complex samples. In terms of baseline (Figure 5C), the sensors showed an average drift of 6–11 mV per day (mV/d) in the first week and have stabilized since then with the drift typically below 3 mV/d. However, even with a drift of 3 mV/d, the error could be as high as 13% for daily measurement. Therefore, the sensors here would need regular calibration in the long term, especially for the first 2 weeks. For calibration-free sensors, drift on the order of μV/h was suggested [10].

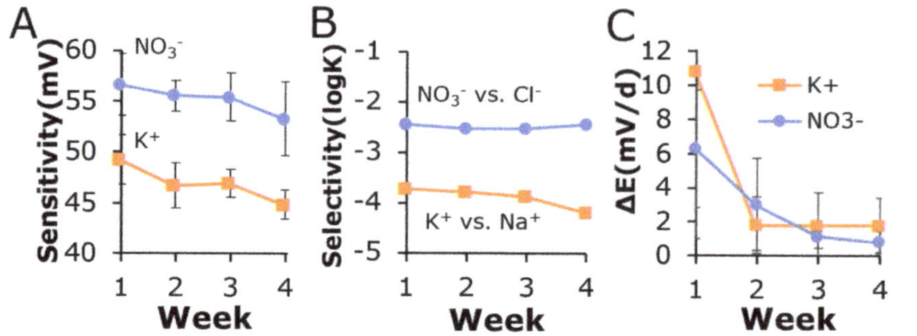

Figure 5. Long-term sensor performance: (**A**) Sensitivity of ISEs. The sensitivities were measured by two-point calibration between 1 and 10 mM KNO_3. (**B**) Selectivity of ISEs. For K^+ electrodes, selectivity was measured against Na^+ ions; for NO_3^- electrodes, selectivity was against Cl^- ions. (**C**). Baseline drift of ISEs at 1 mM KNO_3 (in mV per day, mV/d) over a 4-week period. Reference: electroplated Ag/AgCl. For all measurements, 10 mM NaCl was added as background. At least 3 electrodes were used for each measurement. Sensors were stored dry in air between measurements.

3.4. Self-Calibration and Plant Sap Test

To evaluate the sensor system for in situ measurements, the response of the sensor was measured in an automated two-point calibration (Figure 6). At each step, the sensor was measured for 60 s at default followed by solution switching for 25 s. These parameters were chosen to ensure complete rinsing of the channel while minimizing reagent consumption. The sample volume used for each measurement was ~187.5 μL. As shown in Figure 6, after about two calibrating cycles, both K^+ and NO_3^- sensors showed a highly reproducible response, with the K^+ ISEs exhibiting 20.4 ± 0.45 and 68.4 ± 0.45 mV for the last five cycles at 1 and 10 mM KNO_3, respectively, and the NO_3^- ISEs showing 299.1 ± 0.18 and 244.9 ± 0.36 mV. The high reproducibility indicates that the sensor system can robustly perform self-calibration.

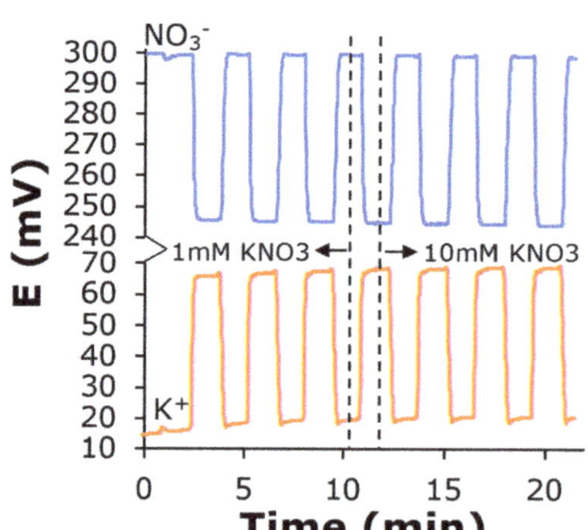

Figure 6. Automated two-point calibration. Potentials were recorded for 1 min at stopped-flow and then the channel was rinsed with a new solution for 25 s. Electrolyte: KNO_3 + 10 mM NaCl. Reference: electroplated Ag/AgCl.

The system was further applied to detect ionic nutrients in plant sap (*Z. mays*). As shown in Figure 7A, a program of pump and valve actions was implemented so that multiple sap samples were analyzed in a continuous manner and the sensors were calibrated before and after each sample. Based on the calibrations, the concentrations of K^+ and

NO_3^- in the original sap were calculated to be 1–20 mM (Figure 7B, Table 1), which were consistent with reported values in maize [15,31]. The low NO_3^- levels in some samples indicate that plants were most likely suffering from nitrogen deficiency. Moreover, the results of our sensors were consistent with that of commercial ISEs, with typical differences of $\leq$3 mM and correlation coefficients >0.95. The mean accuracy of our measurement against commercial sensors was 105.7% for K^+ ions. The NO_3^- sensor showed a small systematic bias (~ +1.3 mM), with respect to commercial sensors, possibly due to interference from Cl^- ions (which were suppressed in commercial ISEs with the use of an ionic strength adjustor). When such bias was corrected by subtraction, the adjusted mean accuracy of measurement was calculated to be 95.2% for NO_3^-.

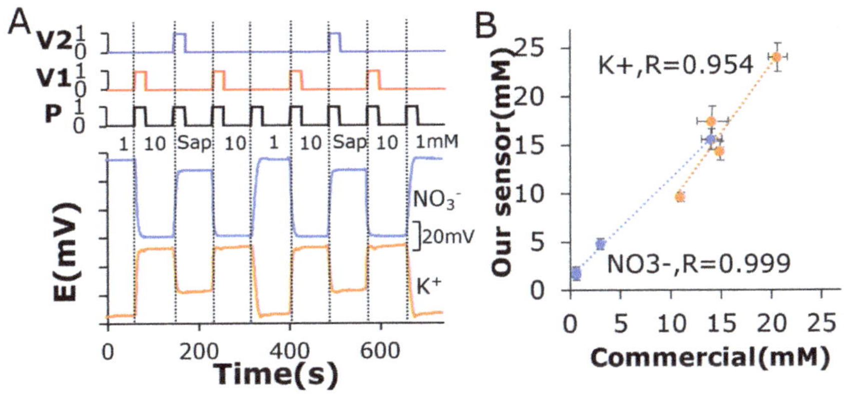

Figure 7. Xylem sap test: (**A**) Real-time sensor response in a programmed sequence of pump (P) and valves (V1,V2). The sensors were calibrated before and after each sap measurement with KNO_3 (1 and 10 mM) + NaCl. (**B**) Comparison of ion concentrations in original sap between our sensors and commercial ISEs. Sap samples were prediluted 10× with 1 mM KNO_3 (+ 10 mM NaCl) for our sensor and 10–20× for commercial ISEs.

Table 1. Concentrations of K^+ and NO_3^- ions in xylem sap (*Z. mays*) measured by our sensors and commercial ISEs (n is the number of measurements).

Sample	K^+_Our Sensor (mM)	K^+_Commercial (mM)	NO_3^-_Our Sensor (mM)	NO_3^-_Commercial (mM)
1	9.59 ± 0.48 (n = 16)	10.98 ± 0.15 (n = 3)	1.63 ± 0.57 (n = 16)	0.62 ± 0.19 (n = 3)
2	14.37 ± 0.98 (n = 16)	14.94 ± 0.34 (n = 3)	4.76 ± 0.56 (n = 16)	3.01 ± 0.2 (n = 3)
3	17.46 ± 1.5 (n = 4)	14.17 ± 1.59 (n = 4)	1.9 ± 0.52 (n = 4)	0.63 ± 0.23 (n = 4)
4	24.02 ± 1.52 (n = 8)	20.64 ± 0.92 (n = 8)	15.59 ± 1.05 (n = 12)	14.07 ± 1.03 (n = 8)

4. Conclusions

This work demonstrated a self-calibrating sensor system by integrating SCISEs with flow-control modules into the same readout circuit. The resulting sensor system is compact, cost-effective, and robust, with the potential for in situ measurements in a variety of agricultural, biomedical, environmental, or geochemical applications. In the future, the capability of the system can be further expanded. For example, the system currently supports multiplexed sensing of three ion analytes via USB-A interface; the number of analytes can be readily increased using an alternative electronic interface such as HDMI (16-pin) or USB-C (24-pin). At the same time, the sensors can be tailored to other ions by simply varying the selective components (ionophores/ion exchangers). In addition, a microcontroller unit can be integrated into the custom PCB circuit to further reduce the footprint of the electronics. Furthermore, chip-based potentiostat circuits can be integrated to enable other sensing schemes (e.g., amperometry or voltammetry). The fluid control modules of the system can also be readily adapted into other analytical systems that require automated sensor calibration.

Author Contributions: Z.Z.: Conceptualization, Methodology, Investigation, Data Curation, Formal analysis, Visualization, Writing—Original Draft. E.B.: Investigation, Writing—Review and Editing. I.P.: Conceptualization, Supervision, Writing—Review and Editing. All authors have read and agreed to the published version of the manuscript.

Funding: This research was funded by the National Science Foundation and the industrial members of the Center for Advanced Design and Manufacturing of Integrated Microfluidics (NSF I/UCRC award IIP-1841473).

Institutional Review Board Statement: Not applicable.

Informed Consent Statement: Not applicable.

Data Availability Statement: The data that support the findings of this study are available from the corresponding author upon reasonable request.

Acknowledgments: We are grateful to Dylan Lynch at UIC Engineering Makerspace for assistance with the laser cutter, and Mark Lehman at 3M for tape samples.

Conflicts of Interest: The authors declare that they have no competing interests.

References

1. Nemiroski, A.; Christodouleas, D.; Hennek, J.W.; Kumar, A.; Maxwell, E.J.; Fernández-Abedul, M.; Whitesides, G.M. Universal mobile electrochemical detector designed for use in resource-limited applications. *Proc. Natl. Acad. Sci. USA* **2014**, *111*, 11984–11989. [CrossRef] [PubMed]
2. Lauks, I.R. Microfabricated Biosensors and Microanalytical Systems for Blood Analysis. *Acc. Chem. Res.* **1998**, *31*, 317–324. [CrossRef]
3. Sheikh, M.; Qassem, M.; Triantis, I.F.; Kyriacou, P.A. Advances in Therapeutic Monitoring of Lithium in the Management of Bipolar Disorder. *Sensors* **2022**, *22*, 736. [CrossRef] [PubMed]
4. Lynch, A.; Diamond, D.; Leader, M. Point-of-need diagnosis of cystic fibrosis using a potentiometric ion-selective electrode array. *Analyst* **2000**, *125*, 2264–2267. [CrossRef]
5. Gao, W.; Emaminejad, S.; Nyein, H.Y.Y.; Challa, S.; Chen, K.; Peck, A.; Fahad, H.M.; Ota, H.; Shiraki, H.; Kiriya, D.; et al. Fully integrated wearable sensor arrays for multiplexed in situ perspiration analysis. *Nature* **2016**, *529*, 509–514. [CrossRef]
6. Alizadeh, A.; Burns, A.; Lenigk, R.; Gettings, R.; Ashe, J.; Porter, A.; McCaul, M.; Barrett, R.; Diamond, D.; White, P.; et al. A wearable patch for continuous monitoring of sweat electrolytes during exertion. *Lab Chip* **2018**, *18*, 2632–2641. [CrossRef]
7. Huang, S.; Shih, W.; Chen, Y.; Wu, Y.; Chen, L. Ion composition profiling and pattern recognition of vegetable sap using a sol-id-contact ion-selective electrode array. *Biosens. Bioelectron.* **2021**, *9*, 100088.
8. Cuartero, M.; Crespo, G.A.; Cherubini, T.; Pankratova, N.; Confalonieri, F.; Massa, F.; Tercier-Waeber, M.-L.; Abdou, M.; Schaefer, J.; Bakker, E. In Situ Detection of Macronutrients and Chloride in Seawater by Submersible Electrochemical Sensors. *Anal. Chem.* **2018**, *90*, 4702–4710. [CrossRef]
9. Cuartero, M.; Pankratova, N.; Cherubini, T.; Crespo, G.A.; Massa, F.; Confalonieri, F.; Bakker, E. In Situ Detection of Species Relevant to the Carbon Cycle in Seawater with Submersible Potentiometric Probes. *Environ. Sci. Technol. Lett.* **2017**, *4*, 410–415. [CrossRef]
10. Rousseau, C.R.; Bühlmann, P. Calibration-free potentiometric sensing with solid-contact ion-selective electrodes. *TrAC Trends Anal. Chem.* **2021**, *140*, 116277. [CrossRef]
11. Calvo-López, A.; Puyol, M.; Casalta, J.; Alonso-Chamarro, J. Multi-parametric polymer-based potentiometric analytical microsystem for future manned space missions. *Anal. Chim. Acta* **2017**, *995*, 77–84. [CrossRef] [PubMed]
12. Ceresa, A.; Sokalski, T.; Pretsch, E. Influence of key parameters on the lower detection limit and response function of solvent polymeric membrane ion-selective electrodes. *J. Electroanal. Chem.* **2001**, *501*, 70–76. [CrossRef]
13. Lindner, E.; Gyurcsányi, R.; Buck, R. Tailored Transport Through Ion-Selective Membranes for Improved Detection Limits and Selectivity Coefficients. *Electroanalysis* **1999**, *11*, 695–702. [CrossRef]
14. Zhang, Z.; Papautsky, I. Solid Contact Ion-selective Electrodes on Printed Circuit Board with Membrane Displacement. *Electroanalysis* **2022**. [CrossRef]
15. Goodger, J.Q.D.; Sharp, R.E.; Marsh, E.L.; Schachtman, D.P. Relationships between xylem sap constituents and leaf conductance of well-watered and water-stressed maize across three xylem sap sampling techniques. *J. Exp. Bot.* **2005**, *56*, 2389–2400. [CrossRef]
16. Thermo Fisher Scientific. Potassium Ion Selective Electrode User Guide. Available online: www.thermofisher.com (accessed on 31 December 2022).
17. Laser, D.; Santiago, J. A review of micropumps. *J. Micromech. Microeng.* **2004**, *14*, R35–R64. [CrossRef]
18. Oh, K.; Ahn, C. A review of microvalves. *J. Micromech. Microeng.* **2006**, *16*, R13–R39. [CrossRef]

19. Zhang, Z.; Papautsky, I. Miniature Ion-selective Electrodes with Mesoporous Carbon Black as Solid Contact. *Electroanalysis* **2021**, *33*, 2143–2151. [CrossRef]
20. Merkel, T.; Graeber, M.; Pagel, L. A new technology for fluidic microsystems based on PCB technology. *Sens. Actuators A Phys.* **1999**, *77*, 98–105. [CrossRef]
21. Li, J.; Wang, Y.; Dong, E.; Chen, H. USB-driven microfluidic chips on printed circuit boards. *Lab Chip* **2013**, *14*, 860–864. [CrossRef]
22. Kontakis, K.; Petropoulos, A.; Kaltsas, G.; Speliotis, T.; Gogolides, E. A novel microfluidic integration technology for PCB-based devices: Application to microflow sensing. *Microelectron. Eng.* **2009**, *86*, 1382–1384. [CrossRef]
23. Burdallo, I.; Jimenez-Jorquera, C.; Fernández-Sánchez, C.; Baldi, A. Integration of microelectronic chips in microfluidic systems on printed circuit board. *J. Micromech. Microeng.* **2012**, *22*, 105022. [CrossRef]
24. Ghanim, M.H.; Abdullah, M.Z. Design of disposable DNA biosensor microchip with amperometric detection featuring PCB substrate. *BioChip J.* **2013**, *7*, 51–56. [CrossRef]
25. Marshall, L.A.; Wu, L.L.; Babikian, S.; Bachman, M.; Santiago, J.G. Integrated Printed Circuit Board Device for Cell Lysis and Nucleic Acid Extraction. *Anal. Chem.* **2012**, *84*, 9640–9645. [CrossRef] [PubMed]
26. Nath, P.; Fung, D.; Kunde, Y.A.; Zeytun, A.; Branch, B.; Goddard, G. Rapid prototyping of robust and versatile microfluidic components using adhesive transfer tapes. *Lab Chip* **2010**, *10*, 2286–2291. [CrossRef]
27. Bhattacharjee, N.; Urrios, A.; Kang, S.; Folch, A. The upcoming 3D-printing revolution in microfluidics. *Lab Chip* **2016**, *16*, 1720–1742. [CrossRef]
28. Moschou, D.; Trantidou, T.; Regoutz, A.; Carta, D.; Morgan, H.; Prodromakis, T. Surface and Electrical Characterization of Ag/AgCl Pseudo-Reference Electrodes Manufactured with Commercially Available PCB Technologies. *Sensors* **2015**, *15*, 18102–18113. [CrossRef]
29. Papamatthaiou, S.; Zupancic, U.; Kalha, C.; Regoutz, A.; Estrela, P.; Moschou, D. Ultra stable, inkjet-printed pseudo reference electrodes for lab-on-chip integrated electrochemical biosensors. *Sci. Rep.* **2020**, *10*, 17152. [CrossRef]
30. Bühlmann, P.; Amemiya, S.; Yajima, S.; Umezawa, Y. Co-Ion Interference for Ion-Selective Electrodes Based on Charged and Neutral Ionophores: A Comparison. *Anal. Chem.* **1998**, *70*, 4291–4303. [CrossRef]
31. Bahrun, A.; Jensen, C.; Asch, F.; Mogensen, V. Drought-induced changes in xylem pH, ionic composition, and ABA concentration act as early signals in field-grown maize (*Zea mays* L.). *J. Exp. Bot.* **2002**, *53*, 251–263. [CrossRef]

chemosensors

MDPI

Article

Potentiometric Surfactant Sensor for Anionic Surfactants Based on 1,3-dioctadecyl-1*H*-imidazol-3-ium tetraphenylborate

Nikola Sakač [1,*], Dubravka Madunić-Čačić [1,2], Dean Marković [3], Lucija Hok [4], Robert Vianello [4], Valerije Vrček [5], Bojan Šarkanj [6], Bojan Đurin [7], Bartolomeo Della Ventura [8], Raffaele Velotta [8] and Marija Jozanović [9,*]

1 Faculty of Geotechnical Engineering, University of Zagreb, 42000 Varaždin, Croatia
2 Saponia Chemical, Pharmaceutical and Foodstuff Industry, Inc., 31000 Osijek, Croatia
3 Department of Biotechnology, University of Rijeka, 51000 Rijeka, Croatia
4 Laboratory for the Computational Design and Synthesis of Functional Materials, Division of Organic Chemistry and Biochemistry, Ruđer Bošković Institute, 10000 Zagreb, Croatia
5 Department of Organic Chemistry, Faculty of Pharmacy and Biochemistry, University of Zagreb, A. Kovačića 1, 10000 Zagreb, Croatia
6 Department of Food Technology, University North, 48000 Koprivnica, Croatia
7 Department of Civil Engineering, University North, 42000 Varaždin, Croatia
8 Department of Physics "E. Pancini"—Università Di Napoli Federico II, 80126 Napoli, Italy
9 Department of Chemistry, University of Osijek, 31000 Osijek, Croatia
* Correspondence: nikola.sakac@gfv.unizg.hr (N.S.); mjozanovic@kemija.unios.hr (M.J.); Tel.: +385-915830336 (N.S.); +385-996865716 (M.J.)

Citation: Sakač, N.; Madunić-Čačić, D.; Marković, D.; Hok, L.; Vianello, R.; Vrček, V.; Šarkanj, B.; Đurin, B.; Della Ventura, B.; Velotta, R.; et al. Potentiometric Surfactant Sensor for Anionic Surfactants Based on 1,3-dioctadecyl-1*H*-imidazol-3-ium tetraphenylborate. *Chemosensors* **2022**, *10*, 523. https://doi.org/10.3390/chemosensors10120523

Academic Editors: Iulia Gabriela David and Dana Elena Popa

Received: 14 November 2022
Accepted: 7 December 2022
Published: 8 December 2022

Abstract: As anionic surfactants are used as cleaning agents, they pose an environmental and health threat. A novel potentiometric sensor for anionic surfactants based on the 1,3-dioctadecyl-1*H*-imidazol-3-ium tetraphenylborate (DODI–TPB) ionophore is presented. The newly developed approach for DODI–TPB synthesis is faster and simpler than the currently used strategies and follows the green chemistry principles. The DODI–TPB ionophore was characterized by computational and instrumental techniques (NMR, LC–MS, FTIR, elemental analysis) and used to produce a PVC-based DODI–TPB sensor. The sensor showed linear response to dodecylbenzenesulfonate and dodecyl sulfate in concentration ranges of 6.3×10^{-7}–3.2×10^{-4} M and 5.9×10^{-7}–4.1×10^{-3} M, for DBS and SDS, respectively. The sensor exhibits a Nernstian slope (59.3 mV/decade and 58.3 mV/decade for DBS and SDS, respectively) and low detection limits (7.1×10^{-7} M and 6.8×10^{-7} M for DBS and SDS, respectively). The DODI–TPB sensor was successfully tested on real samples of commercial detergents and the results are in agreement with the referent methods. A computational analysis underlined the importance of long alkyl chains in $DODI^+$ and their C–H$\cdots\pi$ interactions with TPB^- for the ionophore formation in solution, thereby providing guidelines for the future design of efficient potentiometric sensors.

Keywords: anionic surfactants; computational analysis; green chemistry; quaternary ammonium compounds; synthetic ionophore; potentiometric sensor

1. Introduction

Surfactants are organic molecules which decrease the tension of the surface. Their structure usually consists of a hydrophilic and a hydrophobic part. Depending on the charge, surfactants are grouped into anionic, cationic, amphoteric, and nonionic (with no charge). Surfactants can be classified as zwitterionics, depending on their charge. Cationic surfactants are used as disinfecting agents and as preservatives [1,2], while anionic surfactants are used for cleaning and washing [3]. Nonionic surfactants are employed for enhancing surfactant activity, reducing foaming, and enhancing cleaning. Anionic surfactants participate in 70% of the global surfactant market, which is expected to increase by 4.5% in the period from 2020 to 2025 due to high demand and standard growth [4].

Anionic surfactants are used in many products of daily life and in the industry sector (recently for electric vehicles [5]), which is why they are also likely to be found in wastewater, rivers, and lakes. Apart from the foaming effect, the negative consequences on the environment include their aptitude to prevent oxygen and gas exchange at the water surface; moreover, they easily degrade the cell membrane—as they have a similar nature to lyophilic molecules in the phospholipid bilayer of the cell, cause skin irritation, etc. [6,7]. Cationic surfactants are the most toxic, compared to the other types of surfactants.

Anionic surfactants in water are usually quantified by highly manual procedures, such as methylene blue active substances (MBASs), two-phase titration [8], or instrumental methods, such as liquid chromatography and gas chromatography [3]. These manual procedures are time-consuming, slow, low in reproducibility, and require specialized personnel, while instrumental methods are faster, but need expert personnel, toxic solvents, and expensive instrumentation, as well as they can often be performed only in confined spaces.

Compared to the above analytical procedures, the ion-selective electrodes (ISEs) for the surfactants are robust, simple, fast, and do not use toxic solvents. They can also be portable and used in field measurements [9]. In addition, ISEs are not affected by color changes or transparency problems. ISEs are also capable of measuring concentrations of anions and cations, depending on which ionophore is used [10–14]. All the above advantages are used to fabricate surfactant sensors for measuring cationic surfactants, nonionic surfactants, and anionic surfactants in aqueous media [1,15]. There are mainly two types of surfactant sensors: solid-state surfactant sensors [16,17] and PVC-based liquid membrane-type surfactant sensors [18]. Membranes based on PVC consist of a high molecular weight PVC and a plasticizer in a 2:1 weight ratio, and usually a 1% ionophore, which is an active substance of the sensing membrane [19]. The plasticizer softens the PVC and serves as a solvent for the highly lipophilic ionophore. Ionophores are poorly soluble and possess high molecular weight, associates (ion-pairs) of the cationic surfactant and anionic surfactant, or some high lipophilic counter ions. The role of the ionophore, as a sensing element, is mainly determined by its potential and ability to handle complex interactions with the analyte in its binding sites [20]. Synthetic ionophores designed according to their complexation constants can provide improved electrochemical ionophores leading to the development of optimized sensors with higher selectivity, reproducibility, and longer sensor lifetime [21]. Different commercially available quaternary ammonium compounds (QACs) have been used as a part of the ionophores for anionic surfactants detection, namely dodecyltrimethylammonium [22], cetyltrimethylammonium [23], hyamine [24], 1,3-dihexadecyl-1*H*-benzo[*d*]imidazol-3-ium recently described by our group [1], and others. The introduction of a designed QAC for potentiometric sensor provides a sensing material with planned and improved properties compared to classical ion exchangers.

The aim of this work is to present a new strategy for the direct synthesis of a new ionophore, namely 1,3-dioctadecyl-1*H*-imidazol-3-ium tetraphenylborate (DODI–TPB), and to use it as a sensing component in the preparation of a new DODI–TPB-based surfactant sensor for the quantification of anionic surfactants by potentiometric titrations in real samples of commercial products. The ionophore was extensively studied by computational and instrumental techniques such as NMR, LC–MS, FTIR-ATR, and elemental analysis. The fabricated DODI–TPB surfactant sensor was characterized analytically in terms of response, pH influence, selectivity, and accuracy.

2. Materials and Methods

2.1. Reagents

Chemical reagents used for organic synthesis were all analytical grade chemicals: 1-bromooctadecane, 1*H*-imidazole, and $NaHCO_3$ (Sigma Aldrich, Darmstadt, Germany) and utilized without further purification.

For measurements of the direct potentiometric response of anionic surfactant, analytical grade sodium dodecylsulfate (SDS) and technical grade sodium dodecylbenzenesulfonate (DBS) (all from Fluka, Buchs, Switzerland) were used. Titrations were performed using benzethonium chloride (Hyamine 1622), hexadecyltrimethylammonium bromide (CTAB), cetylpyridinium chloride (CPC), and 1,3-didecyl-2-methylimidazolium chloride (DMIC) (all analytical grade from Merck, Munich, Germany). Ultrapure deionized water was used for all measurements.

For sensing membrane preparation, a high molecular weight PVC and an analytical grade plasticizer *o*-nitrophenyloctylether (*o*-NPOE) (both from Sigma Aldrich, Darmstadt, Germany) were used. The solvent was tetrahydrofuran (THF) (Merck, Munich, Germany).

For interference measurements a series of organic and inorganic anions were used; acetate, borate, benzoate, bromide, chloride, carbonate, dihydrogenphosphate, EDTA, hydrogen sulfate, nitrate, sulfate, all analytical grade sodium salts (all from Kemika, Zagreb, Croatia).

2.2. Ionophore Synthesis and Characterization

Bisalkylated imidazole product (1) was synthesized according to the literature's procedure with some modifications. Alkylation reaction was conducted by adding 0.25 g 1*H*-imidazole (3.37 mmol) and 200 mg $NaHCO_3$ (2.38 mmol) in anhydrous acetonitrile (10 mL) and stirring at 90 °C for 1 h. Then, the solution was cooled and 5 g 1-bromooctadecane (15.00 mmol) was slowly added to the mixture and continued to be vigorously stirred at 120 °C under an inert nitrogen atmosphere for 48 h. The product was washed with methanol and hexane three times. Flash column purification was conducted in DCM/methanol = 10:0.25. The product was dried under a vacuum at 40 °C for 4 h. The desired product (1) (2.00 g, 3.06 mmol) was obtained in a 90.75% yield.

To prepare the sensing ion-pair DODI–TPB, 1.50 g (2.29 mmol) of 1,3-dioctadecyl-1*H*-imidazol-3-ium bromide (1) was mixed for 6 h with 1.50 g (4.38 mmol) of sodium tetraphenylborate, in 100 mL of acetonitrile. This mixture was filtered, washed with dichloromethane, cooled in a freezer for 24 h, and then again filtered. Dichloromethane was removed by rotary evaporation for 3 h. A total of 3.39 g (mmol) of DODI–TPB ionophore was produced as a white solid (yield: 86.44%). After the evaporation of dichloromethane, the crude DODI–TPB (2) was dried at 80 °C to the constant mass. Such prepared ion-pair was used for sensor membrane fabrication.

The DODI–TPB ion-pair was analyzed by FTIR–ATR spectrometer Spectrum Two (Perkin Elmer, Waltham, MA, USA).

The ^{1}H NMR spectra were recorded at 400 MHz and ^{13}C NMR spectra at 100.613 MHz by a VARIAN INOVA 400 (Varian, Crawley, United Kingdom).

MS spectra were recorded in q1 ms scan mode at API 2000 LC-ESI-MS/MS (Applied Biosystems, Foster City, CA, USA).

PerkinElmer 2400 CHNS/O Series II System was used for the elemental analysis (PerkinElmer Inc., Waltham, MA, USA).

2.3. Computational Details

RESP charges at the HF/6–31G(d) level were used to describe the DODI-cation, while the TPB^- anion was parameterized in line with the literature's recommendations. [25] Both ions were positioned in a 15 Å rectangular box and solvated with 8.553 water molecules and supplemented with Na^+ and Br^- counterions to assure electrical neutrality. Such a system underwent the geometry optimization in AMBER 16 [26] with periodic boundary conditions in all directions, followed by a 30 ps equilibration under NTV conditions and a gradual temperature increase from 0 to 300 K. This was, then, submitted to 300 ns of productive and unconstrained MD simulations, with a time step of 2 fs at a constant pressure of 1 atm and temperature of 300 K, employing a threshold of 11.0 Å to truncate the nonbonded interactions. The binding energies between sensor components, ΔG_{BIND}, were obtained through the MM-PBSA analysis [27,28], in line with our recent reports on similar systems [29–31], utilizing every second snapshot, 75.000 in total, recorded from the entire MD trajectory.

2.4. Preparation of Surfactant SENSOR

The surfactant sensor membrane was prepared by dissolving a high molecular weight PVC in a plasticizer *o*-NPOE (1:2) by sonication. Then, 1% of the ion-pair was added. Next, 2 mL of THF was added and sonicated for 15 minutes. After this, the cocktail was poured into the glass mold and left to dry for 1 day. The membrane was planted at the bottom of the Philips electrode body IS-561 (Supelco, Bellefonte, PA, USA). NaCl (3 M) was used as an inner electrolyte.

2.5. Potentiometric Measurements

Surfactant sensor characterizations of the DODI–TPB surfactant sensor by direct potentiometry were performed by Metrohm 794 Basic Titrino with a stirred paired with Metrohm 781 pH meter (Metrohm, Herisau, Switzerland) and a reference electrode silver/silver (I) chloride electrode with potassium chloride (3 M) electrolyte (Metrohm, Herisau, Switzerland). To obtain the response characteristics, selected cationic surfactants (Hyamine 1622, CPC, CTAB, and DMIC) were incrementally added to the deionized water to reach a concentration range from 1×10^{-8} to 1×10^{-2} M. Response time was set to 90 seconds. An interference study was performed on selected organic and inorganic anions usually found in commercial product formulations or in wastewaters. Solution of interfering anion (0.01 M) was placed in a beaker and CPC was incrementally added (0.5 mM) and measured in 90 seconds intervals. IUPAC fixed interference method was used to calculate the potentiometric selectivity coefficients. The effect of pH change was observed for CPC (0.5 mM) at pH levels from 2 to 12.

Potentiometric titrations were performed by Metrohm 808 Titrando with a stirrer, a reference electrode silver/silver (I) chloride electrode with potassium chloride (3 M) electrolyte, and a Metrohm Tiamo software (all from Metrohm, Herisau, Switzerland). SDS (0.4 mM) and DBS (0.4 mM) were titrated with the corresponding concentration of the selected cationic surfactants Hyamine 1622, CTAB, CPC or DMIC, and a DODI–TPB surfactant sensor as an endpoint indicator. The anionic surfactant contents in 12 commercial products were measured by potentiometric titration with CPC in corresponding concentrations and a DODI–TPB surfactant sensor as an endpoint indicator.

3. Results

3.1. Ionophore Synthesis and Characterization

A new, faster, and simpler way for the direct synthesis of ion-pairs was presented (Scheme 1). The common synthesis of ion-pairs employs anionic and cationic surfactants, or large lipophilic ions [32,33]. In the presented approach, the preparation of the ion-pair was achieved by the direct addition of the counter ion (TPB) into the reaction mixture containing the newly synthesized cationic surfactant. The latter two-steps one-pot process is much faster, simpler, and employs less toxic chemicals, which is consistent with the green chemistry principles. A detailed characterization of DODI-Br by MS, ^{1}H NMR, and ^{13}C NMR spectroscopy and elementary analysis and the detailed data is offered in the supplementary materials (Figures S1–S3). The C2 symmetry of DODI-Br and DODI–TPB simplified their NMR spectra. As seen from the ^{1}H NMR spectrum of DODI-Br, the newly formed triplets of methylene signals of $-CH_2N=$ arrived at 4.37 ppm, while the other methylene signals of alkyl chains range from 1.95 to 1.25 ppm. The ^{1}H NMR signals of ^{1}H-imidazole were identified at 10.57 and 7.41 ppm. The characteristic signal of $-CH_2N=$ in the ^{13}C NMR spectrum arrived at 50.1 ppm. The other carbons of the methylene chain groups appeared between 31.9 and 22.7 ppm. The signals at 137.3 and 121.8 can be attributed to the ^{1}H-imidazole unit. In the IR spectrum C = N, C-N stretching was assigned as strong adsorptions at 1650 and 1150 cm^{-1}. The presence of bromine was confirmed by MS and elemental analysis.

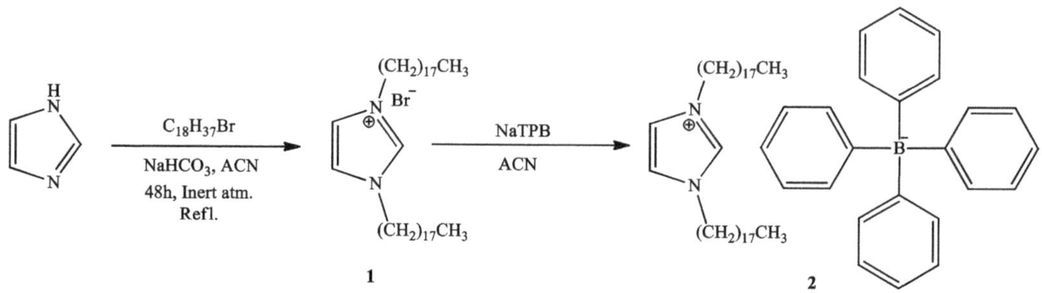

Scheme 1. Synthesis of DODI–TPB ion-pair sensing complex (2) via quaternary alkyl ammonium salt 1,3-dioctadecyl-1*H*-imidazol-3-ium bromide (1).

Characterization of DODI–TPB was also performed. Its ^{1}H NMR spectrum confirmed the presence of the TPB anion. The phenyl -CH = signals were identified in the range of 7.57–7.46 and at 7.00 and 6.83 ppm. Other ^{1}H NMR signals were also consistent with the proposed DODI–TPB structure. Due to the carbon–boron coupling in the ^{13}C NMR spectrum, the signal at 164.0 ppm was a quartet, while the other C atoms of TPB and ^{1}H imidazole were identified at 135.9, 134.9, 125.9, 121.9, and 120.7. ^{13}C-signals of alkyl chains appeared at 49.4, 31.9, and in the range between 29.8 and 14.1 ppm. Similar to DODI-Br, DODI–TPB showed the IR-stretching of C = N, C-N at 1650 and 1150 cm^{-1}. The MS spectrum and elemental analysis were also consistent with the proposed DODI–TPB structure.

3.2. Computational Analysis

Computational analysis was employed to inspect the dynamics and features of the investigated DODI$^+$ cation in the water solution and elucidate intermolecular interactions responsible for its complexation with TPB$^-$. We mainly focused on characterizing the DODI–TPB adduct through electronic, geometric, and thermodynamic features.

Despite containing a rigid imidazole skeleton, DODI$^+$ has two unconstrained C-18 chains, whose flexibility over the anionic TPB$^-$ is clearly evident in the obtained RMSD plots (Figure S4). Interestingly, it appears that it is precisely this feature of DODI$^+$ that is essential for its effective TPB$^-$ recognition through a range of favorable C–H···π interactions with the aromatic phenyl rings in the latter.

DODI–TPB complexation is a favorable, yet reversible, event, due to the adduct formation and its dissociation exchange during the MD simulation. To illustrate that, the matching distance graphs reveal that the relevant N(DODI$^+$)···B(TPB$^-$) distances extend much beyond 40 Å, while taking values below 12 Å in around 48% of the recorded structures (Figure S5). Since DODI$^+$ is not a globular molecule, we identified the latter as the upper limit beyond which no appreciable interactions were observed. In order to quantify these interactions, the calculated MM-PBSA binding free energy shows that the DODI–TPB complex formation is indeed exergonic at $\Delta G_{BIND} = -5.0$ kcal mol^{-1}, which confirms its feasibility (Table 1). To put the latter number in an appropriate context, and given the highly favorable sensing features of the DODI–TPB system, let us mention that precisely the same ΔG_{BIND} value was obtained for the benzimidazole analogue of DODI$^+$ with a slightly shorter C-16 alkyl chain. [29] This indicates that the calculated ΔG_{BIND} value falls into an optimal range as it needs to ensure two opposing aspects, namely the stability of the adduct and the reversibility of its formation, the latter being essential for its sensing features through the potential exchange with other anionic analytes.

Table 1. MM-PBSA calculated binding affinities (ΔG_{BIND}, in kcal mol^{-1}) among the TPB^- anion and selected cations following the molecular dynamics simulation in water.

Cation Component	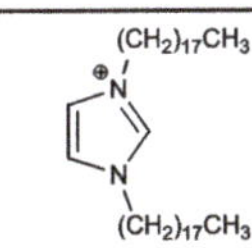	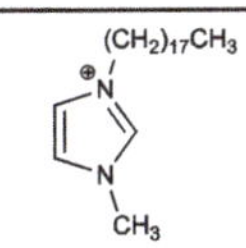	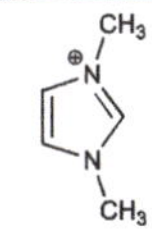	$CH_3(CH_2)_{34}CH_3$
ΔG_{BIND}	−5.0	−2.4	−0.5	−3.2

The representative adduct structure (Figure 1) shows three kinds of favorable interactions, namely (i) the π···π stacking contacts among charged imidazole in $DODI^+$ and a phenyl ring in TPB^-, (ii) the C–H···π interactions involving the central imidazole carbon and the other phenyl ring in TPB^-, and (iii) a range of C–H···π interactions that both C-18 chains form with the remaining phenyl groups in TPB^-. Interestingly, in the elucidated structure, the matching $B(TPB^-)\cdots N(DODI^+)$ distances are 5.4 and 5.7 Å, which is in excellent agreement with the calculated RDF plot, which indicates the largest number of these contacts assuming values around 5.7 Å (Figure 1), and therefore validating the identified structure as representative.

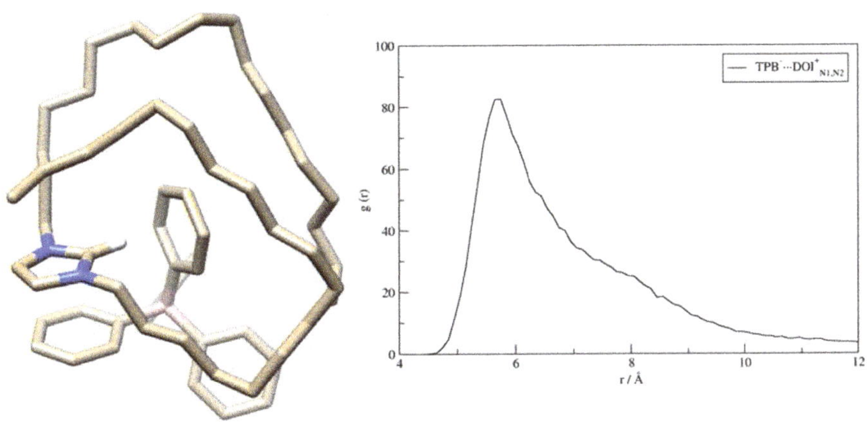

Figure 1. Representative structure of the DODI–TPB complex in the aqueous solution ((**left**); hydrogen atoms omitted for the clarity) as elucidated from the 300 ns molecular dynamics simulation, and the matching RDF graph considering $N(DODI^+)\cdots B(TPB^-)$ distances with the peak value located at 5.7 Å (**right**).

Such vicinity between boron and nitrogen atoms could lead to the assumption that the electrostatic attraction determines the binding among components. Yet, this can be excluded as it follows from the analysis of atomic charges prior and after the adduct formation (Figure S6). The results show that, before binding, the imidazole unit in $DODI^+$ accommodates only one-third of the excess positive charge (0.33 |e|), while the rest is accumulated within its C-18 chains, which is, interestingly, not changed in the formed adduct. In fact, once DODI–TPB is formed, only 2% of the total charge density is exchanged between components, as is evident in the sum of all atomic charges on $DODI^+$ and TPB^- being +1.02 and −0.98 |e|, respectively, which eliminates electrostatic interactions as predominantly responsible for the adduct stability.

In addition, the extent of the π···π stacking interactions was estimated by analyzing distances between the center of mass of the imidazole ring in $DODI^+$ and that of each phenyl group in TPB^-. Adopting the value of 4 Å as the recommended threshold for these contacts [29], it resulted that these interactions take place in only 2.2% of structures occurring during MD simulations. Along these lines, the frequency of the C–H···π interactions involving the central imidazole C–H group was analogously inspected and we counted 5.7% of structures where this was identified. Therefore, both sets of values suggest only a very limiting significance of both of these interactions for the complex formation.

Lastly, due to a large number of C–H groups within both C-18 chains potentially engaging in the C–H$\cdots\pi$ interactions with TPB^-, in evaluating their frequency we have repeated the MD simulations with several model cationic components, having either one or both C-18 chains replaced by a methyl group (Table 1). The results show that when one of the C-18 chains is already substituted by a smaller methyl unit, the binding affinity is reduced in half ($\Delta G_{BIND} = -2.4$ kcal mol^{-1}), while it exhibits an even larger further decrease when no C-18 chains are present ($\Delta G_{BIND} = -0.5$ kcal mol^{-1}). This evidently underlines the crucial effect of the long alkyl chains and highlights this type of interaction as dominant in recognizing and binding the TPB^- in solution. Such a conclusion is further supported by analyzing the situation with a simple alkane, with its length, C-36, equaling the sum of both chains in $DODI^+$. Interestingly, despite being formally uncharged, the affinity of $C_{36}H_{74}$ towards TPB^- surpasses those from both model cationic imidazoles and achieves $\Delta G_{BIND} = -3.2$ kcal mol^{-1} (Table 1), thereby reaching 64% of the affinity of the full $DODI^+$. We believe all of this clearly illustrates the importance of these hydrophobic fragments and their flexibility for the efficient TPB^- binding, and it provides useful guidelines for the future design of efficient potentiometric sensors based on selected ionophores.

3.3. *Sensor Characterization*

3.3.1. Sensor Response to Anionic Surfactants

According to the modified Nernst equation, the electromotive force of the surfactant sensor membrane in the presence of anionic surfactant is the following:

$$E = E^0 - Slog a_{AS^-}$$

where E^0 is a constant potential term, S is a slope in the linear part of the response curve, and a_{AS^-} is an activity of the investigated surfactant anion.

The first step in DODI–TPB surfactant sensor characterization was to observe the direct potentiometric response of the sensor for two anionic surfactants, namely DBS and SDS, over a wide range of concentrations in deionized water. In this way, it is possible to determine the response of the sensor for the desired anions. The response curves and response characteristics for DBS and SDS are presented in Figure 2 and Table 2. For the DBS anion, the linear concentration range was from 6.3×10^{-7} M to 3.2×10^{-4} M with an estimated correlation coefficient (R^2) of 0.9997 in the linear part of the response curve. For the SDS anion, the linear concentration ranged from 5.9×10^{-7} M to 4.1×10^{-3} M with an R^2 of 0.9998 in the linear part of the response curve. The limits of detection (LOD) were calculated according to IUPAC recommendations. [34] For DBS, the calculated LOD were 6.1×10^{-7} M and 5.5×10^{-7} M for SDS. The estimated slope values in the linear response region for DBS were 59.29 ± 0.5 mV/decade and for SDS anion 58.31 ± 0.4 mV/decade.

Table 2. Response characteristics of DODI–TPB surfactant sensor to anionic surfactants DBS and SDS in H_2O, given together with ±95% confidence limits.

Parameters	Anionic Surfactants	
	DBS	**SDS**
Slope (mV/decade)	59.3 ± 0.5	58.3 ± 0.4
Correlation coefficient (R^2)	0.9997	0.9998
Limit of detection (M)	7.1 ± 10^{-7}	6.8 ± 10^{-7}
Useful linear concentration range (M)	6.3×10^{-7} to 3.2×10^{-4}	5.9×10^{-7} to 4.1×10^{-3}

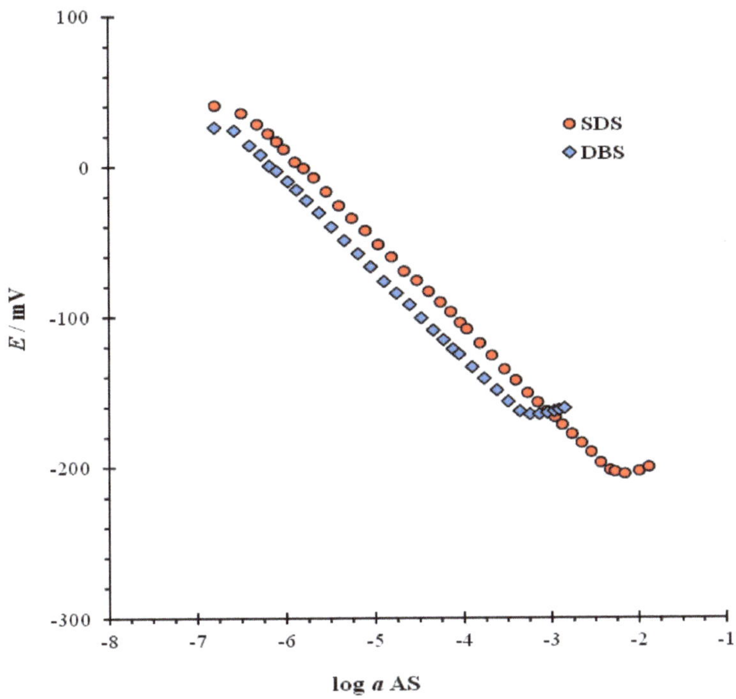

Figure 2. Direct potentiometric response characteristics of DODI–TPB surfactant sensor to anionic surfactants SDS and DBS in H_2O at a wide concentration range, with mean values at ±95% confidence limits.

3.3.2. Selectivity and pH

An interference study was performed to investigate the influence of some common organic and inorganic anions (present in products or waters and wastewaters) on the response characteristics of the DODI–TPB surfactant sensor in the presence of an anionic surfactant SDS in the broad concentration range from 5×10^{-6} to 5×10^{-3} M. The selected interfering anion concentration was 0.01 M. To calculate the selectivity coefficient, a fixed interference method [34] was applied. The selectivity was tested to observe the ability of the sensor to discriminate certain anions in the presence of interfering ions. The results of the calculated selectivity coefficient ($logK^{pot.}_{An_i^-}$) for different inorganic and organic interfering sodium salts are presented in Table 3. The DODI–TPB surfactant sensor showed good selectivity toward SDS in the presence of common interfering anions.

The response of the DODI–TPB surfactant sensor to SDS (0.5 mM) was tested in the range of pH 2 to 13 and is presented in Figure 3. The measured electromotive force was stable in the pH range from 2 to 9, and this pH region could be used in further investigations. At pH 10 to 13, the DODI–TPB surfactant sensor showed higher fluctuation in the signal change, and this region was not suitable for measurements. This likely comes as a result of the fact that both sensor components lack notably acidic/basic sites [35], which is why lowering the pH conditions exerts only a moderate effect, while a significant increase in the latter possibly disintegrates TBP^- due to the introduced OH^- anions that either compete with phenyl ligands for the boron coordination or help with the hydrolysis of either component.

Table 3. Calculated selectivity coefficient ($logK^{pot.}_{An^-_i}$) for different inorganic and organic interfering sodium salts (10 mM) for SDS response with the DODI–TPB surfactant sensor for SDS.

Anion	$logK^{pot.}_{An^-_i}$
Acetate	−4.27
Borate	−4.75
Benzoate	−4.35
Bromide	−4.00
Chloride	−3.92
Carbonate	−3.92
Dihydrogenphophate	−3.92
EDTA	−3.92
Hydrogen sulfate	−3.30
Nitrate	−3.43
Sulfate	−3.48
Toluensulfonate	−3.93

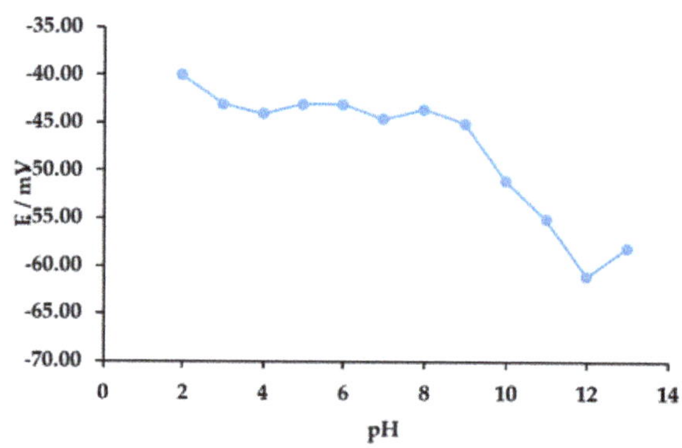

Figure 3. The response characteristics of DODI–TPB surfactant sensor towards SDS (0.5 mM) in deionized water at different pH levels.

3.4. Potentiometric Titrations

3.4.1. Titration of DBS with Selected Cationic Surfactants

Hyamine 1622, CTAB, CPC, and DMIC were used for the potentiometric titrations of DBS (4 mM) with selected cationic surfactants of analytical grade (4 mM). Titration conditions were the same for all measurements and the DODI–TPB surfactant sensor was used as the endpoint indicator. Potentiometric titration curves are shown in Figure 4. All titration curves had a sigmoidal shape with a well-defined signal change in the endpoint region. The DODI–TPB surfactant sensor showed the best signal change in the titrations with DMIC, where the signal change was 382.5 ± 2.4 mV. The signal change for CPC was 350.4 ± 3.1 mV, for CTAB 335.6 ± 1.9 mV, while the lowest signal change was obtained for Hyamine 1622, at 316.6 ± 3.2 mV. From the titration curves, the corresponding first derivatives were calculated ($\Delta E/\Delta V$) and included in Figure 4. Analogous to the titration curves data, the first derivative values for DMIC showed the highest change in the endpoint, 79.4 mV/mL, with a sharp peak. The first derivative values for the other three cationic surfactants, CPC, CTAB, and Hyamine1622, were 77.7, 77.3, and 62.1 mV/mL, respectively, with well-defined peaks. The falling properties of the cationic titrants for potentiometric titrations of DBS using the DODI–TPB surfactant sensor as the endpoint indicator were DMIC > CPC > CTAB > Hyamine 1622. All the studied cationic surfactants were suitable for DBS detection using the DODI–TPB surfactant sensor.

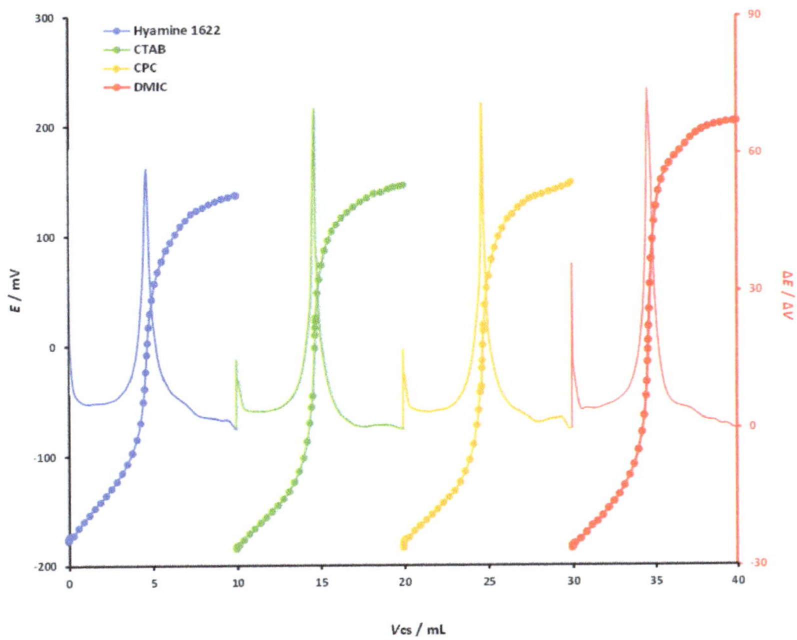

Figure 4. Potentiometric titration curves, and first derivatives, of DBS (0.4 mM) with four different cationic surfactants (0.4 mM) used as a titrant: Hyamine 1622, CTAB, CPC, and DMIC. The titration curves and their first derivatives were rearranged for the sake of clarity.

3.4.2. Titration of SDS with Selected Cationic Surfactants

For the potentiometric titrations of SDS, the same approach was used as for DBS. Hyamine 1622, CTAB, CPC, and DMIC were used for the potentiometric titrations of SDS (4 mM) with selected cationic surfactants of analytical grade (4 mM). Titration conditions were the same for all measurements, and the DODI–TPB surfactant sensor was used as the endpoint indicator. Potentiometric titration curves are shown in Figure 5. All titration curves had a sigmoidal shape with well-defined signal changes in the area of the endpoint. The DODI–TPB surfactant sensor showed the best signal change in the titrations with DMIC where the signal change was 376.5 ± 3.8 mV. The signal change for CPC was 319.3 ± 3.7 mV, for CTAB 303.9 ± 2.5 mV, while the lowest signal change was obtained for Hyamine 1622 with 272.4 ± 3.1 mV. From the titration curves, the corresponding first derivatives were calculated ($\Delta E/\Delta V$) and included in Figure 5. Analogous to the titration curve data, the first derivative values for DMIC showed the largest change in the endpoint, 79.6 mV/mL, with a sharp peak. The first derivative values for the three other cationic surfactants, CPC, CTAB, and Hyamine1622, were 75.7, 75.4, and 53.5 mV/mL, respectively, with well-defined peaks. The falling properties of the cationic titrants for potentiometric titrations of DBS using the DODI–TPB surfactant sensor as an endpoint indicator were DMIC > CPC > CTAB > Hyamine 1622. All of the studied cationic surfactants were suitable for SDS detection with the DODI–TPB surfactant sensor.

The standard addition of SDS and DBS at two concentration levels (0.1 and 0.01 mM) was used to estimate the accuracy of the DODI–TPB surfactant sensor. Selected cationic surfactants (Hyamine 1622, CTAB, CPC, and DMIC) were used as titrants. The measured and found concentrations of DBS and SDS, as well as the corresponding recoveries, are listed in Table 4. The recoveries for titrations of DBS ranged from 97.3 to 102.0%, with the best recoveries for DMIC being 99.3% (for 0.1 mM) and 99.7% (for 0.01 mM). The recoveries for the titrations of SDS ranged from 98.1 to 99.8%, with the best recovery for DMIC being 99.7% (for 0.1 mM) and 99.6% (for 0.01 mM). The proposed DODI–TPB surfactant sensor showed good accuracy against the studied anionic surfactants.

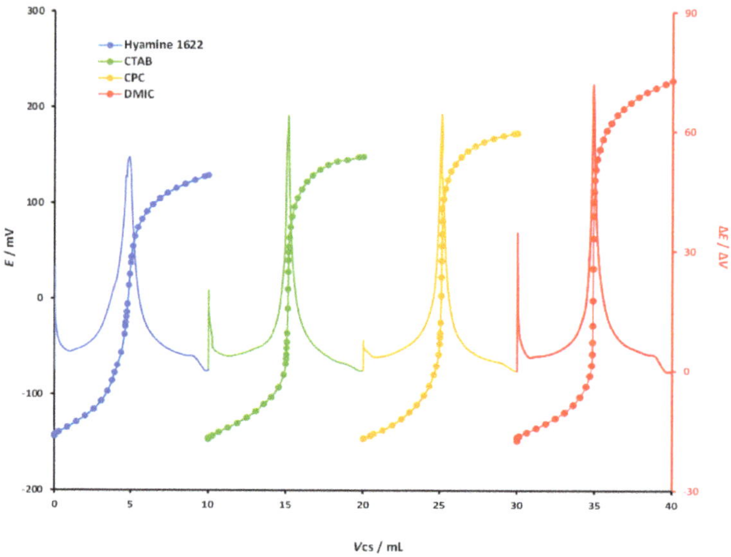

Figure 5. Potentiometric titration curves, and first derivatives, of SDS (0.4 mM) with four different cationic surfactants (0.4 mM) used as a titrant: Hyamine 1622, CTAB, CPC, and DMIC. The titration curves and their first derivatives were rearranged for the sake of clarity.

Table 4. Potentiometric titration results of SDS and DBS anionic surfactants with selected cationic surfactants (4 mM) as a titrant and the DODI–TPB surfactant sensor as an endpoint indicator.

		Analyte					
		DBS			**SDS**		
		Taken (mM)	**Found (mM)**	**Recovery (%)**	**Taken (mM)**	**Found (mM)**	**Recovery (%)**
Titrant	CTAB	0.1	0.0973	97.3	0.1	0.0989	98.9
		0.01	0.00102	102.0	0.01	0.00981	98.1
	Hyamine 1622	0.1	0.0972	97.2	0.1	0.0988	98.8
		0.01	0.00975	97.5	0.01	0.00989	98.9
	CPC	0.1	0.0993	99.3	0.1	0.0991	99.1
		0.01	0.00985	98.5	0.01	0.00987	98.9
	DMIC	0.1	0.0993	99.3	0.1	0.0998	99.7
		0.01	0.00997	99.7	0.01	0.00996	99.6

3.4.3. Titrations of Commercial Samples

The DODI–TPB surfactant sensor was used as an endpoint indicator in potentiometric titration of 12 commercial products containing anionic surfactants. These products were detergents intentionally divided into three groups: powder, liquid gel, and hand wash, to cover all potential product types. They were all declared to contain anionic surfactants. An appropriate concentration of CPC was used as a titrant. The studies with CPC showed good analytical properties, just below those of DMIC, but CPC was selected because the price of DMIC is much higher. Five independent titrations were performed for each detergent sample (Table 5). Hand detergents were found to have the highest levels of anionic surfactants ranging from 14.89 to 15.98%. Powder detergent samples contained 4.25 to 6.01% of anionic surfactants, while gel liquid detergents contained the lowest amounts of anionic surfactants, ranging from 2.11 to 3.21%. The data were compared with the previously published ISE DMI–TPB surfactant sensor [13] and a two-phase titration referent method [8], which revealed good agreement between all three methods.

Table 5. Potentiometric titration results of detergent commercial products with the DODI–TPB surfactant sensor as an endpoint indicator and CPC as a titrant, compared with the ISE surfactant sensor and a referent two-phase titration method.

Detergent Sample		% Anionic Surfactant		
		DODI–TPB	**DMI–TPB [13]**	**Two-Phase Titration [8]**
Solid/powder	Sample 1	4.25 ± 0.08	4.21	4.71
	Sample 2	5.21 ± 0.09	5.22	5.01
	Sample 3	5.66 ± 0.07	5.61	5.51
	Sample 4	6.01 ± 0.09	6.09	6.11
Liquid/gel	Sample 5	2.11 ± 0.07	2.15	2.27
	Sample 6	2.45 ± 0.11	2.41	2.85
	Sample 7	2.25 ± 0.09	2.26	2.41
	Sample 8	3.21 ± 0.09	3.27	3.28
Handwashing	Sample 9	15.22 ± 0.11	15.21	15.32
	Sample 10	14.89 ± 0.09	14.85	15.07
	Sample 11	15.42 ± 0.08	15.48	15.55
	Sample 12	15.98 ± 0.12	15.99	16.12

4. Conclusions

The new procedure for the direct synthesis of the 1,3-dioctadecyl-1*H*-imidazol-3-ium tetraphenylborate (DODI–TPB) ionophore was successfully applied. The DODI–TPB ionophore was characterized and confirmed by NMR, ATR FT-IR, LC–MS, and elemental analysis.

A computational analysis confirmed the feasibility of the DODI–TPB complex formation in solution and elucidated a range of C–H···π interactions among flexible C-18 alkyl chains in $DODI^+$ and phenyl groups in TPB^- as dominant for their recognition in solution. The calculated MM-PBSA binding affinity among components of $\Delta G_{BIND} = -5.0$ kcal mol^{-1} appears optimal, as it efficiently combines the ionophore stability with its ability to exchange anions in solution, thereby allowing for favorable analytic and sensing responses.

The DODI–TPB ionophore was used to fabricate the PVC-based liquid membrane-type potentiometric surfactant sensor for anionic surfactants. The DODI–TPB surfactant sensor exhibited good response characteristics, a wide linear response range (5.9×10^{-7}–4.1×10^{-3} M for SDS), a low detection limit (6.8×10^{-7} M for SDS), a Nernstian slope, high selectivity toward interfering anions, high accuracy (98.1–99.8% for SDS), and a wide operational pH range (2–9). Both anionic surfactants, DBS and SDS, were successfully titrated with four cationic surfactants, resulting in sigmoid curves with high potential changes (up to 382.5 mV), well-defined inflexions, and sharp endpoint peaks for the first derivatives. The DODI–TPB surfactant sensor was successfully used to quantify anionic surfactants in 12 commercially available detergents (powder, hand wash, and liquid gel detergents). The results show good agreement with the ISE surfactant sensor and the two-phase titration method. The DODI–TPB surfactant sensor could be used in quality control in the industry. Even though the sensor showed high stability and a broad response range, the sensor seems to be more suitable to observe surfactant concentrations in wastewaters than lakes and rivers. Further investigations will be conducted using the DODI–TPB surfactant sensor to quantify anionic surfactants in water and wastewater.

Supplementary Materials: The following supporting information can be downloaded at: https://www.mdpi.com/article/10.3390/chemosensors10120523/s1, Figure S1: 1H NMR (400 MHz; CDCl3) of 1,3-dioctadecyl-1H-imidazol-3-ium tetraphenylborate (2); Figure S2: 13C NMR (100.613 MHz; CDCl3) of 1,3-dioctadecyl-1H-imidazol-3-ium tetraphenylborate (2); Figure S3: ATR-FTIR spectra of powder 1,3-dioctadecyl-1H-imidazol-3-ium tetraphenylborate (2); Figure S4: RMSD graphs during the molecular dynamics simulation in the aqueous solution; Figure S5: Time dependence of the distance between the boron atom in TPB− and the nitrogen atoms in DODI+ during the molecular dynamics simulation; Figure S6: Charge distribution within the DODI+ cation and the TPB− anion.

Author Contributions: Conceptualization, M.J. and N.S.; Methodology, M.J., N.S.; D.M.-Č. and R.V. (Robert Vianello); Formal Analysis, D.M.-Č., N.S., D.M., L.H., V.V. and M.J.; Investigation, M.J., N.S., D.M.-Č. and L.H.; Resources, M.J., N.S., B.Š., R.V. (Robert Vianello) and B.Đ.; Data Curation, N.S., D.M., D.M.-Č., L.H., V.V., B.D.V., R.V. (Raffaele Velotta) and M.J.; Writing—Original Draft Preparation, N.S., M.J., D.M.-Č., B.Š. and R.V. (Robert Vianello); Writing—Review and Editing, N.S., D.M., R.V. (Robert Vianello) and M.J.; Visualization, L.H., B.Š., B.D.V., R.V. (Raffaele Velotta) and B.Đ.; Supervision, M.J., N.S. and R.V. (Robert Vianello); Project Administration, M.J. and N.S. All authors have read and agreed to the published version of the manuscript.

Funding: Croatian Science Foundation, projects IP-2019-04-8846, IP-2020-02-8090, and DOK-2020-01-3482.

Institutional Review Board Statement: Not applicable.

Informed Consent Statement: Not applicable.

Data Availability Statement: The data presented in this study are available in the Supplementary Materials.

Acknowledgments: N.S., B.Š. and M.J. would like to thank the EuChemS Sample Preparation Task Force and Network. N.S., B.D.V. and R.V. would like to thank the COST Action PortASAP—European network for the promotion of portable, affordable and simple analytical platforms. D.M. would like to thank to Croatian Science Foundation for funding the research project IP-2019-04-8846 and the University of Rijeka for funding the research grant UNIRI-prirod-18-102. L.H. and R.V. (Robert Vianello) would like to thank the Croatian Science Foundation for a financial support (IP-2020-02-8090) and the Zagreb University Computing Centre (SRCE) for granting computational resources on the ISABELLA cluster. L.H. wishes to thank the Croatian Science Foundation for a doctoral stipend through the Career Development Project for Young Researchers (DOK-2020-01-3482).

Conflicts of Interest: The authors declare no conflict of interest.

References

1. Sakač, N.; Marković, D.; Šarkanj, B.; Madunić-Čačić, D.; Hajdek, K.; Smoljan, B.; Jozanović, M. Direct Potentiometric Study of Cationic and Nonionic Surfactants in Disinfectants and Personal Care Products by New Surfactant Sensor Based on 1,3-Dihexadecyl−1H-Benzo[d]Imidazol−3-Ium. *Molecules* **2021**, *26*, 1366. [CrossRef] [PubMed]
2. Fizer, O.; Filep, M.; Pantyo, V.; Elvira, D.; Fizer, M. Structural Study and Antibacterial Activity of Cetylpyridinium Dodecyl Sulfate Ion Pair. *Biointerface Res. Appl. Chem.* **2022**, *12*, 3501–3512. [CrossRef]
3. Jozanović, M.; Sakač, N.; Karnaš, M.; Medvidović-Kosanović, M. Potentiometric Sensors for the Determination of Anionic Surfactants—A Review. *Crit. Rev. Anal. Chem.* **2019**, *51*, 115–137. [CrossRef] [PubMed]
4. Research and Markets Global Surfactants Market (2019 to 2025)—Drivers, Restraints, Opportunities and Challenges. Available online: https://www.prnewswire.com/news-releases/global-surfactants-market-2019-to-2025---drivers-restraints-opportunities-and-challenges-301086922.html (accessed on 13 November 2022).
5. *Global Anionic Surfactants Market—Industry Trends and Forecast to 2027*; Data Bridge Market Research: Pune, India, 2020.
6. Effendy, I.; Maibach, H.I. Surfactants and Experimental Irritant Contact Dermatitis. *Contact Dermat.* **1995**, *33*, 217–225. [CrossRef] [PubMed]
7. Venhuis, S.H.; Mehrvar, M. Health Effects, Environmental Impacts, and Photochemical Degradation of Selected Surfactants in Water. *Int. J. Photoenergy* **2004**, *6*, 115–125. [CrossRef]
8. *ISO 2271*; International Organization for Standardization Surface Active Agents, Detergents, Determination of Anionic-Active Matter by Manual or Mechanical Direct Two-Phase Titration Procedure. ISO: Geneva, Switzerland, 1989.
9. Dimeski, G.; Badrick, T.; John, A.S. Ion Selective Electrodes (ISEs) and Interferences—A Review. *Clin. Chim. Acta* **2010**, *411*, 309–317. [CrossRef]
10. Fizer, M.; Fizer, O.; Sidey, V.; Mariychuk, R.; Studenyak, Y. Experimental and Theoretical Study on Cetylpyridinium Dipicrylamide—A Promising Ion-Exchanger for Cetylpyridinium Selective Electrodes. *J. Mol. Struct.* **2019**, *1187*, 77–85. [CrossRef]
11. Fizer, O.; Fizer, M.; Sidey, V.; Studenyak, Y. Predicting the End Point Potential Break Values: A Case of Potentiometric Titration of Lipophilic Anions with Cetylpyridinium Chloride. *Microchem. J.* **2021**, *160*, 105758. [CrossRef]
12. Najafi, M.; Maleki, L.; Rafati, A.A. Novel Surfactant Selective Electrochemical Sensors Based on Single Walled Carbon Nanotubes. *J. Mol. Liq.* **2011**, *159*, 226–229. [CrossRef]

13. Sakač, N.; Karnaš, M.; Jozanović, M.; Medvidović-Kosanović, M.; Martinez, S.; Macan, J.; Sak-Bosnar, M. Determination of Anionic Surfactants in Real Samples Using a Low-Cost and High Sensitive Solid Contact Surfactant Sensor with MWCNTs as the Ion-to-Electron Transducer. *Anal. Methods* **2017**, *9*, 2305–2314. [CrossRef]
14. Seguí, M.J.; Lizondo-Sabater, J.; Martínez-Máñez, R.; Pardo, T.; Sancenón, F.; Soto, J. Ion-Selective Electrodes for Anionic Surfactants Using a New Aza-Oxa-Cycloalkane as Active Ionophore. *Anal. Chim. Acta* **2004**, *525*, 83–90. [CrossRef]
15. Samardžić, M.; Budetić, M.; Széchenyi, A.; Marković, D.; Živković, P.; Šarkanj, B.; Jozanović, M. The Novel Anionic Surfactant Selective Sensors Based on Newly Synthesized Quaternary Ammonium Salts as Ionophores. *Sens. Actuators B Chem.* **2021**, *343*, 130103. [CrossRef]
16. Kovács, B.; Csóka, B.; Nagy, G.; Ivaska, A. All-solid-state surfactant sensing electrode using conductive polymer as internal electric contact. *Anal. Chim. Acta* **2001**, *437*, 67–76. [CrossRef]
17. Matysik, S.; Matysik, F.M.; Einicke, W.D. A disposable electrode based on zeolite-polymer membranes for potentiometric titrations of ionic surfactants. *Sens. Actuators B Chem.* **2002**, *85*, 104–108. [CrossRef]
18. Madunić-Čačić, D.; Sak-Bosnar, M.; Matešić-Puač, R. A New Anionic Surfactant-Sensitive Potentiometric Sensor with a Highly Lipophilic Electroactive Material. *Int. J. Electrochem. Sci.* **2011**, *6*, 240–253.
19. Sakač, N.; Madunić-Čačić, D.; Karnaš, M.; Đurin, B.; Kovač, I.; Jozanović, M. The Influence of Plasticizers on the Response Characteristics of the Surfactant Sensor for Cationic Surfactant Determination in Disinfectants and Antiseptics. *Sensors* **2021**, *21*, 3535. [CrossRef]
20. Olkowska, E.; Polkowska, Z.; Namieśnik, J. Analytical Procedures for the Determination of Surfactants in Environmental Samples. *Talanta* **2012**, *88*, 1–13. [CrossRef]
21. Pires, A.R.; Araújo, A.N.; Montenegro, M.C.B.S.M.; Chocholous, P.; Solich, P. New Ionophores for Vitamin B1 and Vitamin B6 Potentiometric Sensors for Multivitaminic Control. *J. Pharm. Biomed. Anal.* **2008**, *46*, 683–691. [CrossRef]
22. Mahajan, R.K.; Shaheen, A. Effect of Various Additives on the Performance of a Newly Developed PVC Based Potentiometric Sensor for Anionic Surfactants. *J. Colloid Interface Sci.* **2008**, *326*, 191–195. [CrossRef]
23. Devi, S.; Chattopadhyaya, M.C. Determination of Sodium Dodecyl Sulfate in Toothpastes by a PVC Matrix Membrane Sensor. *J. Surfactants Deterg.* **2013**, *16*, 391–396. [CrossRef]
24. Issa, Y.M.; Mohamed, S.H.; Baset, M.A. El Chemically Modified Carbon Paste and Membrane Sensors for the Determination of Benzethonium Chloride and Some Anionic Surfactants (SLES, SDS, and LABSA): Characterization Using SEM and AFM. *Talanta* **2016**, *155*, 158–167. [CrossRef] [PubMed]
25. Arooj, M.; Arrigan, D.W.M.; Mancera, R.L. Characterization of Protein-Facilitated Ion-Transfer Mechanism at a Polarized Aqueous/Organic Interface. *J. Phys. Chem. B* **2019**, *123*, 7436–7444. [CrossRef] [PubMed]
26. Case, D.A.; Betz, R.M.; Cerutti, D.S.; Cheatham, T.E., III; Darden, T.A.; Duke, R.E.; Giese, T.J.; Gohlke, H.; Goetz, A.W.; Homeyer, N.; et al. *AMBER 2016*; University of California: San Francisco, CA, USA, 2016.
27. Genheden, S.; Ryde, U. The MM/PBSA and MM/GBSA Methods to Estimate Ligand-Binding Affinities. *Expert Opin. Drug Discov.* **2015**, *10*, 449–461. [CrossRef] [PubMed]
28. Hou, T.; Wang, J.; Li, Y.; Wang, W. Assessing the Performance of the MM/PBSA and MM/GBSA Methods. 1. The Accuracy of Binding Free Energy Calculations Based on Molecular Dynamics Simulations. *J. Chem. Inf. Model.* **2011**, *51*, 69–82. [CrossRef] [PubMed]
29. Sakač, N.; Madunić-Čačić, D.; Marković, D.; Hok, L.; Vianello, R.; Šarkanj, B.; Đurin, B.; Hajdek, K.; Smoljan, B.; Milardović, S.; et al. Potentiometric Surfactant Sensor Based on 1,3-Dihexadecyl-1H-Benzo[d]Imidazol-3-Ium for Anionic Surfactants in Detergents and Household Care Products. *Molecules* **2021**, *26*, 3627. [CrossRef]
30. Hok, L.; Mavri, J.; Vianello, R. The Effect of Deuteration on the H2 Receptor Histamine Binding Profile: A Computational Insight into Modified Hydrogen Bonding Interactions. *Molecules* **2020**, *25*, 6017. [CrossRef]
31. Hok, L.; Rimac, H.; Mavri, J.; Vianello, R. COVID-19 Infection and Neurodegeneration: Computational Evidence for Interactions between the SARS-CoV-2 Spike Protein and Monoamine Oxidase Enzymes. *Comput. Struct. Biotechnol. J.* **2022**, *20*, 1254–1263. [CrossRef]
32. Galović, O.; Samardžić, M.; Petrušić, S.; Sak-Bosnar, M. A New Sensing Material for the Potentiometric Determination of Anionic Surfactants in Commercial Products. *Int. J. Electrochem. Sci.* **2014**, *9*, 3802–3818.
33. Mao, C.; Robinson, K.J.; Yuan, D.; Bakker, E. Ion–ionophore interactions in polymeric membranes studied by thin layer voltammetry. *Sens. Actuators B Chem.* **2022**, *358*, 131428. [CrossRef]
34. Buck, R.P.; Lindner, E. Recomendations for Nomenclature of Ion-Selective Electrodes (IUPAC Recommendations 1994). *Pure Appl. Chem.* **1994**, *66*, 2527–2536. [CrossRef]
35. Tshepelevitsh, S.; Kütt, A.; Lõkov, M.; Kaljurand, I.; Saame, J.; Heering, A.; Plieger, P.G.; Vianello, R.; Leito, I. On the Basicity of Organic Bases in Different Media. *Eur. J. Org. Chem.* **2019**, *2019*, 6735–6748. [CrossRef]

chemosensors

MDPI

Article

Ultrathin PANI-Decorated, Highly Purified and Well Dispersed Array Cncs for Highly Sensitive HCHO Sensors

Qingmin Hu [1,2,†], Zhiheng Ma [1,2,†], Jie Yang [3], Tiange Gao [1,2], Yue Wu [1,2], Zhe Dong [1,2], Xuyi Li [1,2], Wen Zeng [4], Shichao Zhao [3] and Jiaqiang Xu [1,2,*]

1 NEST Lab, Department of Physics, College of Science, Shanghai University, Shanghai 200444, China; huqingmin@shu.edu.cn (Q.H.); Marquerzh@shu.edu.cn (Z.M.); gaotiange@shu.edu.cn (T.G.); mrtun@shu.edu.cn (Y.W.); ss996584188@shu.edu.cn (Z.D.); xuyili@shu.edu.cn (X.L.)
2 NEST Lab, Department of Chemistry, College of Science, Shanghai University, Shanghai 200444, China
3 State Key Laboratory of Coal Conversion, Institute of Coal Chemistry, Chinese Academy of Sciences, Taiyuan 030001, China; yangjie16@mails.UCAS.ac.cn (J.Y.); zhaoshichao@sxicc.ac.cn (S.Z.)
4 School of Materials Science and Engineering, Chongqing University, Chongqing 400044, China; wenzeng@cqu.edu.cn
* Correspondence: xujiaqiang@shu.edu.cn
† These authors contribute equally to this work.

Abstract: The flocculation of small surficial groups on pristine CNCs (carbon nanocoils) bundles limit their application. In this study, we designed and fabricated novel array CNCs with a surficial decoration of polyaniline (PANI) using in situ methods. Atomic layer deposition (ALD) and chemical vapor deposition (CVD) methods were employed to fabricate the highly pure array CNCs. The array CNCs decorated with ultra-thin PANI were confirmed by different characterizations. Furthermore, this material displayed a good performance in its detection of formaldehyde. The detection results showed that the CNCs coated with PANI had a low limit of detection of HCHO, as low as 500 ppb, and the sensor also showed good selectivity for other interfering gases, as well as good repeatability over many tests. Furthermore, after increasing the PANI loading on the surface of the CNCs, their detection performance exhibited a typical volcanic curve, and the value of the enthalpy was extracted by using the temperature-varying micro-gravimetric method during the process of detection of the formaldehyde molecules on the CNCs. The use of array CNCs with surficial decoration offers a novel method for the application of CNCs and could be extended to other applications, such as catalysts and energy conversion.

Keywords: array carbon nanocoils; atomic layer deposition (ALD); QCM sensors; surficial decoration; PANI

Citation: Hu, Q.; Ma, Z.; Yang, J.; Gao, T.; Wu, Y.; Dong, Z.; Li, X.; Zeng, W.; Zhao, S.; Xu, J. Ultrathin PANI-Decorated, Highly Purified and Well Dispersed Array Cncs for Highly Sensitive HCHO Sensors. *Chemosensors* **2021**, *9*, 276. https://doi.org/10.3390/chemosensors9100276

Academic Editor: Boris Lakard

Received: 23 August 2021
Accepted: 20 September 2021
Published: 28 September 2021

1. Introduction

As an important carbon material, carbon nanocoils (CNCs) have attracted increasing attention in the study of catalysis templates, wave-absorption and energy conversion-storage due to their unique geometrical and mechanical elegance, light weight, good stability, and electrical properties [1–4]. For example, carbon nanocoils have exhibited their application potential in terms of the advantages they offer in the design of electromagnetic wave absorbers, flat-panel displays, and catalysis templates [5]. Given the advantages of CNCs, it is interesting to further explore their application in sensors for environmental monitoring. However, there are few studies about their application in gas sensors. Quartz crystal microbalance (QCM), which offers advantages such as high sensitivity, low working temperature, high portability and low power consumption, offers a novel direction for the application of CNCs in gas sensors [6,7].

Sensing materials with large specific surface and surficial groups, such as metal organic framework (MOF) [8], covalent organic framework (COF) [9] and Santa Barbara Amorphous-15 (SBA-15)/ordered mesoporous carbon material FDU-15 (FDU-15) [10,11]

offer high QCM detection performance. However, although the CNCs have highly specific surfaces, pristine CNCs haves little surficial groups and could be flocculated into bundles, like carbon nanotubes (CNTs), through van der Waals interaction. It is interesting and important to develop highly dispersed CNCs with surficial decoration of functional groups. Efforts have been made to improve the dispersion of CNCs and to decorate surficial functional groups. The direct fabrication of CNCs on different matrices can lead to good dispersion of CNT, resulting in high mechanicity and a wide range of potential applications [12–14]. The surficial decoration of polymers has been widely applied in the form of active gas materials due to its high sensitivity, low working temperature and unique physicochemical properties [15–20].

Formaldehyde (HCHO) is widely used in the chemical industry because it is an economical chemical. However, formaldehyde as gaseous pollution from indoor decoration and industrial production, described as a carcinogen by the US National Toxicology Program, is harmful and toxic to human beings [21–24]. Different methods, including gas/liquid chromatography (GC/LC), colorimetric analysis and mass spectrometry (MS) have been developed to mitigate this. However, these methods are always time-consuming and costly [21,22].

Efforts have been made by researchers to develop portable, sensitive and selective sensors. Functional materials applied as QCM sensors offer a novel direction for the detection of formaldehyde. The reactions between $-NH_2$ and R_2CO are always used in the design of formaldehyde-sensing materials [25–27]. Polyaniline (PANI) has been widely studied and successfully applied in the surficial decoration of gas sensors (such as H_2S, NH_3, NO_2, CO, acetone, benzene etc.) because of its easy fabrication, good environmental stability and other unique physicochemical properties [15,28–36]. The abundant –NH– and $-NH_2$ groups in PANI offer active sites for the adsorption of formaldehyde. Inspired by previous studies, it is interesting to fabricate well dispersed CNCs with PANI surficial decoration for the detection of formaldehyde. Moreover, the hydrophobicity of polyimide can improve the resistance of QCM to humidity [37–40].

In this work, we employed ALD technology to deposit uniform Cu-based catalysts on graphene. Subsequently, low-temperature CVD was employed to maintain the uniformity of the sites and fabricate in situ array CNCs. The surficial decoration of PANI on the CNCs was carried out by the electrodeposition method. The novel PANI-decorated array CNCs displayed an excellent ability to detect formaldehyde. The formaldehyde sensor coated with CNCs-PANI exhibited the highest sensitivity (70 Hz@0.5 ppm) and good selectivity. The value of the adsorbed enthalpy was extracted by an adsorption isotherm method that we have reported previously. The Action Level and TVA for formaldehyde were 0.5 and 0.75 ppm, respectively. Our sensor met the required LOD. As far as we know, these were the first CNCs to have been used as a mass-type sensing material for the detection of HCHO.

2. Experimental Method

2.1. Preparation of Cu-Based Catalysts

Firstly, grapheme was ultrasonically dispersed for thirty minutes before the ALD. Next, the 2 mL solution was dropped onto quartz wafers and dried in air. Subsequently, the Cu catalysts were deposited by ALD (Chongqing Nuotu Loc., Chongqing, China) in a hot-wall closed-chamber reactor. Bis (2,2,6,6-tetramethyl-3,5-heptanedionato) copper (III) (99%, Alfa Aesar, Shanghai, China) and ozone (O_3) were used as precursors in the preparation of the Cu nanoparticles, with N_2 (99.999%) as the carrier gas. O_3 acts as an oxidant and enables growth over a wide temperature range, as shown in Scheme 1. The deposition temperature was 270 °C. In this work, the Cu-based catalysts with different numbers of cycles by ALD were designated Cu_x, where x is the number of Cu oxide ALD cycles (100, 150, 200, 400, respectively).

2.2. Synthesis of Array CNCs

After Cu-based catalysts were processed by ALD, the Cu catalysts were transferred into a tube furnace equipped with a quartz tube (60 mm × 1100 mm). The CNCs' growth temperature was set as 250 °C, at a rate of 10 °C/min, in an H_2/Ar (5%/95%) atmosphere. The CNCs' growth time was 30 min under atmospheric pressure, as shown in Scheme 1. The obtained CNCs were transferred to the HNO_3 (68 wt%) solution at 100 °C for about 2 h in order to remove the Cu catalysts. Subsequently, the CNCs were filtered and dried. Finally, the CNCs were dispersed in ethanol for further study.

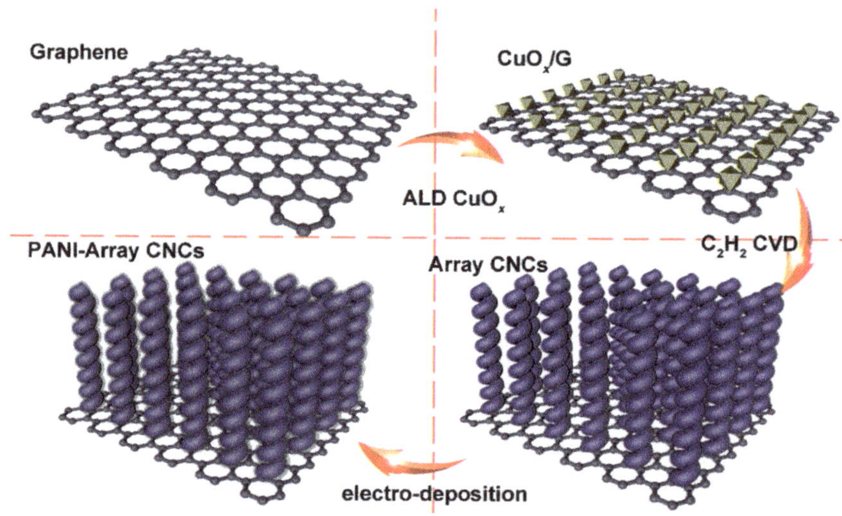

Scheme 1. Illustration for the fabrication of the PANI-CNC.

2.3. Synthesis of Array PANI-CNCs

For the preparation of the PANI-CNCs, CNC was ground with several drops of N-methylpyrrolidone for 30 min, then the black slurry was printed onto the FTO and placed in a vacuum oven at 60 °C for 24 h. The obtained sample was then cooled naturally.

The PANI-CNCs were fabricated through the electrodeposition method, featuring a 100 mL mixed solution containing 100 μL of aniline monomer, and 1 M H_2SO_4 was used as the electrolyte. The electrochemical deposition was performed in a three-electrode-cell using FTO coated with CNT as the working electrode, a Pt foil as the counter electrode and Ag/AgCl as the reference electrode, at a constant voltage of 0.8 V for 10 min. After electrodeposition, the PANI-CNCs was washed with deionized water and ethanol for several times and dried at 60 °C for 24 h.

2.4. Characterization

The transmission electron micrographs (TEM) and high-resolution TEM (HRTEM) images for the morphologies of the materials were acquired by JEOL-2100F microscope (JEOL, Tokyo, Japan). Field emission scanning electron microscopy (FE-SEM, JSM-6700 F, JEOL, Tokyo, Japan) was also employed to observe the surface morphologies of the materials. The atomic force microscope (AFM) analysis of the samples was performed using Nanosurf Naio AFM (Nanosurf, Liestal, Switzerland).

The adsorption–desorption isotherms for the different samples were carried out by Micromeritics ASAP 2020 (Microm, San Jose, CA, USA). The XRD patterns were collected on a DX-2700 diffractometer, using a Cu Kα1 source (Haoyuan, Dandong, China). The surface area was analyzed by Micromeritics Tristar 3000 at 77 K. The X-ray photoelectron spectroscopy (XPS) data were obtained with a WSCLAB X-ray photoelectron spectroscopy system (Thermal Fisher, Agawam, MA, USA), using an Al K_α source. The attenuated total reflection spectra (ATR) data for the samples were collected on an IS10 spectrometer (Thermal Fisher, IS10, Agawam, MA, USA) with an MCT detector.

2.5. Fabrication of QCM Sensor and Sensing Performance

Piezoelectric quartz crystal QCM resonators (WestSensor Co., Chengdu, China) with silver electrodes on both sides were used as transducers. The change of QCM frequency was related to a change in the silver electrodes' mass, and the relationship was described by Sauerbrey equation $\Delta F = -2.26 \times 10^{-6} f_0^2 \Delta m/A$, where f_0 is the fundamental resonant frequency, which depends on the nature of the QCM chip, and A is the area of the silver plates coated in quartz crystal. Apart from the invariants mentioned above, the frequency shift (ΔF) was proportional to the mass change of adsorption (Δm) in the electrode surface of the QCM.

Prior to their fabrication, the materials were dispersed in a solution of 25% ethanol and 75% ethanediol by ultrasonic agitation for 2 min. Subsequently, the suspension was dropped onto the silver electrodes of the QCM resonators, which were marked 1-PANI-CNC, 2-PANI-CNC, 3-PANI-CNC, 4-PANI-CNC sensors. Subsequently, the resonators were dried at 50 °C in a vacuum drying oven for 20 min.

The prepared QCM sensors were tested in a laboratory-made sealed chamber with different concentrations of volatile organic compounds (VOC) gas. The analyte VOCs introduced by injection were steadily obtained with a dynamic gas-mixing apparatus.

Nitrogen was used as the carrier gas because there was little difference between the oxygen and nitrogen in the air, and they did not have a special adsorption effect in this experiment. However, the nitrogen test is generally capable of obtaining a more stable baseline, so nitrogen was used in most of our tests [41]. At the start of the measurement, a nitrogen stream was flowed into the chamber until a stable baseline was obtained for the QCM sensors. Next, the analyte vapor was introduced by injection. After obtaining a stable response from the analyte, the reference N_2 was re-introduced into the sampling chamber to re-establish the baseline. All the detection experiments were performed in an air-conditioned room at 25 °C.

3. Results and Discussion

3.1. Fabrication and Characterization of Array CNCs

The reducibility of Cu-based catalysts was analyzed by temperature-programmed reduction (TPR), as shown in Figure 1. There was a strong peak at 245 °C which could have resulted from the reduction of CuO [42]. Therefore, the Cu-based catalyst could have reduced at 250 °C in the H_2/Ar atmosphere during the CNCs' fabrication. The growth of the CNCs could be attributed to the Cu^0 or the Cu^+. The Cu^+ could be attributed to the active species for the growth of CNCs because the Cu^+ will have been further reduced to Cu^0 by the hydrogen reduction resulting from the acetylene's decomposition at the high temperature [2].

The morphologies of the Cu-based catalysts after reduction at 250 °C were also characterized by the SEM. As shown in Figure 2, the decorated Cu particles were deposited uniformly on the graphene. The Cu nanoparticles were about 50 nm in diameter for the Cu200 catalysts. Furthermore, the diameter of the Cu nanoparticles increased with the increase of the ALD Cu cycles. After 200 cycles of ALD Cu, the Cu nanoparticles increased in number instead of diameter. After further ALD Cu cycles, the diameter of the Cu nanoparticles increased. Furthermore, the nanoparticles maintained their uniform dispersion after 400 ALD cycles. The Cu800 catalysts displayed a non-homogenous dispersion. This could be attributed to the high content of the Cu catalysts and the agglomeration of the Cu nanoparticles. The uniformly deposited Cu catalysts lead to the well dispersed growth of the CNCs.

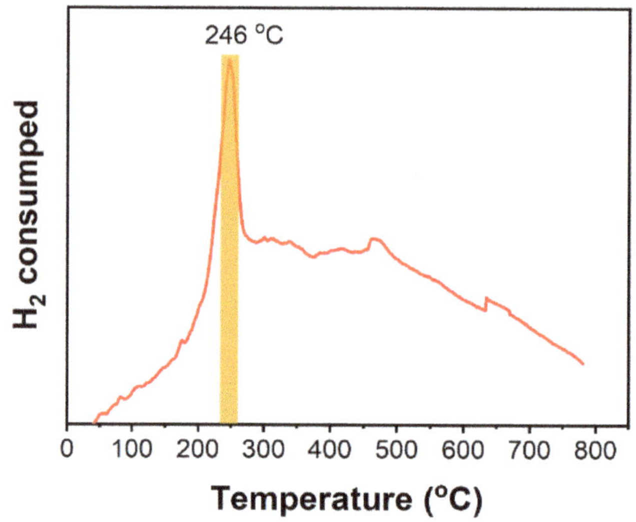

Figure 1. TPR profile for the Cu200 catalyst.

Figure 2. The SEM images for different Cux catalysts after reduction.

The AFM was employed to further confirm the morphology of the Cu200 catalyst. The bright spots in the images in Figure 3 were ascribed to the Cu nanoparticles. The AFM results confirmed that the Cu nanoparticles (<80 nm) were uniformly deposited on the graphene. This was consistent with the SEM results shown in Figure 2, and further confirmed the uniform deposition of Cu nanoparticles.

The morphologies of CNCs were characterized by SEM, as shown in Figure 4. As the ALD Cu cycles increased, the products of the fibers and the CNCs transformed into purified CNCs, and then back into fibers and CNCs. The application of the Cu100 catalyst lead to a mixture of long fibers and, soon afterwards y, CNCs, as revealed by the SEM images (Figure 4a,c). For the Cu400 catalysts, there were long fibers and CNCs on the graphene. By contrast, the highly pure CNCs were fabricated by Cu200 catalysts, as straight fibers were rarely shown on the SEM images, as shown in Figure 4b,d. The CNCs fabricated by Cu200 exhibited uniform coil diameters, demonstrating the uniformity of the Cu200 catalyst and stable growth during the CVD. Most importantly of all, the highly pure CNCs fabricated by Cu200 lead to a nearly upright array on the graphene instead of the array lying on the graphene, as shown by the fibers fabricated by Cu100 and Cu400, which allowed the quick adsorption and desorption of the gas.

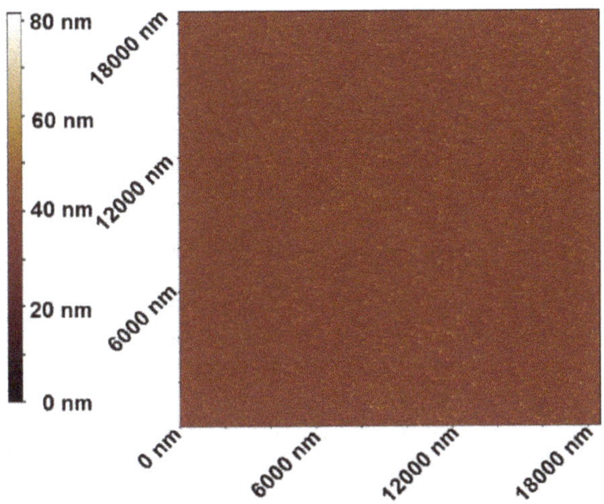

Figure 3. AFM images for the Cu200 catalysts after reduction.

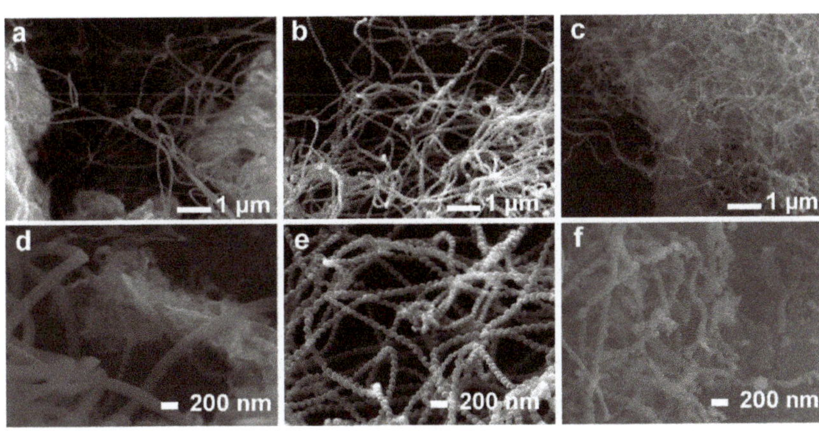

Figure 4. SEM images for CNCs. (**a**,**d**) CNCs synthesized by Cu100, (**b**,**e**) CNCs synthesized by Cu200, (**c**,**f**) CNCs synthesized by Cu 400.

The N_2 adsorption–desorption isotherms of the CNCs before carbonization, the CNCs, and the 3-PANI-CNCs are shown in Figure 5. They exhibited typical type IV-shaped isotherms for CNC-based samples, according to the International Union of Pure and Applied Chemistry (IUPAC)'s classification. These were attributed to the physical adsorption of N_2 on the CNCs with nonporous nanotube walls. The Brunauer-Emmett-Teller (BET) model was employed for the analysis of the surface areas. The BET surface areas were calculated to be 104.8, 248.4 and 312.3 m^2g^{-1} for the CNCs before carbonization, the CNCs and the 3-PANI-CNCs, respectively. After the decoration of the PANI, the BET surface area of the 3-PANI-CNCs increased compared with that of the CNCs. This was attributed to the mesoporous structure of the ultra-thin layer of the PANI.

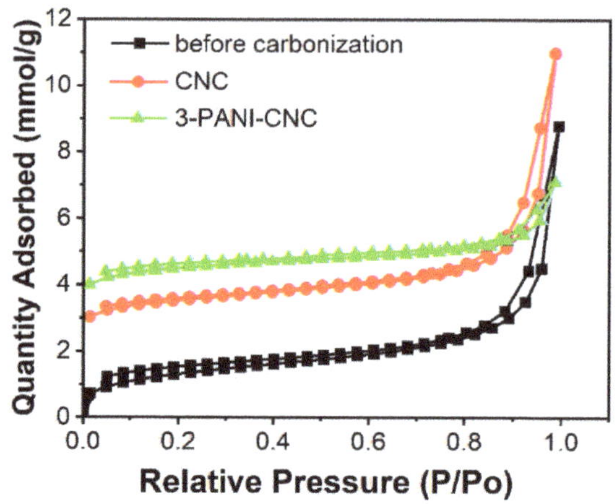

Figure 5. The BET surface of the different samples.

3.2. Confirmation of Surficial Decoration of PANI on Array CNCs

TEM was employed to further investigate the morphology of these CNCs and the decoration of the PANI on the CNCs' surfaces. As shown in Figure 6, the morphology of the CNCs with a diameter of 50–80 nm was consistent with the SEM results. From the HRTEM image, as shown in Figure 6b, there was no obvious lattice fringe. This could be attributed to the low degree of crystallinity of the CNCs or the amorphous CNCs. As shown in Figure 6c, the HAADF-STEM and EDX (energy-dispersive X-ray spectroscopy, JEOL-2100F microscope, JEOL, Tokyo, Japan) mapping were employed to confirm the decoration of the PANI on the CNCs surface. It was clear that the distribution of the N element was uniform on the surface of the CNCs, confirming the successful decoration of the PANI on the CNCs' surfaces. The N maintained the morphology of the pristine CNCs. The pale distribution of the N elements belonged to the ultra-thin layer of the PANI.

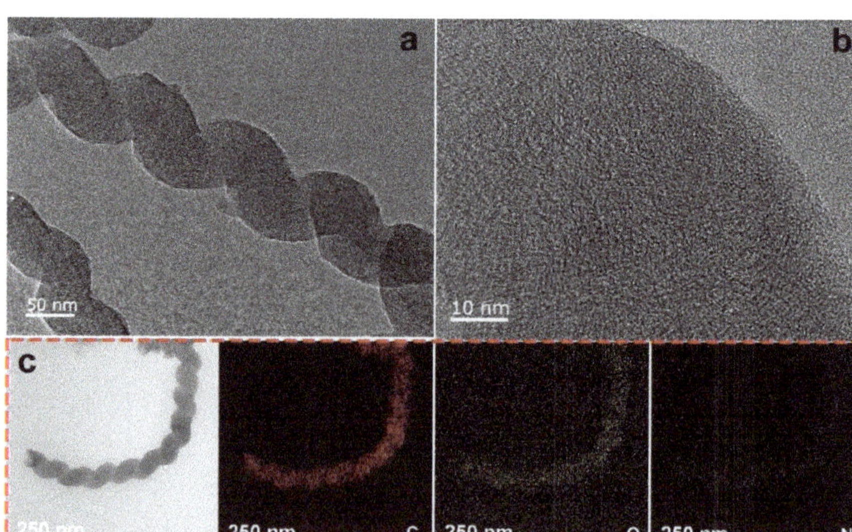

Figure 6. (**a**) TEM for PANI-CNCs, (**b**) HRTEM images for PANI-CNCs, (**c**) HAADF-STEM and EDS mapping for the PANI-CNCs.

Further confirmation of the decoration of the PANI layer on the CNCs was obtained through the performance of X-ray photoelectron spectra (XPS) to. As shown in Figure 7, the XPS spectra revealed that there were N, C and O on the surface of the 3-PANI-CNCs. The XPS results indicated the successful decoration of the PANI on the surface of the CNCs. As shown in the surface elements analysis spectrum, there were three peaks for O_{1s}, N_{1s} and C_{1s}. There was no Cu peak in the surface elements analysis spectrum, revealing the successful coating of the PANI and the complete removal of the catalysts.

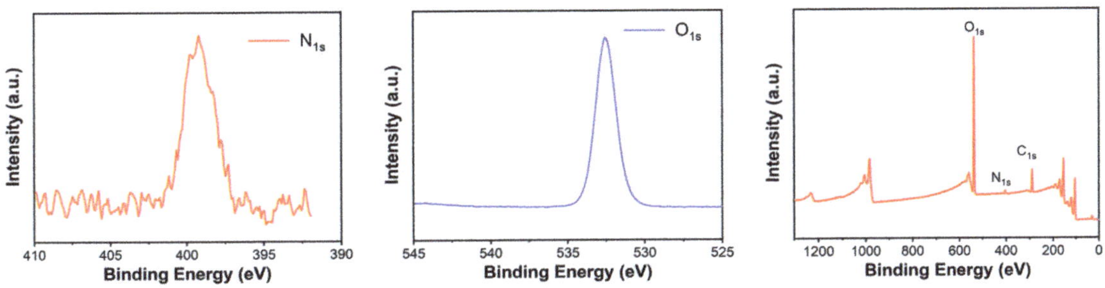

Figure 7. XPS spectra for 3-PANI-CNCs. (**a**) N_{1s} spectrum, (**b**) O_{1s} spectrum and (**c**) surface elements analysis spectrum.

The ATR-IR spectra for the bare CNCs and the 3-PANI-CNCs are shown in Figure 8. There were no obvious peaks on the bare CNCs spectrum. There were peaks on the 3-PANI-CNCs spectrum. The peaks at 1486, 1415, 1304, 1213, 1160 and 822 cm^{-1} were the characteristic peaks of the PANI. The peak centered at 1486 and 1411 cm^{-1} was attributed to the stretching of the C=C group in the benzenoid or quinoid ring. The peaks located at 1304 cm^{-1} and 1213 cm^{-1} were attributed to the stretching of the C–N group on the secondary aromatic amines [43]. The broad peak located at 1160 cm^{-1} was attributed to the stretching of N=Q=N (where Q means the quinone-type rings), which is a characteristic peak of the "electronic-like band". The peak at 822 was attributed to the stretching of the C–H in the benzenoid ring. The presence of the N=Q group and the C–N group offered active sites for the adsorption of the formaldehyde because of the preferred adsorption of R_2CO and the Schiff-base reaction between the R_2CO and the N=Q C–N groups [25,44].

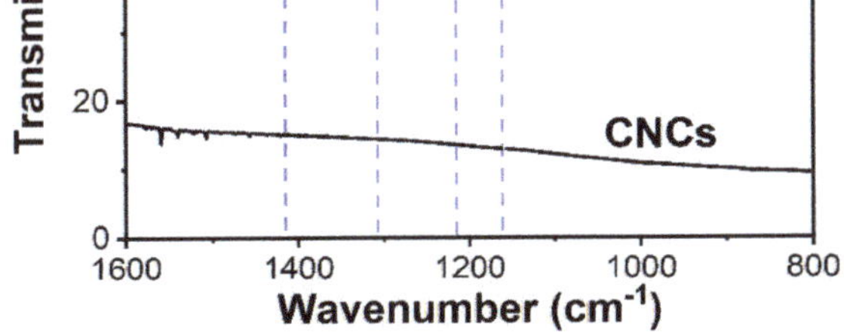

Figure 8. ATR-IR spectra for the bare CNCs and the 3-PANI-CNCs.

Further confirmation of the crystal structure of the coating of the CNC and PANI-CNC samples was obtained through XRD. As shown in Figure 9, there were two broad peaks in the CNCs' XRD curves, which was caused by the amorphous structure of the CNCs. The

XRD results of the CNCs further verified the HRTEM results. For the 3-PANI-CNC, there were new peaks located at about 12° and 25°, which were attributed to (001) and (002) of the graphitized prism-shaped hollow carbon [43]. This confirmed that with the deposition of PANI, the 3-PANI-CNCs' surfaces increased as the new graphitized prism-shaped hollow carbon emerged.

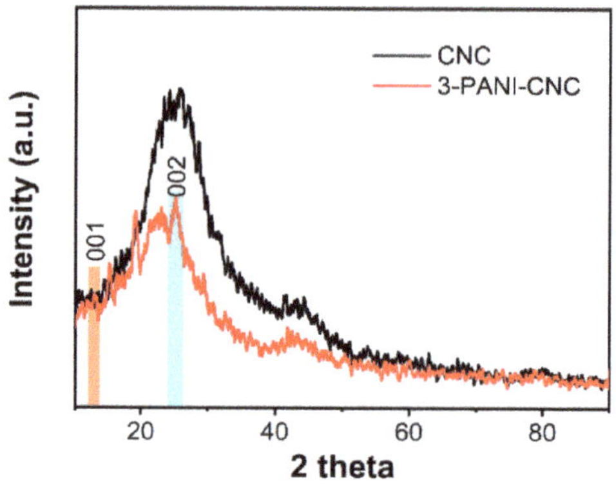

Figure 9. XRD Profiles for the CNC and 3-PANI-CNCs.

The ATR-IR further proved the deposition of the PANI on the CNCs, which was consistent with the EDS mapping and XPS results. Furthermore, the functional group of the C–N group confirmed by IR could be the active sites for the adsorption of the HCHO. The XRD further confirmed the graphitized crystal of the PANI.

3.3. Detection Performance of Array PANI-CNC-Based QCM Sensors

It is well known that there is a special interaction between the amino group and the formaldehyde molecule. Therefore, the QCM sensors coated with polyaniline were employed to investigate its response to formaldehyde.

From Figure 10a, it can be seen that the 3-PANI-CNC samples had a higher response to the formaldehyde compared with the other three samples at the different concentrations. The reason for this result is that, on the one hand, the amino loading on the surface of the four carbon helical structures gradually increased, which was conducive to the improvement of gas-sensing performance. On the other hand, the greater polyaniline loading lead to a decrease in specific surface area and pore size, which is a critical factor for gas response. Under the influence of these two aspects, the gas sensitivity of the four samples finally showed a volcanic change. Comparing the samples of 1-PANI-CNC, 2-PANI-CNC and 4-PANI-CNC, it can be clearly seen that the response value of the 4-PANI-CNC was greater than those of the 1-PANI-CNC and the 2-PANI-CNC due to the increase in amino loading.

Figure 10b exhibits the response curve of the 3-PANI-CNC to formaldehyde at various concentrations. It is obvious that the frequency shift of the 3-PANI-CNC sample changed more and more with the increase of the gas concentration form 0.5 ppm to 10 ppm. When the gas concentration was 0.5 ppm, the response value was about 75 Hz, and the signal-to-noise ratio was larger than 5 in this experiment; therefore, the logical limit of detection (LOD) of the PANI-CNC-modified QCM sensor for HCHO was 260 ppb. This was calculated by using the calibration curve ($y = 57.897x + 24.489$, $R^2 = 0.99978$) with data from Figure 4b, through the equation: $LOD = 3\sigma/S$. The square of R represents the correlation of the fitted straight line, and the value 0.9998 proves that the fitting of the straight line was reliable. The error variance of the baseline was less than 5; here, we used the maximum value of 5, so the logical limit of detection of this sensor can be inferred as 260 ppb. In addition, the response/recovery time of the QCM sensor was also satisfied. Based on the results in Figure 10a, the response time was about 12 s and then reached an equilibrium state of adsorption and desorption. When the formaldehyde vapor was blown away, the desorption of formaldehyde vapor was very quick and there was little drift in the baseline during continuous testing.

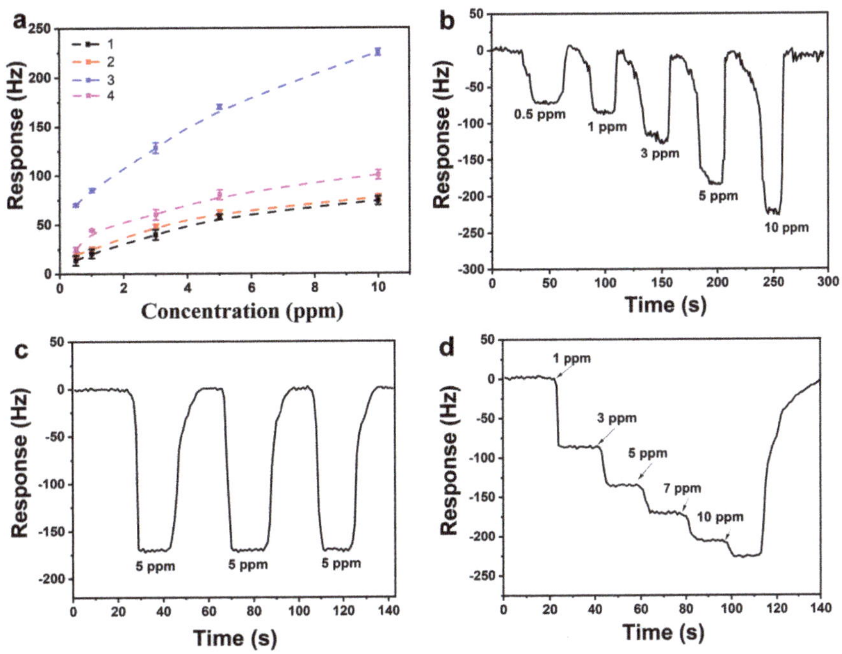

Figure 10. (**a**) The frequency responses of x-PANI-CNCS (x = 1, 2, 3 and 4) with 0.5–10 ppm formaldehyde gas, (**b**) response of the PANI-CNC-based QCM sensors to HCHO at different concentrations in the range of 0.5–10 ppm, (**c**) cycling stability of PANI-CNC-based QCM to 5 ppm HCHO vapor, (**d**) dynamic responses of PANI-CNC-based QCM to 1–10 ppm HCHO vapor.

Another key factor is the repeatability of the QCM sensor. As shown in Figure 10c, the sensor was tested three times for its ability to detect 5 ppm formaldehyde. The differences between the response values across the three tests was less than 5 Hz, which reflected the good repeatability of the sensor.

Figure 10d is a successive response curve of the 3-PANI-CNC samples to 1–10 ppm formaldehyde. With the continuous increase of the gas concentration, the response value also gradually increased, however, the frequency shift caused by the increase of the same concentration (2 ppm) became smaller and smaller. This result meant that the sensitivity of the material decreased as the amount of adsorption increased. This was consistent with the Langmuir sorption theory. The results could be attributed to the decrease of the PANI surficial active sites. Furthermore, the 3-PANI-CNC showed a high level of long-term stability: after three months of testing, the error range did not exceed 8 Hz, as shown in Figure S1.

Compared with previous research into formaldehyde gas sensors (Table 1), this QCM sensor not only showed a satisfactory response and recovery speed, but also high sensitivity. The high detection performance of the PANI-CNCs was attributed to the array CNC, which was beneficial for the diffusion and the ultra-thin decoration of the PANI offering the active sites. Considering the sensitivity and stability of the HCHO sensor, it is more reliable to use this sensor when the sensitivity requirement is not particularly high.

Table 1. Room-temperature detection properties of various sensing materials based on QCM.

Material	Target Gas	Response/ Recovery Time	Limit of Detection	Reference
PAN-PVA	HCHO	120 s/Not given	0.5 ppm	[45]
Pristine PANI	HCHO	Not given	12.87 ppm	[27]
PLY-PANI	HCHO	Not given	1.11 ppm	[27]
LYS-PANI	HCHO	Not given	0.4 ppm	[27]
A polyaniline/silver	HCHO	Not given	1.24 ppm	[46]
PANI/Fluoral-P	HCHO	42 s/Not given	3.7 ppm	[47]
PVAm	HCHO	120 s/Not given	0.5 ppm	[45]
Graphene oxide	HCHO	Not given	0.06 ppm	[18]
PANI–Cd^{2+}	HCHO	25 s/30 s	Not given	[48]
PAAm/MWCNTs	HCHO	80 s/100 s	0.5 ppm	[49]
MIL-101(Cr)	HCHO	24 s/75 s	1.79 ppm	[50]
PANI-CNC	HCHO	12 s/14 s	260 ppb	This work

It is known that the enthalpy of adsorption (ΔH) is a key parameter in adsorption-induced mass-type detection; according to classical physical-chemical adsorption theories, weak reversible physical adsorption without selectivity occurs when the ΔH value is greater than −40 kJ/mol but less than 0 kJ/mol. Irreversible chemistry reactions occur when the value of ΔH is less than −80 kJ/mol [51]. If selectivity and reversibility are considered carefully, reversible chemical adsorption is ideal for meeting the needs of the gas sensor when the value of ΔH is in the range of −80 kJ/mol to −40 kJ/mol. Therefore, we used a temperature-varying micro-gravimetric method to extract the ΔH. Combined with Figure 10b,d, two adsorption isotherms were fitted; they are shown in Figure 11a. The ΔH was calculated to be −64.76 kJ/mol, according to the Clausius–Clapeyron equation. The moderate ΔH indicated a reversible Schiff base addition reaction and an –NH_2 functional group of PANI.

Besides sensitivity, response/recovery time, and repeatability, selectivity for target molecules (i.e., cross-sensitivity) is also a critical parameter of sensors. Herein, seven different gases were used to evaluate the selectivity of the HCHO sensor. As shown in Figure 11b, the 3-PANI-CNC sensor displayed the highest response to the HCHO, of up to 140 Hz@3 ppm; with the responses of the other interfering gases were all lower than 70 Hz at the same concentration. The negligible interfering signals of these other gases may be ascribed to the inevitable physical/nonspecific adsorption.

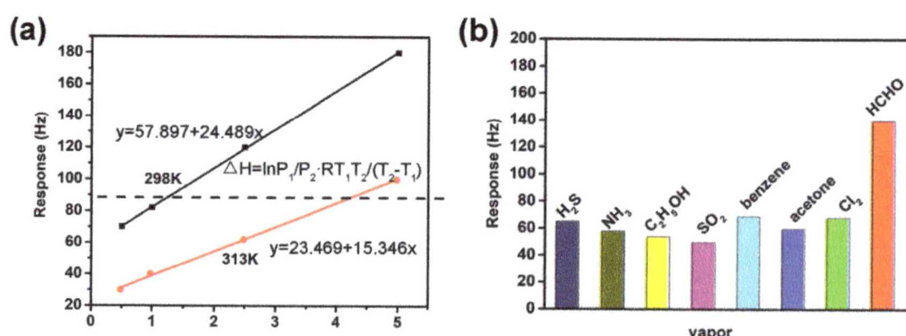

Figure 11. (**a**) Based on the experimental results in Figure 10a,c, two isotherms were plotted to calculate the adsorption heat (ΔH), (**b**) response of PANI-CNC-based QCM to different vapors in 3 ppm (25 °C).

3.4. Discussion

Based on the results, the successful decoration of the ultra-thin PANI layer was confirmed. The PANI-CNC sensors' sensitivity to formaldehyde increased at first and then decreased along with the increasing of the PANI layer. The reason for the increase in the sensor's performance was attributed to the gradually increase of the amino loading on the surface of the four carbon helical structures, which was conducive to the improvement of the gas sensor's performance. The reason for this increase in detection performance was a decrease in the specific surface area and pore size, which is a critical factor for gas response, along with a further increase in polyaniline loading.

The high response/recovery performance was attributed to the fact that the formaldehyde molecules quickly diffused in the array PANI-CNCs. Moreover, the helical PANI-CNCs reinforced the Knudsen diffusion, which increased the probability between the formaldehyde molecules and the PANI layer, which in turn lead to the high response and low detection limits. As shown in Figure 12, the –NH– group and the $-NH_2$ group acted as active sites in the Schiff-base reaction and caused the formaldehyde molecules to be absorbed on the surface of the PANI-CNCs. Both the quick diffusion of the formaldehyde molecules between the array PANI-CNCs and the Schiff-base reaction between the –NHx (–NH– and $-NH_2$) and the formaldehyde molecules lead to a good detection performance.

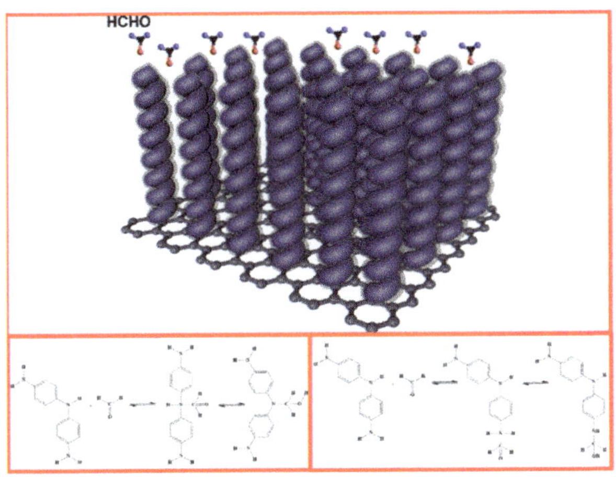

Figure 12. The mechanism of the 3-PANI-CNCs.

4. Conclusions

In summary, we fabricated highly purified array CNCs through the optimization of ALD Cu-based catalysts. The ultra-thin PANI layer was decorated on the surfaces of the CNCs through electrodeposition. The element mapping, XPS, ATR-IR and XRD results confirmed the successful decoration of the ultra-thin layer of the PANI on the CNCs. The array PANI-CNC showed a high HCHO detection performance. The detection limit of the sensor for formaldehyde was lower than 500 ppb. The thermodynamic experiments indicated that CNCs-PANI material is suitable for formaldehyde detection due to the moderate value of adsorption ΔH (-64.76 kJ/mol). The good detection performance was attributed to the reversible Schiff base reaction between the –NH– group and the HCHO. The high response/recovery performance was due to the array-distributed CNC. This study offers a novel direction for the design and application of CNCs in gas detection, which demonstrate good selectivity, sensitivity and repeatability with formaldehyde. Furthermore, the surficial decoration of ultra-thin PANI on nanomaterial also offers a novel strategy for the functional decoration of materials in order to increase the adsorption of target gas.

Supplementary Materials: The following are available online at https://www.mdpi.com/article/10.3390/chemosensors9100276/s1. Figure S1: The long-term stability of the 3-PANI-CNC sensor.

Author Contributions: Conceptualization and data curation, Q.H., Z.M. and J.Y.; methodology, T.G., Y.W., Z.D. and X.L.; writing—original draft preparation, Q.H., Z.M.; writing—review, editing and supervision, W.Z., S.Z. and J.X. All authors have read and agreed to the published version of the manuscript.

Funding: This work was supported by the Shanghai Natural Science Foundation (19ZR1418900) and the Key Basic Research Program of the Science and Technology Commission of Shanghai Municipality (20JC1415300). We also gratefully appreciate the financial support from China Scholarship Council (CSC).

Institutional Review Board Statement: Not applicable.

Informed Consent Statement: Not applicable.

Data Availability Statement: The data presented in this study are available on request from the corresponding author.

Conflicts of Interest: The authors declare no conflict of interest.

References

1. Wang, G.; Gao, Z.; Tang, S.; Chen, C.; Duan, F.; Zhao, S.; Lin, S.; Feng, Y.; Zhou, L.; Qin, Y. Microwave Absorption Properties of Carbon Nanocoils Coated with Highly Controlled Magnetic Materials by Atomic Layer Deposition. *ACS Nano* **2012**, *6*, 11009–11017. [CrossRef] [PubMed]
2. Wang, G.; Ran, G.; Wan, G.; Yang, P.; Gao, Z.; Lin, S.; Fu, C.; Qin, Y. Size-Selective Catalytic Growth of Nearly 100% Pure Carbon Nanocoils with Copper Nanoparticles Produced by Atomic Layer Deposition. *ACS Nano* **2014**, *8*, 5330–5338. [CrossRef] [PubMed]
3. Ge, H.; Zhang, B.; Gu, X.; Liang, H.; Yang, H.; Gao, Z.; Wang, J.; Qin, Y. A Tandem Catalyst with Multiple Metal Oxide Interfaces Produced by Atomic Layer Deposition. *Angew. Chem.* **2016**, *55*, 7081–7085. [CrossRef] [PubMed]
4. Gao, Z.; Dong, M.; Wang, G.; Sheng, P.; Wu, Z.; Yang, H.; Zhang, B.; Wang, G.; Wang, J.; Qin, Y. Multiply Confined Nickel Nanocatalysts Produced by Atomic Layer Deposition for Hydrogenation Reactions. *Angew. Chem.* **2015**, *54*, 9006–9010. [CrossRef] [PubMed]
5. Gao, Z.; Qin, Y. Design and Properties of Confined Nanocatalysts by Atomic Layer Deposition. *Acc. Chem. Res.* **2017**, *50*, 2309–2316. [CrossRef] [PubMed]
6. Wang, L.; Zhu, Y.; Xiang, Q.; Cheng, Z.; Chen, Y.; Xu, J. One novel humidity-resistance formaldehyde molecular probe based hydrophobic diphenyl sulfone urea dry-gel: Synthesis, sensing performance and mechanism. *Sens. Actuators B Chem.* **2017**, *251*, 590–600. [CrossRef]
7. Wang, L.; Wang, Z.; Xiang, Q.; Chen, Y.; Duan, Z.; Xu, J. High Performance Formaldehyde Detection Based on a Novel Copper (II) Complex Functionalized QCM Gas Sensor. *Sens. Actuators B Chem.* **2017**, *248*, 820–828. [CrossRef]
8. Wang, L. Metal-organic frameworks for QCM-based gas sensors: A review. *Sens. Actuators A Phys.* **2020**, *307*, 111984. [CrossRef]
9. Zhang, S.; Yang, Q.; Xu, X.; Liu, X.; Li, Q.; Guo, J.; Torad, N.L.; Alshehri, S.M.; Ahamad, T.; Hossain, M.S.A.; et al. Assembling Well-Arranged Covalent Organic Frameworks on MOF-Derived Graphitic Carbon for Remarkable for-Maldehyde Sensing. *Nanoscale* **2020**, *12*, 15611–15619. [CrossRef]
10. Znamenskaya, Y.; Björklund, S.; Kocherbitov, V.; Alfredsson, V. Effect of Hydration and Dehydration on the Properties of SBA-15 Layer Studied by Humidity Scanning QCM-D. *Microporous Mesoporous Mater.* **2016**, *230*, 58–65. [CrossRef]
11. Lou, H.; Shen, C.; Xiang, Q.; Xu, J.; Lou, T. FDU-12 Mesoporous Materials Detection Hg (II) Ions by QCM. *Nano* **2016**, *11*, 1650094. [CrossRef]
12. Zhang, Q.; Zhao, M.; Liu, Y.; Cao, A.; Qian, W.; Lu, Y.; Wei, F. Energy-Absorbing Hybrid Composites Based on Alternate Car-Bon-Nanotube and Inorganic Layers. *Adv. Mater.* **2009**, *21*, 2876–2880. [CrossRef]
13. Zhao, M.-Q.; Zhang, Q.; Jia, X.-L.; Huang, J.-Q.; Zhang, Y.-H.; Wei, F. Hierarchical Composites of Single/Double-Walled Carbon Nanotubes Interlinked Flakes from Direct Carbon Deposition on Layered Double Hydroxides. *Adv. Funct. Mater.* **2010**, *20*, 677–685. [CrossRef]
14. Lv, R.; Cui, T.; Jun, M.-S.; Zhang, Q.; Cao, A.; Su, D.S.; Zhang, Z.; Yoon, S.-H.; Miyawaki, J.; Mochida, I.; et al. Open-Ended, N-Doped Carbon Nanotube-Graphene Hybrid Nanostructures as High-Performance Catalyst Support. *Adv. Funct. Mater.* **2011**, *21*, 999–1006. [CrossRef]
15. Fratoddi, I.; Venditti, I.; Cametti, C.; Russo, M.V. Chemiresistive polyaniline-based gas sensors: A mini review. *Sens. Actuators B Chem.* **2015**, *220*, 534–548. [CrossRef]
16. Choi, H.H.; Lee, J.; Dong, K.-Y.; Ju, B.-K.; Lee, W. Gas Sensing Performance of Composite Materials Using Conducting Polymer/Single-Walled Carbon Nanotubes. *Macromol. Res.* **2012**, *20*, 143–146. [CrossRef]
17. Wu, Z.; Chen, X.; Zhu, S.; Zhou, Z.; Yao, Y.; Quan, W.; Liu, B. Enhanced Sensitivity of Ammonia Sensor Using Graphene/Polyaniline Nanocomposite. *Sens. Actuators B Chem.* **2013**, *178*, 485–493. [CrossRef]

18. Yang, M.; He, J. Graphene Oxide as Quartz Crystal Microbalance Sensing Layers for Detection of Formaldehyde. *Sens. Actuators B Chem.* **2016**, *228*, 486–490. [CrossRef]
19. Wang, Z.; Peng, X.; Huang, C.; Chen, X.; Dai, W.; Fu, X. CO Gas Sensitivity and Its Oxidation Over TiO_2 Modified by PANI under UV Irradiation at Room Temperature. *Appl. Catal. B Environ.* **2017**, *219*, 379–390. [CrossRef]
20. Tanrıverdi, E.E.; Uzumcu, A.T.; Kavas, H.; Demir, A.; Baykal, A. Conductivity Study of Polyaniline-Cobalt Ferrite (PANI-$CoFe_2O_4$) Nanocomposite. *Nano-Micro Lett.* **2011**, *3*, 99–107. [CrossRef]
21. Im, D.; Kim, D.; Jeong, D.; Park, W.I.; Chun, M.; Park, J.-S.; Kim, H.; Jung, H. Improved Formaldehyde Gas Sensing Properties of Well-Controlled Au Nanoparticle-Decorated In_2O_3 Nanofibers Integrated on Low Power MEMS platform. *J. Mater. Sci. Technol.* **2019**, *38*, 56–63. [CrossRef]
22. Yan, D.; Xu, P.; Xiang, Q.; Mou, H.; Xu, J.; Wen, W.; Li, X.; Zhang, Y. Polydopamine Nanotubes: Bio-Inspired Synthesis, Formaldehyde Sensing Properties and Thermodynamic Investigation. *J. Mater. Chem. A* **2016**, *4*, 3487–3493. [CrossRef]
23. Xu, L.; Ge, M.; Zhang, F.; Huang, H.; Sun, Y.; He, D. Nanostructured of SnO_2/NiO Composite as a Highly Selective Formaldehyde Gas Sensor. *J. Mater. Res.* **2020**, *35*, 3079–3090. [CrossRef]
24. Li, N.; Fan, Y.; Shi, Y.; Xiang, Q.; Wang, X.; Xu, J. A Low Temperature Formaldehyde Gas Sensor Based on Hierarchical SnO/SnO_2 Nano-Flowers Assembled from Ultrathin Nanosheets: Synthesis, Sensing Performance and Mechanism. *Sens. Actuators B Chem.* **2019**, *294*, 106–115. [CrossRef]
25. Liu, X.-Y.; Yin, X.-M.; Yang, S.-L.; Zhang, L.; Bu, R.; Gao, E.-Q. Chromic and Fluorescence-Responsive Metal–Organic Frameworks Afforded by N-Amination Modification. *ACS Appl. Mater. Interfaces* **2021**, *13*, 20380–20387. [CrossRef] [PubMed]
26. Akbar, S.A.; Mardhiah, A.; Saidi, N.; Lelifajri, D. The Effect of Graphite Composition on Polyaniline Film Performance for Formalin Gas Sensor. *Bull. Chem. Soc. Ethiop.* **2021**, *34*, 597–604. [CrossRef]
27. Srinives, S.; Sarkar, T.; Mulchandani, A. Primary Amine-Functionalized Polyaniline Nanothin Film Sensor for Detecting Formal-Dehyde. *Sens. Actuators B Chem.* **2014**, *194*, 255–259. [CrossRef]
28. Lim, J.-H.; Phiboolsirichit, N.; Mubeen, S.; Deshusses, M.A.; Mulchandani, A.; Myung, N.V. Electrical and Gas Sensing Properties of Polyaniline Functionalized Single-Walled Carbon Nanotubes. *Nanotechnology* **2010**, *21*, 75502. [CrossRef] [PubMed]
29. Parmar, M.; Balamurugan, C.; Lee, D.-W. PANI and Graphene/PANI Nanocomposite Films-Comparative Toluene Gas Sensing Behavior. *Sensors* **2013**, *13*, 16611–16624. [CrossRef] [PubMed]
30. Wang, Q.; Dong, X.; Pang, Z.; Du, Y.; Xia, X.; Wei, Q.; Huang, F. Ammonia Sensing Behaviors of TiO_2-PANI/PA_6 Composite Nanofibers. *Sensors* **2012**, *12*, 17046–17057. [CrossRef]
31. Huang, J.; Yang, T.; Kang, Y.; Wang, Y.; Wang, S. Gas Sensing Performance of Polyaniline/ZnO Organic-Inorganic Hybrids for Detecting VOCs at Low Temperature. *J. Nat. Gas. Chem.* **2011**, *20*, 515–519. [CrossRef]
32. Kim, M.-J.; Kim, K.H.; Yang, X.; Yu, Y.; Lee, Y.-S. Improvement in NO Gas-Sensing Properties Using Heterojunctions between Polyaniline and Nitrogen on Activated Carbon Fibers. *J. Ind. Eng. Chem.* **2019**, *76*, 181–187. [CrossRef]
33. Bai, H.; Shi, G. Gas Sensors Based on Conducting Polymers. *Sensors* **2007**, *7*, 267–307. [CrossRef]
34. Reddy, K.R.; Sin, B.C.; Ryu, K.S.; Kim, J.-C.; Chung, H.; Lee, Y. Conducting Polymer Functionalized Multi-Walled Carbon Nanotubes with Noble Metal Nanoparticles: Synthesis, Morphological Characteristics and Electrical Properties. *Synth. Met.* **2009**, *159*, 595–603. [CrossRef]
35. Bai, H.; Sheng, K.; Zhang, P.; Li, C.; Shi, G. Graphene Oxide/Conducting Polymer Composite Hydrogels. *J. Mater. Chem.* **2011**, *21*, 18653–18658. [CrossRef]
36. Zhang, T.; Nix, M.B.; Yoo, B.-Y.; Deshusses, M.A.; Myung, N.V. Electrochemically Functionalized Single-Walled Carbon Nano-tube Gas Sensor. *Electroanalysis* **2006**, *18*, 1153–1158. [CrossRef]
37. Zhao, Y.; Zhang, Z.; Yu, L.; Jiang, T. Hydrophobic Polystyrene/Electro-Spun Polyaniline Coatings for Corrosion Protection. *Synth. Met.* **2017**, *234*, 166–174. [CrossRef]
38. Ding, H.; Zhu, C.; Zhou, Z.; Wan, M.; Wei, Y. Hydrophobicity of Polyaniline Microspheres Deposited on a Glass Substrate. *Macromol. Rapid Commun.* **2006**, *27*, 1029–1034. [CrossRef]
39. Leng, W.; Zhou, S.; Gu, G.; Wu, L. Wettability Switching of SDS-Doped Polyaniline from Hydrophobic to Hydrophilic Induced by Alkaline/Reduction Reactions. *J. Colloid Interface Sci.* **2012**, *369*, 411–418. [CrossRef]
40. Fan, H.; Wang, H.; Guo, J.; Zhao, N.; Xu, J. SDBS-Assisted Preparation of Novel Polyaniline Planar-Structure: Morphology, Mechanism and Hydrophobicity. *J. Colloid Interface Sci.* **2013**, *414*, 46–49. [CrossRef] [PubMed]
41. Ma, Z.; Yuan, T.; Fan, Y.; Wang, L.; Duan, Z.; Du, W.; Zhang, D.; Xu, J. A Benzene Vapor Sensor Based on a Metal-Organic Framework-Modified Quartz Crystal Microbalance. *Sens. Actuators B Chem.* **2020**, *311*, 127365. [CrossRef]
42. Luo, M.-F.; Fang, P.; He, M.; Xie, Y.-L. In Situ XRD, Raman, and TPR Studies of CuO/Al_2O_3 Catalysts for CO Oxidation. *J. Mol. Catal. A Chem.* **2005**, *239*, 243–248. [CrossRef]
43. Li, X.; Yu, L.; Zhao, W.; Shi, Y.; Yu, L.; Dong, Y.; Zhu, Y.; Fu, Y.; Liu, X.; Fu, F. Prism-Shaped Hollow Carbon Decorated with Polyaniline for Microwave Absorption. *Chem. Eng. J.* **2020**, *379*, 122393. [CrossRef]
44. Ariyageadsakul, P.; Vchirawongkwin, V.; Kritayakornupong, C. Determination of Toxic Carbonyl Species Including Acetone, Formaldehyde, and Phosgene by Polyaniline Emeraldine Gas Sensor Using DFT Calculation. *Sens. Actuators B Chem.* **2016**, *232*, 165–174. [CrossRef]
45. Huang, W.; Wang, X.; Jia, Y.; Li, X.; Zhu, Z.; Li, Y.; Si, Y.; Ding, B.; Wang, X.; Yu, J. Highly Sensitive Formaldehyde Sensors Based on Polyvinylamine Modified Polyacrylonitrile Nanofibers. *RSC Adv.* **2013**, *3*, 22994–23000. [CrossRef]

46. Zhang, J.; Guan, P.; Li, W.; Shi, Z.; Zhai, H. Synthesis and Characterization of a Polyaniline/Silver Nanocomposite for the Deter-Mination of Formaldehyde. *Instrum. Sci. Technol.* **2015**, *44*, 249–258. [CrossRef]
47. Carquigny, S.; Redon, N.; Plaisance, H.; Reynaud, S. Development of a Polyaniline/Fluoral-P Chemical Sensor for Gaseous Formaldehyde Detection. *IEEE Sens. J.* **2011**, *12*, 1300–1306. [CrossRef]
48. Zhang, L.; Li, X.; Mu, Z.; Miao, J.; Wang, K.; Zhang, R.; Chen, S. A Novel Composite of CdS Nanorods Growing on a Polyaniline-Cd^{2+} Particles Surface with Excellent Formaldehyde Gas Sensing Properties at Low Temperature. *RSC Adv.* **2018**, *8*, 30747–30754. [CrossRef]
49. Feng, L.; Feng, L.; Li, Q.; Cui, J.; Guo, J. Sensitive Formaldehyde Detection with QCM Sensor Based on PAAm/MWCNTs and PVAm/MWCNTs. *ACS Omega* **2021**, *6*, 14004–14014. [CrossRef] [PubMed]
50. Haghighi, E.; Zeinali, S. Formaldehyde Detection Using Quartz Crystal Microbalance (QCM) Nanosensor Coated by Nanoporous MIL-101(Cr) Film. *Microporous Mesoporous Mater.* **2020**, *300*, 110065. [CrossRef]
51. Cao, Y.; Fan, Y.; Ma, Z.; Cheng, Z.; Xiang, Q.; Duan, Z.; Xu, J. Urea-Functionalized SBA-15 Hybrids: Post-grafting Synthesis, High-Performance Organophosphorus Sensing and Their Response Mechanism. *Sens. Actuators B Chem.* **2018**, *273*, 1162–1169. [CrossRef]

chemosensors

MDPI

Review

Electrochemical Sensors for Liquid Biopsy and Their Integration into Lab-on-Chip Platforms: Revolutionizing the Approach to Diseases

Salma Umme [1,*], Giulia Siciliano [1], Elisabetta Primiceri [1], Antonio Turco [1], Iolena Tarantini [2], Francesco Ferrara [1,*] and Maria Serena Chiriacò [1]

[1] CNR NANOTEC–Institute of Nanotechnology, Via per Monteroni, 73100 Lecce, Italy; giulia.siciliano@nanotec.cnr.it (G.S.); elisabetta.primiceri@nanotec.cnr.it (E.P.); antonio.turco@nanotec.cnr.it (A.T.); mariaserena.chiriaco@nanotec.cnr.it (M.S.C.)

[2] Department of Mathematics and Physics "Ennio De Giorgi", University of Salento, Via per Arnesano, 73100 Lecce, Italy; iolena.tarantini@unisalento.it

* Correspondence: salma.umme@nanotec.cnr.it (S.U.); francesco.ferrara@nanotec.cnr.it (F.F.)

Abstract: The screening and early diagnosis of diseases are crucial for a patient's treatment to be successful and to improve their survival rate, especially for cancer. The development of non-invasive analytical methods able to detect the biomarkers of pathologies is a critical point to define a successful treatment and a good outcome. This study extensively reviews the electrochemical methods used for the development of biosensors in a liquid biopsy, owing to their ability to provide a rapid response, precise detection, and low detection limits. We also discuss new developments in electrochemical biosensors, which can improve the specificity and sensitivity of standard analytical procedures. Electrochemical biosensors demonstrate remarkable sensitivity in detecting minute quantities of analytes, encompassing proteins, nucleic acids, and circulating tumor cells, even within challenging matrices such as urine, serum, blood, and various other body fluids. Among the various detection techniques used for the detection of cancer biomarkers, even in the picogram range, voltammetric sensors are deeply discussed in this review because of their advantages and technical characteristics. This widespread utilization stems from their ability to facilitate the quantitative detection of ions and molecules with exceptional precision. A comparison of each electrochemical technique is discussed to assist with the selection of appropriate analytical methods.

Keywords: liquid biopsy; electrochemical sensors; lab-on-a-chip; miniaturization; sensor integration; microfabrication

Citation: Umme, S.; Siciliano, G.; Primiceri, E.; Turco, A.; Tarantini, I.; Ferrara, F.; Chiriacò, M.S. Electrochemical Sensors for Liquid Biopsy and Their Integration into Lab-on-Chip Platforms: Revolutionizing the Approach to Diseases. *Chemosensors* **2023**, *11*, 517. https://doi.org/10.3390/chemosensors11100517

Academic Editor: Gerd-Uwe Flechsig

Received: 30 August 2023
Revised: 22 September 2023
Accepted: 26 September 2023
Published: 1 October 2023

1. Introduction

A biopsy is a technique in which tissue samples are taken from the body and examined under a microscope to see if cancer (though the concept is applicable to many other diseases) or abnormal cells are present. Biopsies can be classified into the following categories based on the sample being taken (Figure 1).

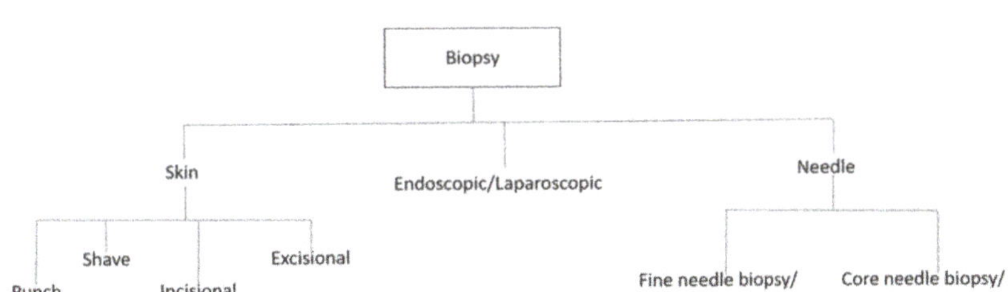

Figure 1. Classification of biopsies: understanding the different categories.

In the last decades, the liquid biopsy, namely, the possibility to have a diagnosis from body fluids without resorting to a tissue biopsy, has been increasingly investigated. The possibility to detect and classify tumors or other diseases, even at a very early stage, in a minimally invasive and repeatable way could have a significant clinical impact, and significant progress has been made in the development of devices able to do this in a smarter manner compared to standard analytical methods. Despite the great advantages for patients' compliance and the minimally invasive features, this approach has not yet attained the status of a conventional tool in the armory of clinical oncologists [1].

Biosensors are considered to be promising tools for the quantitative or semi-quantitative detection of analytes [2]. In this type of sensor, a biological molecule interacts with the analyte, previously immobilized on the biosensor, producing a physicochemical signal that is detected by the transducer. Biosensors can be divided into two categories: catalytic-based, which produce a substance starting from substrate's compounds, and affinity-based, which directly bind the analyte. According to the type of signal being transmitted, biosensors can be classified as electrochemical, optical [3], magnetic, or piezoelectric, just to name a few [4].

Electrochemical techniques excel among these methods by offering rapid, sensitive, selective, and cost-effective detection and monitoring of a wide range of biological molecules associated with diverse diseases. Additionally, their seamless integration into portable systems enables the implementation of point-of-care diagnostic approaches [5]. An electrochemical biosensor is a compact device that utilizes both biorecognition processes and electrochemical transducers to convert biological information into electrical signals. This conversion provides either quantitative or semi-quantitative information about the analyte being detected [6].

Electrochemical biosensors were recommended by the international union of pure and applied chemistry (IUPAC). Which states that an electrochemical biosensor is a self-contained integrated device that is capable of providing specific quantitative or semi-quantitative analytical information using a biological recognition element (biochemical receptor) which is kept in direct spatial contact with an electrochemical transduction element.

Electrochemical methods, such as electrochemical impedance spectroscopy (EIS), differential pulse voltammetry (DPV), and cyclic voltammetry (CV), play a crucial role in both the development of biosensors and the evaluation of their performance [7]. These methods are highly valuable approaches in the field of liquid biopsy (Figure 2). For research purposes, the CV technique is widely used in biosensor development because it provides valuable information such as the types of redox processes present in the analysis and the reversibility of reactions. A sensing system that can identify a cell's location within a microfluidic channel was designed by Rapier et al. The results from electrochemical impedance spectroscopy (EIS) show that cells in microfluidic channels can be positioned between different pairs of electrodes at varied locations along the device's length. Impedance spectra distinguish among confluent, sparse, and empty microfluidic channels. A huge boost to the development of this kind of sensors as well as to their application in the liquid biopsy was given by their high suitability to miniaturization. Electrodes' architecture, indeed, can be easily achieved through micro- and nanofabrication methods, increasing the number of sensing elements per area and allowing high-throughput performances [8].

Also, lab-on-a-chip integration is directly linked to miniaturization. The possibility to perform multiple assays by lowering the dimensions of sensing elements has led to the necessity to differently functionalize and use them for the detection of different biomarkers [9,10].

The utilization of impedance-based methods significantly facilitated the simplification of cellular assays, providing quantitative and highly sensitive results that are amenable to automation and scalability in multi-well formats. An extensive review by W. Gamal et al. shed light on their efficacy. Moreover, in the context of the real-time monitoring of a three-dimensional cell culture, dielectric spectroscopy and electrical impedance tomography emerged as promising alternatives to two-dimensional impedance sensing [11]. These impedance-based cellular assays (IBCAs) serve as label-free phenotyping assays and are gaining increasing interest in the field of regenerative medicine applications [12]. To determine the flow in a microfluidic chip, Evangelos S. et al. developed a strain-sensing module based on microfluidic and lab-on-a-chip systems that offers simple integration with most microfluidic systems. The sensor consists of interconnected platinum nanoparticles that self-assemble on flexible polyolefin substrates, which also serve as the sealing layer for the microfluidic channels. These nanoparticle networks are formed using a modified sputtering approach and are implemented on printed circuit board substrates (PCBs) through milling and computer numerical control machining. The resulting module exhibits a competitive limit of detection (LOD), cost-effectiveness, low power requirements, and seamless integration with existing microfluidic systems. It can be utilized as an independent unit or integrated into the sealing material, enabling the detection of flow rates as low as 5 μL/min (equivalent to a strain of 0.00337%). The sensor demonstrates a sensitivity of 0.021 μL per minute [13].

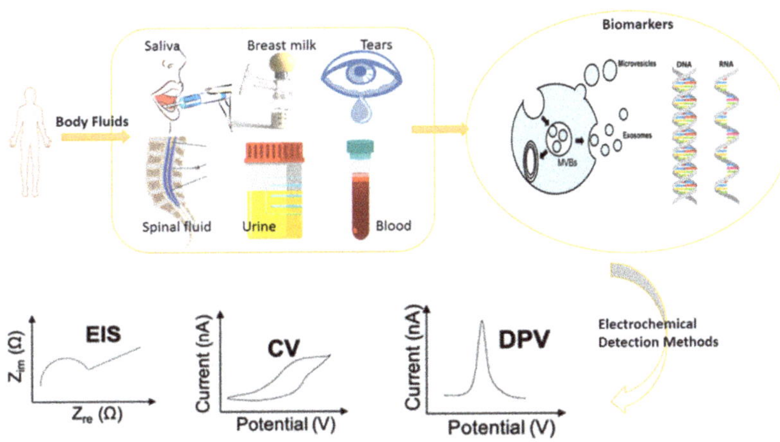

Figure 2. Overview of the electrochemical biosensors' operation.

Lucile et al. reported another novel microfluidic method that can selectively extract, preconcentrate, and fluorescently detect IL-6 directly on the chip by the fluidization of magnetic beads. The ability to switch between packed and fluidized states allowed the authors to evaluate how the physical characteristics of the beads could be altered to increase mass transport, lessen non-specific binding, and triple the detection signal. A high dynamic range (10 pg/mL to 2 ng/mL) and a twofold reduction in LOD compared to traditional approaches were demonstrated by integrating the entire ELISA protocol into a single microfluidic chamber [14].

In this review, an overview about electrochemical methods and their applications as transduction techniques in the development of biosensors is provided. Particular attention is paid to the liquid biopsy and to the use of a miniaturized platform, allowing for the spread of point-of-care devices in this field.

Essential Biomarkers for Liquid Biopsy Detection and Monitoring

The liquid biopsy is increasingly used for the detection, analysis, and monitoring of circulating tumor cells (CTCs), circulating tumor DNA, and circulating extracellular nucleic acids [15], in blood or other body fluids such as urine, with the main advantage of diagnosing cancer at an early stage. In contrast to a traditional invasive biopsy, which includes cells or tissues from the lesion, the liquid biopsy is a non-invasive procedure that can also be used during treatment planning as well as in the follow-up of the disease. In addition to blood, other body fluids are under consideration for the liquid biopsy:

urine [16] is already used as a source of biomarkers (PCA3 in prostate and pancreatic cancer detection), and several examples of state-of-the-art detection from saliva, seminal fluid [17], or stool were described [18].

The liquid biopsy is a non-invasive method for detecting, analyzing, and monitoring cancer cells, DNA, and other nucleic acids in body fluids such as blood and urine. It offers several advantages over traditional biopsy methods, including the ability to diagnose cancer at an early stage without the need for invasive procedures. This technique can be used not only for the initial diagnosis of cancer but also during treatment planning and in follow-up procedures. Other body fluids such as urine, saliva, seminal fluid, and stool are also being studied as potential sources of biomarkers for the liquid biopsy.

Biomarkers are recognized to have a critical role in the early detection of cancer, the creation of personalized treatments, and the identification of the disease-related underlying processes. An ideal cancer biomarker would have high clinical sensitivity and specificity, rapid release in the blood for early detection, high concentration in the blood for prolonged periods of time, and the possibility to be quantified.

The detection targets of the liquid biopsy primarily encompass cells and nucleic acids found in body fluids, including circulating tumor cells (CTCs) in blood, cell-free DNA (cfDNA), circulating tumor DNA (ctDNA), exosomes, micro-RNA (miRNA), and proteins.

Circulating tumor cells are released either from primary tumors or from secondary metastatic sites. Identifying CTCs among blood cells and enumerating them has significant prognostic value [19]. Investigating their presence in blood at the early stage of tumor pathologies means that a highly sensitive method should be used, since a very low number of cells/mL should be detected. To this aim, the presence of a specific membrane biomarker can be used as the target for probes aiming to immobilize or label CTCs [20].

Protein biomarkers are widely used in the liquid biopsy, more often but not only in blood and serum. As an example, urine is already being used to detect PCA3 in prostate and pancreatic cancer, and several state-of-the-art detection methods were developed for other body fluids [21]. Typically, in the development of an electrochemical sensor, a protein biomarker can be identified using antibody/antigen probes immobilized on the surface of the electrodes, aptamers specifically designed to bind the protein, and molecularly imprinted polymers (MIP) directly synthesized on the electrodes' surface, producing artificial antibodies [22–24].

Nucleic acid biomarkers include several kinds of molecules, ranging from fragments of DNA circulating in blood (i.e., cell-free DNA) to miRNA-associated (or not) to extracellular vesicles (EVs) to circulating DNA, also in this case associated with CTCs or as the cargo of EVs. The interest in these last entities is strongly increasing since EVs are identified as important messengers of information among cells and body districts. Indeed, they are physiologically produced and up taken by all types of cells and released into all biofluids. EVs are commonly divided into two classes: large EVs, including microvesicles originating from membrane budding, and small EVs, including exosomes originating from intracellular vesicular bodies. Capturing, counting, and deeply characterizing the surface and the cargo of these membrane bodies is one of the biggest challenges in molecular and cellular biology, as they could be very informative as diagnostic and prognostic biomarkers. The standard methods for the analysis of EVs suffer from strong limitations, and some biochip-based devices were developed to overcome them [25,26].

An innovative way of using the liquid biopsy, going in the direction of point-of-care medicine, is the integration of smart assays into portable devices, which need a small amount of the sample, low power to work, and minimal equipment [27]. Electrochemical methods are particularly suitable for integration into the near-the-bed platform, since most of them are label-free methods, and all the components can be easily miniaturized into lab-on-a-chip platforms. A list of examples from literature including biomarkers investigated, biofluids and electrochemical methods with related LODs is reported in Table 1.

Table 1. Some common biomarkers used for liquid biopsy, the biofluid in which they are detected, and electrochemical method used.

Biomarker	Biofluid/Sample	Electrochemical Method	LOD	Ref.
Exosomes	Plasma	Potentiometric	20 pM 106 mL^{-1} 43 particles μL^{-1} <105 vesicles/10 μL	[28] [29] [30] [31]
Circulating nucleic acids	Human serum	DPV	3.9×10^{-22} g/mL	[32] [33]
• **Circulating tumor DNA (ctDNA)** • **Circulating microRNA (miRNA)**	Serum	DPV and EIS	0.45 fM	[15] [34]
Circulating tumor cells (CTCs)	Blood Blood Peripheral blood	Amperometry DPV DPV	5 cells/mL 27 cells/mL 3 cells/ml	[35] [36] [37]
Proteins	Human serum and saliva	CV and EIS	3.3 fg m/L	[38]

2. Miniaturization Strategies for Biosensing

Miniaturization is the key challenge and research trend currently pursued in the field of biosensing. Moore's law, a theory in the field of microelectronics, postulates the continuous advancement of the industry by the doubling of the number of on-chip transistors every two years. This exponential growth in the transistor count has led to a reduction in the cost per function. Miniaturizing the transducer as well as the biosensing element means a boost in the enhancing sensitivity and specificity features. This advancement will contribute to the achievement of three key performance metrics in biosensors: enhancing the limit of detection, reducing the response time, and lowering production costs.

One of the practical and technological advantages of miniaturizing the biosensing system, indeed, is the improved sensitivity of electrochemical and electronic sensors by increasing the system's signal-to-noise ratio, reaching dimensions of the micro-structured electrodes and nanomaterials that are comparable with entities of interest (as examples: cells in the case of microstructures and protein and nucleic acids in the case of nanomaterials). Signal enhancement in this condition is achieved by the presence of nanogaps and/or nanostructured electrodes that obtain the noise reduction by making available a high surface area for biosensing interactions. This holds true for systems based on field-effect transistors as well as nanoscale electrochemical biosensors. Additionally, reducing the interelectrode spacing at the nanoscale can be also considered as a strategy to amplify the redox current and, in this way, to obtain an electrochemical single-molecule analysis by generating a significate signal-to-noise ratio [39].

The kinetics of transport reactions in biosensing are closely linked to the time needed for biorecognition events to take place. In this context, the background current associated with the charging of the double layer (capacitive current) varies in proportion to the electrode's conductive area. In miniaturized electrochemical cells, the resistive drop is minimized by shortening the ionic current pathway. As a result, the capacitance is reduced, leading to a decreased time constant for the system. This enables faster electron-transfer kinetics compared to macroscale systems.

The possibility of reducing dimensions brings several side advantages. Among these, first of all, is adding portability and integration into a complex platform, which allows for the realization of a multiplexed analysis, including sample treatment tools and parallelized functions. This not only enables the creation of compact yet robust devices but also has the added benefit of reducing manufacturing costs by minimizing material and fabrication expenses per device. This efficiency in mass production contributes to overall cost reduction.

2.1. Micro- and Nanofabrication Methods

Several methods for the micro- and nanofabrication of electrodes are available, allowing for the production of the desired geometries of transducers and sensing elements for electrochemical biosensing. In addition to the consolidated optical and electron-beam lithography technologies, some innovative tools are gaining popularity thanks to the possibility they provide to achieve a better resolution and higher customizable processes in a relatively short time with respect to the standard techniques. These tools offer wide-range opportunities to innovate alongside Moore scaling without requiring high investment levels but offering large-feature, low-end production and high-end performances [40].

Among these technologies, maskless lithography methods based both on a direct laser writing source (for example, a femtosecond laser or UV laser) and two-photon systems are widely used for the fabrication of microelectrodes, ensuring a very high resolution of lithographic patterns even over a large surface area and in a three-dimensional operating mode [41,42]. Recently, several groups were working on these kinds of processes. Zhu and coworkers dealt with the realization of planar electrodes onto flexible substrates and realized direct laser writing on the stacked graphene multilayer of a large-area micro-supercapacitor onto a polyaniline substrate, demonstrating the possibility to fabricate pressure/gas sensors with high sensitivity for multiple applications [43]. Dotan et al. implemented a novel approach for the development of soft and flexible microelectrode structures used in electric and electrochemical sensing. Their method involved the combination of the supersonic cluster beam deposition (SCBD) of gold nanoparticles onto Polydimethylsiloxane (PDMS), followed by femtosecond (fs) laser processing. Through this technique, they successfully produced a nanocomposite film with mechanical properties comparable to those of the elastomeric substrate [44]. Two-photon lithography is a very versatile and flexible micro- and nanofabrication technology that allows for the 3D architecture of a lab-on-a-chip and an integrated platform. The possibility to combine planar and multilevel structures into the same chip thanks to two-photon lithography was explored by Luitz and coworkers, who realized complex 3D micro- and nano-objects using a platinum-containing photoresin, which can be structured via direct lithographic two-photon polymerization, paving the way for novel applications like the production of innovative metamaterials for biomedical applications, where high surface areas and the physicochemical properties of Pt are highly desirable. Moreover, with the subsequential steps of lithography, the two-photon lithography method enables the possibility of embedding sensor structures into microfluidic devices, thus obtaining a monolithic platform for on-chip sample preparation and characterization [45–47].

Microfluidic devices offer the ability to control the flow of fluids at the microscale, enabling the rapid and precise detection of biomolecules. When combined with nanomaterial-based sensors, the real-time monitoring of low concentrations of biomolecules in body fluids such as blood [48], urine [49], tears [50], and saliva [51] can be achieved. The available and easy methods for the fabrication of PDMS made this material emerge as the most popular polymer for the realization of microfluidic channels, which is useful for sample preparation or reaction chambers. Its property of sticking onto various substrates and the possibility to mold it in microstructures allow for its application in a wide range of research fields. Moreover, its adhesion on the surface of connection pads often does not interfere with electrical/acoustic/magnetic signals, allowing for the integration of microfluidics with sensing modules [52–54].

On the other hand, despite the large use of PDMS in research contexts, it has several drawbacks. First of all, its reversible hydrophilicity limits long-term experiments with biological materials and hinders the transition into industrial applications. Moreover, PDMS is not suitable for use with most commonly used solvents (ethanol, isopropanol, acetone, etc.), which could be necessary for preliminary sample treatment. The resistance of PDMS microchannels to high pressures is also limited in the case of stand-alone devices, thus forcing operators to use screws and clamps to secure and keep the watertight sealing. To overcome these aspects, some other polymers are on the rise due to their properties that are more suitable for the realization of robust monolithic devices. Among them, the list includes thermoplastic polymers (plastics) like poly(methyl methacrylate) (PMMA), polystyrene (PS), cyclic olefin copolymers (COC), and polycarbonate (PC), which allow for easy surface treatments, are biocompatible and transparent, and meet some of the industrial requirements for the LOC market [55,56].

2.2. Nanomaterials in Electrochemical Sensors Integrated in LOC Device: From 2D to 3D Electrodes

Nanomaterials have emerged as a promising class of materials for sensing applications due to their unique physicochemical properties [57–59]. The exceptional properties exhibited by nanomaterials, including high surface area, excellent electrical and thermal conductivity, and unique optical characteristics, make them highly advantageous for seamless integration into lab-on-a-chip (LOC) devices as electrochemical sensors. This integration enables the detection of molecules in body fluids with significantly improved sensitivity and accuracy [22].

The incorporation of nanomaterials in electrochemical biosensors holds the potential to bring about a revolutionary transformation in the field of clinical diagnostics. This advancement facilitates the rapid, sensitive, and highly specific detection of biomolecules in body fluids, paving the way for significant advancements in medical diagnosis and patient care. This enhancement is achieved by either promoting electronic transfers or increasing the volume/surface area ratio [58]. This has significant implications for medical diagnosis and treatment, as it can enable the early detection and monitoring of diseases such as cancer, diabetes, and cardiovascular diseases or the timely identification of bacterial infections.

The integration of nanomaterials onto the electrode surface of microfluidic devices plays a crucial role in the advancement of high-performance electrochemical sensors. This is particularly important when electrodes are situated in less accessible locations, as often encountered in lab-on-a-chip systems. By incorporating nanomaterials, the sensitivity and overall performance of the electrochemical sensors can be significantly enhanced, enabling accurate and reliable detection in challenging sample environments. In this kind of device, nanomaterials are typically integrated onto the electrode surface using various techniques, such as in situ synthesis [60], drop casting [61], spin coating [62], electrochemical deposition [63], electrospinning [64], and inkjet printing [65]. Drop casting and spin coating are simple and cost-effective techniques that involve the deposition of nanomaterials onto the electrode surface using a dropper or a spinning device, respectively. Electrochemical deposition involves the deposition of nanomaterials onto the electrode surface by applying a voltage or a current to the electrode in the presence of the nanomaterials in solution. Inkjet printing involves the precise deposition of nanomaterials onto the electrode surface using a specialized printer. Among them, electrochemical deposition and inkjet printing are the techniques that allow for the most precise and controlled deposition of nanomaterials onto the electrode surface, which is critical for the development of high-performance electrochemical sensors.

Various types of organic and inorganic nanomaterials, including carbon nanotubes, graphene, metal and metal oxide, polymer, quantum dots, Prussian blue [66], nanorods, and tubes, are incorporated into electrochemical transducers to enhance electrochemical sensing (Figure 3).

Metallic nanostructures, such as gold [67], silver [68], and platinum [69], are widely used in electrochemical sensors and integrated into microfluidic devices for the detection of molecules in body fluids. Gold nanoparticles (AuNPs), for example, are extensively studied due to their unique electronic, optical, and surface properties, which make them ideal for use in biosensing applications [70]. AuNPs are used in a variety of electrochemical sensors for the detection of different biomolecules, such as glucose, cholesterol [71], and prostate-specific antigen (PSA) [72].

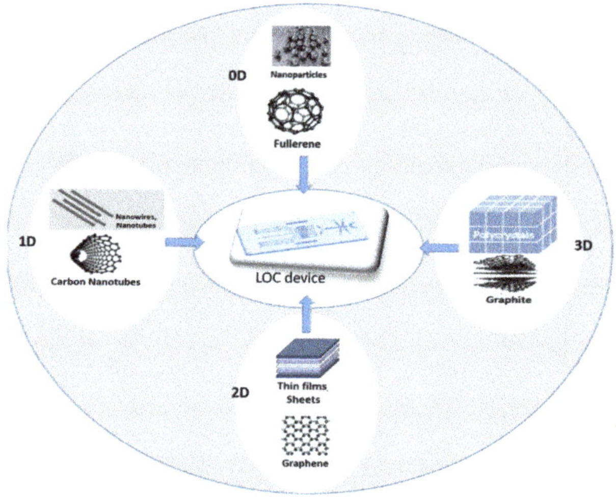

Figure 3. An overview of the nanomaterials used for biosensing.

Magnetic nanoparticles (MNPs), such as iron oxide nanoparticles, are also used in electrochemical biosensors integrated into microfluidic devices for the detection of biomolecules in bodily fluids. MNPs have unique magnetic and surface properties, which make them ideal for use in biosensing applications [73]. These nanomaterials not only improve the limit of detection of the sensors but also enable the separation and transportation of bioanalytes inside the microfluidic device, thereby allowing for the miniaturization of analytical methods [74]. For example, MNPs functionalized with iridium oxide nanoparticles and tyrosinase are used for the detection and quantification of methimazole in microsystems. In the analytical measurements, a permanent magnet was used to immobilize the magnetic complex on the electrode surface. The system was highly sensitive with a low limit of detection (0.004 μM) and demonstrated effectiveness in serum samples. Interestingly, the use of a microfluidic device allows for an improved limit of detection, reusability, automation, the volume of the sample, and response time compared to batch configuration [75].

Among nonmetallic nanomaterials, carbon nanotubes (CNTs) [76], graphene [77], and quantum dots [78,79] (QDs) show high sensitivity toward various analytes, including glucose [77], cholesterol [80], and proteins, [81] with detection limits in the subnanomolar range. Moreover, carbon-based nanomaterials such as graphene and carbon nanotubes are also combined with metallic nanoparticles or polymeric layers to produce nanocomposites with improved performance [82–84] in terms of electronic transfer and selectivity.

One of the challenges associated with the use of nanomaterials in electrochemical sensors integrated into microfluidic devices is the reproducibility and stability of the sensors. Due to the small size of nanomaterials, synthesizing and functionalizing these materials can be challenging, leading to variations in sensor performance. Additionally, the stability of nanomaterial-based sensors can be affected by factors such as temperature, pH, and humidity, resulting in reduced sensor performance over time. Efforts are being made to address these challenges, by developing reproducible synthesis and functionalization methods and optimizing sensor design to enhance stability.

Another challenge related to the use of nanomaterials in electrochemical sensors integrated into microfluidic devices is the integration of these sensors into practical clinical applications. While many studies demonstrated the feasibility of using nanomaterial-based sensors for detecting biomolecules in body fluids, further development and optimization are required before these sensors can be widely adopted in clinical settings. This includes optimizing the sensitivity, selectivity, and stability of the sensors, as well as developing user-friendly and cost-effective instrumentation for their use. In fact, despite the enhanced sensor performance offered by nanomaterials implemented on electrode surfaces, the two-dimensional planar electrodes can still limit component and signal transmission when used in vivo, thereby affecting sensor accuracy and sensitivity [85,86]. This limitation is particularly true for complex samples such as blood or plasma used in point-of-care-devices [87,88]. Furthermore, the planar structure of two-dimensional electrodes poses challenges in achieving the adequate immobilization of active components and an efficient signal transmission, leading to potential issues in sensing accuracy. To address these limitations, the integration of macroscale three-dimensional (3D) porous materials, comprising nanomaterials combined with polymers [89–92], can be employed as electrodes. This approach facilitates expanded microfluidic transport and enables the incorporation of multianalyte detection capabilities, thereby enhancing the overall performance of the sensor system.

The incorporation of porous channels in biosensing systems offers several advantages, including increased surface area, enhanced ion/mass transport pathways, and the improved immobilization and stability of active components. In this context, graphene emerged as a promising avenue for the development of three-dimensional (3D) electrodes. Graphene can be fabricated in the form of aerogel or combined with polymers, providing an excellent opportunity to create highly efficient and versatile 3D electrode structures. Furthermore, the surface of graphene can be easily engineered with other nanomaterials and biorecognition elements. For instance, Xu et al. [93] demonstrated the use of a graphene foam (GF) modified with carbon-doped titanium dioxide nanofibers ($nTiO_2$) as an electrochemical working electrode. The three-dimensional and porous structure of the GF facilitated the penetration and attachment of $nTiO_2$ onto its surface, resulting in enhanced charge-transfer resistance, increased surface area, and improved access of the analyte to the sensing surface. The GF–$nTiO_2$ composite was further functionalized with the ErbB2 antibody for the specific detection of the target ErbB2 antigen, a biomarker for breast cancer. The sensor was employed for quantification of the ErbB2 antigen using differential pulse voltammetry and electrochemical impedance spectroscopy techniques. Remarkably, both methods exhibited high sensitivity across a wide concentration range of the target antigen, demonstrating excellent specificity even in the presence of other members of the EGFR family.

In another study, Zhang et al. [94] conducted a study where they developed an enzymatic electrochemical microfluidic biosensor for glucose detection. The biosensor incorporated a three-dimensional porous graphene aerogel and glucose oxidase (GOx), taking advantage of the aerogel's high electrical conductivity and specific surface area to enhance the immobilization of GOx. The microfluidic system implemented in the biosensor reduced sample consumption during testing. The biosensor exhibited excellent selectivity and stability and successfully monitored glucose levels in serum samples. This innovative biosensor shows promise for clinical applications in diabetes diagnosis, and the method employed for preparing the graphene-aerogel-modified electrode holds potential for broader use in diverse electrochemical sensors.

In addition to high sensitivity, nanomaterial-based sensors are highly selective, enabling the detection of specific molecules in complex biological samples. This selectivity is achieved through the functionalization and modification of nanomaterials, by attaching specific ligands (e.g., antibodies, DNA, RNA, aptamers, and enzymes) [95] through covalent or non-covalent interactions to enhance specificity and electronic transfer in sensors. For example, Fan et al. (2022) developed a smartphone-based electrochemical system composed of CNTs functionalized with gold nanoparticles, thionine, and antibodies for the detection of CA125, a biomarker for prostate cancer [96]. The biosensor exhibited high selectivity toward CA125, with no interference from the other biomolecules present in human serum.

Another approach to obtaining sensors with high specificity is the creation of molecular imprinted polymers (MIPs). MIPs are polymer-based artificial receptors with the ability to recognize different types of target molecules such as amino acids, peptides [97], pesticides [98], drugs [99], and even larger molecules such as proteins [100] and whole cells [101]. The target molecules act as a template and interact with functional monomers to form a complex during polymerization, and then the template can be removed, leaving cavities able to rebind the template molecules thanks to its geometry and chemical moieties. MIPs are largely used as recognition elements in electrochemical sensors and offer great advantages such as improved stability, cost-effectiveness, and a rapid fabrication procedure, overcoming the limitations of natural receptors (antibodies, nucleic acids, and peptides) such as sensitivity to enzymatic digestion, low preservation temperature, etc.

So, by combining the advantages of MIPs and electrochemical transducers, several sensing platforms were realized, joining the sensitivity and ease of use of electrochemical sensors with the high selectivity and stability of MIPs [102]. To realize a high-performance sensing platform based on a MIP as an artificial receptor, it is necessary to consider at least two key aspects: (i) the choice of polymer and (ii) imprinting processes. Electrochemical sensors are compatible with different imprinting approaches such as in situ bulk polymerization, surface imprinting, and electrosynthesis.

The most commonly employed approach for imprinting is bulk imprinting, wherein the transducer surface is coated with a mixture of the template and pre-polymer, which exhibit mutual interaction. Then, after polymerization, template molecules are entrapped inside the polymer matrix and can be removed by a washing step, creating cavities able to recognize the analyte in the subsequent analytic steps. To apply this technique to larger molecules is necessary to realize a very thin layer of polymer, so the imprinted binding sites are near the interface, making template removal and rebinding easier. Another possibility for the recognition of large molecules such as proteins is epitope imprinting, which consists of imprinting only a portion of the target molecules [103].

An alternative approach is based on electrosynthesis, in which polymerization is induced by applying a suitable potential range to a solution containing the monomers with the template molecules without any initiator. The characteristics of resulting films can be tuned by modulating electrochemical parameters. Conductive polymers (CP) and insulators/non-conductive polymers (NCP) can be used with different advantages and disadvantages. Non-conducting MIP films self-limit their growth, to allow a fine control on their thickness, while CPs are more flexible and offer the possibility to tune not only the thickness but also the conductive properties by changing the deposition conditions. The selection of polymers is strictly related to the detection methods: for example, capacitive [104] or impedimetric [105] sensors require nonconductive polymers, while for amperometric detection it is better to use conductive ones [106].

Surface imprinting is one of the most used techniques for the development of MIPs for large molecules, cells, and microorganisms: it consists within the template imprinting only on the MIP surface. Several techniques such as soft lithography, microcontact imprinting, and sacrificial template support methods were exploited to confine the imprinted sites on the MIP interface [105].

The analytical performances of electrochemical MIP-based sensors can be furtherly improved by combining MIP technology with nanomaterials and realizing an imprinted nanocomposite. Different nanomaterials were used for this purpose ranging from carbon-based materials (e.g., nanotubes [107,108] and graphene [109]) to metallic nanoparticles [110]: this transition toward a nanoMIP significantly improved the analytical performances of MIP-based sensors in terms of detection limits and sensitivity, and the nanostructuration of the material allowed for better diffusion of the analyte on the transducer surface, resulting in a faster response time for the sensor. Nevertheless, further efforts are still necessary to have standardized procedures for industrial applications and medical diagnostics.

3. Electrochemical Biosensors

Electrochemical biosensors are interesting because they are simple to miniaturize and enable low-cost mass production. In the following section, the most used techniques in electrochemical biosensors and their working principles are briefly illustrated and discussed.

3.1. Amperometric Method

In amperometric biosensors, the electrode current is measured, and most amperometric electrochemical biosensors are based on an enzyme's redox activity (most often horseradish peroxidase (HRP)). Enzymes enhance biosensing systems by catalyzing chemical reactions, thus improving sensitivity [111]. The performance of enzyme-based biosensors relies on factors such as the electrode surface, enzyme type, substrate, and mediator usage [112]. HRP is commonly employed as a secondary detection reagent, with the TMB/H_2O_2 substrate proving most effective. Amperometric biosensors measure current at a constant potential to detect the analyte. This method provides selectivity, as the potential used is characteristic of the analyte. The current is measured after directly setting the desired potential, enhancing the accuracy of the analysis [113].

Zhang et al. reported the development of a nonenzymatic immunosensor for the detection of SCC-Ag, utilizing rGO-TEPA and AuAg NCs [114]. In another study, a competitive RNA/RNA hybridization-assay-based biosensor was developed using Streptavidin–Horseradish Peroxidase (SA–HRP) and biotinylated capture probes. The biosensor employed H_2O_2 as substrate and hydroquinone (HQ) as a RedOx mediator. Two separate platforms, a screen-printed electrode (SPE) with Au NPs and a GCE with tungsten diselenide and Au NPs, were used in the biosensor [115]. Additionally, a microfluidic amperometric immunosensor was developed for the detection of the cancer biomarker CLD7 in circulating extracellular vesicles (EVs). The immunosensor was validated in colorectal cancer (CC) patients [116].

The amperometric graph in Figure 4 is used to detect the biomarker miRNA-211. The biosensor, made of a gold-nanoparticle-modified electrode with an attached RNA probe, measures the change in electric current when different concentrations of miRNA-21 are added. The current is high when no miRNA-21 is present, as all the biotinylated miRNA-21 can bind to the probe. When miRNA-21 is present, it competes with the biotinylated miRNA-21 for hybridization with the probe, reducing the current. The lower the current is, the higher the concentration of miRNA-21 is in the sample.

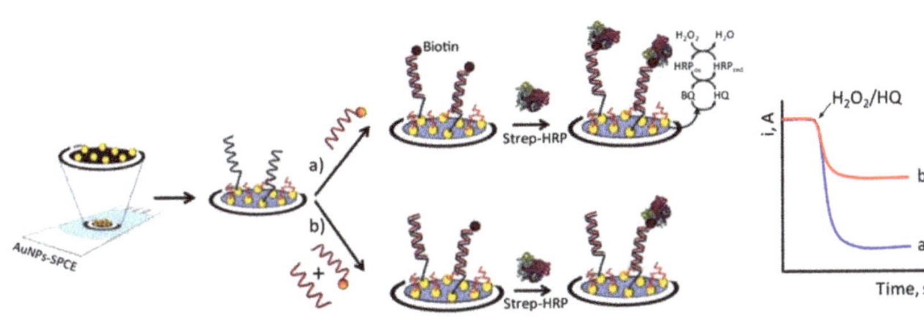

Figure 4. Direct competitive hybridization assay developed for miRNA determination shown schematically (Reprinted with permission from Ref. [117]). The competitive assay is obtained incubating a thiolated RNA complementary to the target miRNA assembled onto an AuNPs-modified SPCE with a biotinylated, short-stranded RNA whose sequence is identical to that of the target miRNA (**a**). Conversely in (**b**), a mix of biotinylated and target miRNA is present. The higher the concentration of the target miRNA the lower amperometric response was measured as a consequence of the smaller number of biotin-miRNA molecules attached to the electrode.

Chronoamperometry is a variation of an amperometric technique that involves monitoring the current generated by the faradaic process at the electrode over time. It involves applying a sufficiently large potential step to the working electrode to initiate a chemical reaction and then observing the current as a function of time.

3.2. Potentiometric Method

Potentiometric biosensors use a tiny amount of current to monitor the potential of an electrochemical cell [118]. Potentiometric sensors, employing the controlled current method, utilize an electrochemical cell containing two reference electrodes to measure the potential across an ion-selective membrane. Enzymes are commonly used to facilitate ion production, which is then detected by the supporting electrode. Controlled current methods offer the advantage of using more affordable measurement instrumentation compared to controlled potential methods. The Nernst equation relates concentration and potential in potentiometric measurements, offering a low limit of detection (LOD) for early-stage cancer diagnosis [117]. Jia et al. developed a Light Addressable Potentiometric Sensor (LAPS) for the detection of the liver cancer biomarker hPRL-3 [119]. Another study [120] utilized surface molecular imprinted self-assembled monolayers (SAM) for a potentiometric biosensor with a linear range of 2.5–250 ng/mL [121]. Label-free potentiometric detection targeted the HAPLN1 protein biomarker in MPM, achieving a pM-range LOD. Goda et al. created a hybridization-based potentiometric microarray for exosomal miRNA identification [28].

Potentiometric methods were employed to target malignant cells, investigating their electrochemistry in microenvironments affected by lactate release and pH fluctuations. Shaibani et al. achieved an LOD of 103 cells per ml, revealing pH flux alterations around cancer cells and their connection to altered cell metabolism [122]. The selective detection of circulating tumor cells (CTCs) in prostate cancer was achieved by utilizing an anti-EpCAM functionalized graphene oxide potentiometric biosensor based on the Light Addressable Potentiometric Sensor (LAPS) approach [123].

Based on pH monitoring, a sensor-integrated microfluidic technique is employed to find cancer cells. In this instance, the CTC's metabolic alteration resulted in a decrease in the pH of the surrounding environment. Potentiometric methods with Ag/AgCl and ZnO electrodes were employed to assess pH variations and cell modifications in vitro, utilizing three cell lines (A549, A7r5, and MDCK) in a microfluidic setup (Figure 5) [124].

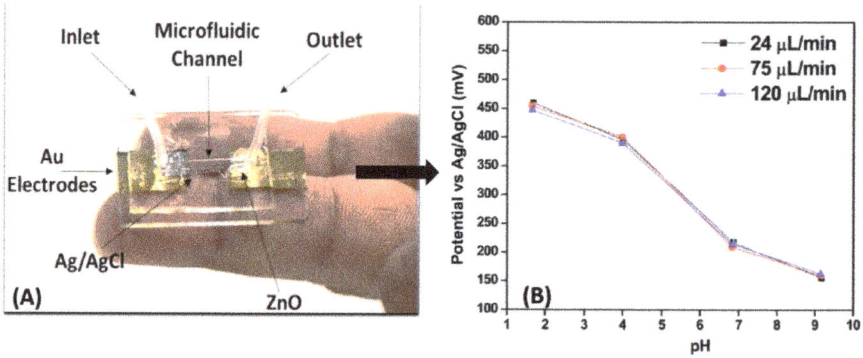

Figure 5. A microfluidics-based pH sensor (**A**) was successfully created with the use of rf sputtered ZnO thin films and Ag/AgCl ink. The potentiometric curve (**B**) obtained using this sensor. (Reprinted with permission from Ref. [124]. Copyright 2017 American Chemical Society.).

The potentiometric curve in Figure 5 shows the effect of the flow rate on the potential difference between the working and reference electrodes in the microfluidic device. The authors tested four different flow rates from 24 to 120 μL/min with four different pH buffer solutions from 1.68 to 9.18. The results indicated that the potential difference was not significantly affected by the flow rate, which implies that the microfluidic device has high stability and reliability over a wide range of flow speeds. The authors also observed a minimal drift of ±3 mV/h for each pH solution, which is acceptable for many applications.

3.3. Impedimetric Method

Electrochemical impedance spectroscopy (EIS) is a technique that examines the resistive and capacitive characteristics of a system by applying an AC excitation signal with varying frequencies. By analyzing the impedance spectra, it is possible to determine the resistive and capacitive components of the system based on the in-phase and out-of-phase current responses. At higher frequencies, the migration rate of redox species can become rate limiting, resulting in a frequency-dependent phase lag when analytes impede access to the electrode surface.

Electrochemical impedance spectroscopy (EIS) evaluates the interfacial characteristics, ion passage, and biomolecule interactions with electrode surfaces. It involves applying the AC potential to an electrochemical cell and measuring the resulting current signal. The resulting frequency-dependent impedance is represented on a Nyquist plot using a Randles circuit, showing semicircles at higher frequencies for electron transfer restrictions and a linear line at lower frequencies indicating the diffusion-limited electron transfer (Figure 6).

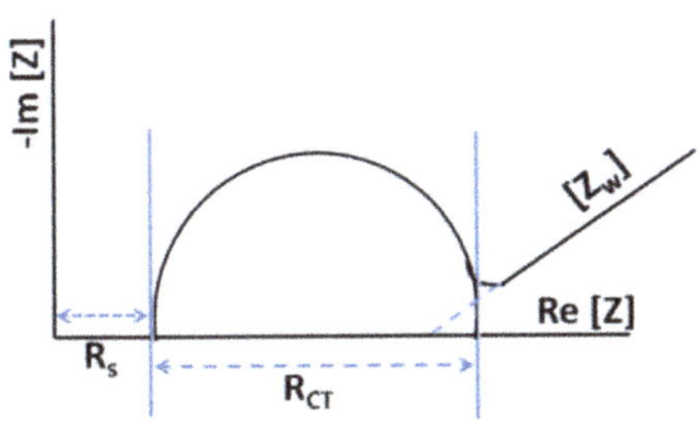

Figure 6. The Nyquist plot exhibits a depressed arc, indicating polarization caused by a combination of kinetic and diffusion processes.

The R_{ct} value reflects electron-transfer kinetics, while RS represents bulk electrolyte characteristics, and ZW represents diffusion. The Nyquist plot helps calculate ZW, represented by a 45-degree sloped straight line intercept. EIS is utilized for the label-free detection of cancer cells.

Elshafey et al. developed a label-free impedimetric biosensor for detecting the cancer biomarker EGFR. The biosensor utilized protein G and gold nanoparticles on a modified gold electrode for efficient immobilization. The calibration curve exhibited a wide dynamic range, from 1 pg/mL to 1 g/mL, with a low detection limit of 0.34 pg/mL in PBS and 0.88 pg/mL in human plasma. Interference from various substances in human plasma led to slight variations in the electrochemical signal during real-world experiments [125]. Han et al. developed a label-free cytosensor for cancer cell identification using phage display technology and EIS, demonstrating rising Rct values with an increasing cell concentration, indicating reduced electron transfer efficiency. The approach employed a specific phage immobilized on a gold electrode, with $[Fe(CN)_6]$ as the redox probe indicator, offering high specificity and repeatability and eliminating the need for complex steps of purification of recognition elements [126].

Hu et al. utilized EIS for the detection of liver cancer cells. They immobilized a mannose-specific lectin (con A) on a gold electrode, leading to changes in the charge-transfer resistance that correlated with the concentration of cancer cells (Bel-7404). This label-free approach directly targeted cancer cells, providing a direct, selective, and sensitive method with a detection limit of 234 cells/mL, eliminating the requirement for probe labeling [127]. Azzouzi et al. developed an impedimetric electrochemical biosensor using a biotinylated DNA/LNA molecular beacon (MB) probe linked with gold nanoparticles (AuNPs) for miRNA-21 detection in blood samples. This biosensor demonstrated high selectivity, good repeatability, and a wide linear detection range of 1–1000 pM, with a low detection limit of 0.3 pM. The use of neutravidin as a recognition element on the electrode surface enhanced biosensing properties, including sensitivity [128]. Kilic et al. introduced a label-free electrochemical sensor to compare the secretion levels of extracellular vesicles (EVs) in hypoxia and normoxic MCF-7 cells. The sensor utilized functionalized gold electrodes and employed differential pulse voltammetry (DPV) and electrochemical impedance spectroscopy (EIS) for EV detection (Figure 7). The sensor exhibited a linear operating range of 10^2–10^9 EVs/mL, with a limit of detection (LOD) of 77 EVs/mL. Selectivity was assessed using the RhD protein, and the results were compared with ELISA and Nanoparticle Tracking Analysis (NTA) [129].

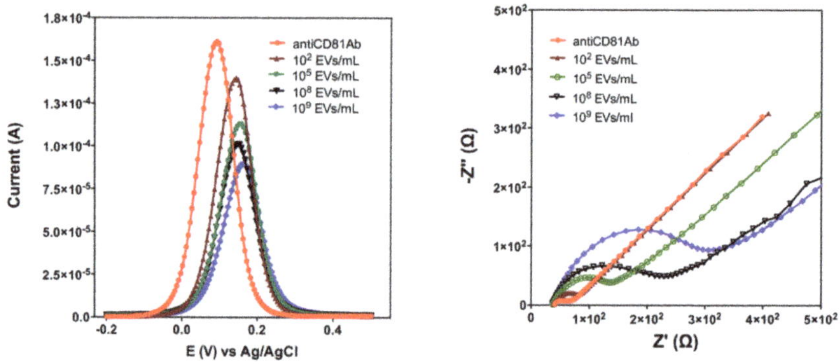

Figure 7. Differential pulse voltammograms recorded for various EVs concentrations (10–10^{10} EVs/mL) (**left**) and EIS measurements in the concentration range of 10^2–5×10^6 EVs/mL (**right**). (Reprinted from Ref. [129]).

3.4. Conductometric Method

Conductometric biosensors offer exciting possibilities for advanced bioanalytical detection. They measure changes in electrical conductivity during chemical reactions, using enzymes to modify the ionic strength of the sample solution. These biosensors are advantageous due to their miniaturization potential, low voltage requirement, and lack of a reference electrode. They are promising for applications in healthcare, environmental monitoring, and food safety [130]. Capacitive biosensors have the benefit of their capacitance measurements being more informative about the biosensor's insulating qualities [131]. Even slight sensor layer desorption generally results in a rise in the capacitance baseline. Furthermore, nonspecific binding is less likely with capacitive sensors.

Liang et al. established a conductometric immunoassay for the detection of alpha-fetoprotein (AFP) in the serum of liver cancer patients. The assay demonstrated robust conductometric responses, achieving a low detection limit of 4.8 pg/mL across a dynamic linear range of 0.01–100 ng/mL [132]. For the quantification of prostate cancer antigen, Bhardwaj et al. [133] introduced a conductometric immunosensing platform utilizing tetracyanoquinodimethane (TCNQ)-doped thin films of copper MOF, $Cu_3(BTC)_2$. The platform exhibited a dynamic linear range of 0.1–100 ng/mL for PSA detection, with a limit of detection as low as 0.06 ng/mL. Lin and coauthors devised conductometric sensors based on silicon nanowires for the detection of apolipoprotein A1, a biomarker associated with bladder cancer. The sensors demonstrated a wide dynamic range spanning from 0.2 ng/mL to 10 μg/mL, with a detection limit of approximately 1 ng/mL [134].

3.5. Voltammetric Method

Voltammetric biosensors measure current intensity by applying potential between a working and a reference electrode, detecting analyte concentration. They offer high sensitivity and multiplexed biomarker detection and serve as valuable point-of-care diagnostic devices [135].

Voltammetry is widely used for characterizing reaction kinetics and obtaining qualitative and quantitative information about analytes. It provides valuable insights into complex electrode reactions through current measurements. Voltammetry offers technical advantages and fewer limitations compared to other techniques for quantitative determination. It is commonly employed for detecting biomarkers during the liquid biopsy. The voltammetric methods broadly used for the detection of biomarkers under the liquid biopsy are described in the following sections.

3.5.1. Cyclic Voltammetry (CV)

Cyclic voltammetry (CV) is a widely used method for investigating redox processes, monitoring reaction intermediates, and assessing reaction product stability. It involves measuring the potential between the working electrode and reference electrode, while measuring the current between the working electrode and counter electrode. CV plots the electrochemical current on the y-axis and the working electrode potential on the x-axis, with the potential cycling back to its starting value. It provides information on the oxidation and reduction of redox species [136]. A partial cycle, a complete cycle, or a series of cycles might be performed depending on the results of the study. The electrons are transferred from the analyte to the WE or from the electrodes to the analyte during the redox reaction.

Kumar et al. utilized nanostructured zirconia ($nZrO_2$) as a transducer surface in a CV-based technique for detecting the oral cancer biomarker CYFRA. The immobilized receptor antibody (anti-CYFRA) on amine-functionalized $nZrO_2$ showed proportional electrochemical current changes, enabling detection in the range of 2–16 ng/mL with a sensitivity of 2.2 mA mL/ng [137]. Later on, Wang et al. developed an electrochemical sensor using a glassy carbon electrode modified with silver hybridized mesoporous silica nanoparticles (Ag@MSNs) to detect the prostate cancer biomarker PSA. The sensor exhibited improved bioreceptor adsorption and electron-transfer rates, utilizing hydroquinone as a redox probe. PSA detection was achieved in a wide concentration range of 0.05 to 50.0 ng/mL, with a detection limit of 15 pg/mL [138]. A study conducted by Kumar et al. was based on using two-dimensional Ti3C2-MXene nanosheets to detect the carcinoembryonic antigen (CEA) biomarker with a detection range of 0.0001–2000 ng/mL [139].

Taleat et al. employed a sandwich technique for detecting the MUC1 protein, a key contributor to tumor development in various cancers. They combined MUC1 monoclonal antibody immobilized on a poly-aminobenzoic-acid-modified graphite screen-printed electrode with a methylene-blue-modified aptamer (specific ss-DNA) [140]. Feng et al. coupled CV and electrochemiluminescence (ECL) techniques to achieve the simultaneous detection of AFP and CEA by tagging the detection antibody and using methylene blue as an electrochemical indicator that binds directly to aptamers' G base for MUC1 protein concentration identification [141].

3.5.2. Differential Pulse Voltammetry (DPV)

Differential pulse voltammetry (DPV) involves the scanning potential with small amplitude pulses while measuring the current at two points before and after each pulse. The difference in current measurements is calculated and plotted as a function of the base potential, allowing for analysis of non-faradaic current decay.

DPV is a widely used electrochemical procedure known for its sensitivity and speed. It involves applying fixed-amplitude electrochemical pulses on a slowly increasing base potential and recording the resulting current differential. DPV is utilized to detect early cancer and study drug performance in cancer. Lin et al. developed a reusable biosensor using a magnetic graphene-oxide-modified gold electrode (MGO-Au) to detect VEGF in human plasma for cancer detection [142]. The DPV-based sensor demonstrated effective sensitivity, a rapid reaction time, and a wide linear detection range, outperforming the ELISA approach.

Amjadi et al. studied the influence of doxorubicin (DOX) and a flavonoid-modified drug (FMD) on lung cancer cells (A549) using the DPV approach, revealing that the FMD had a stronger effect on cancer cells compared to DOX, as demonstrated by the reduction in electrochemical reactivity with increasing drug concentration [143]. Additionally, Pacheco et al. and Wang et al. utilized electrochemical techniques (CV and DPV) with breast cancer cells immobilized on working electrodes to quantify cancer cells in unknown samples using a known cancer cell calibration curve [144,145].

The fabrication and alteration of electrodes are important in electrochemical measurement. The screen printing technique is widely employed in this context for the manufacturing of portable low-cost electronics, particularly disposable electrodes. Compared to conventional electrode fabrication methods, this technology presents numerous advantages, encompassing the precise manipulation of electrode dimensions, a wide range of electrode designs, compact device sizes, reduced production expenses, user-friendly operation, and the capability to create diverse arrays of electrodes [146]. Furthermore, screen-printed electrodes allow for additional customization of the electrode surface by altering it with different nanomaterials, resulting in an increased surface area, increased biomolecule immobilization efficiency, and unique electrochemical characteristics.

Using eight disposable screen-printed microelectrode arrays as the transducer surface, Zani et al. established a sensitive and easy PSA detection technique. Magnetic beads were employed to collect the main PSA antibody in this experiment. The electrochemical measurements were taken using DPV after the collected beads were washed with the antibody-labeled enzyme alkalinephosphatase (AP). The detection range of this biosensor was linear (0–20 ng/mL), with a lower detection limit of 1.4 ng/mL [147]. Erdem et al. described an electrochemical biosensor based on a multichannel screen-printed array of electrodes (MUX-SPE16) for assessing the nucleic acid hybridization of distinct miRNA sequences (miRNA-16, miRNA-15a, and miRNA-660). In this study, streptavidin-coated magnetic beads were placed on the electrode surface before a biotinylated DNA probe was immobilized. Following the hybridization procedure, the electrochemical response was measured using the DPV method on the guanine oxidation signal [148].

Furthermore, due to the limited sensitivity and specificity of a single biomarker, measuring or tracking it is insufficient for reliable cancer diagnosis. Therefore, researchers are interested in the simultaneous detection of numerous tumor markers to obtain more accurate and dependable results. Serum VEGF-C had a specificity of 68% and a sensitivity of 85%, whereas MMP-9 had a specificity of 75% and a sensitivity of 63%. Similarly, VEGF had a specificity of 59% and a sensitivity of 80%, but the combination of these three markers had a greater sensitivity and specificity (83% and 76%, respectively) than the single-biomarker strategy for lung cancer detection [149]. CEA was also demonstrated to enhance cancer prediction when combined with other biomarkers. When CEA was combined with CA 15-3, for example, its sensitivity rose from 89% to 96% [150].

Wu et al. introduced a novel approach for the concurrent detection of CA 19-9 and CA 125 cancer biomarkers, which involved the utilization of a disposable two-throughput immune-electrode array. The researchers applied a cellulose acetate membrane onto the graphite working electrodes (W1 and W2) of a screen-printed chip, followed by the co-immobilization of thionine/CA 19-9 and thionine/CA 125 on separate electrodes. Antibodies labeled with HRP were then detected on these working electrodes. By establishing an electron shuttle mechanism facilitated by the immobilized thionine, the enzymatic reduction of H_2O_2 by HRP resulted in the generation of electrochemical signals, enabling the simultaneous detection of both biomarkers [151,152].

Two separate cancer biomarkers (CEA and AFP) were concurrently identified utilizing DPV technology and a metal-ion-tagged immunocolloidal gold nanocomposite as a signal tag in another manner. The signal antibody was modified with two metal ions (AuNPs/anti-CEA/Cu21 and AuNPs/anti-AFP/Pb21) in this manner. The authors leveraged the intrinsic electrochemical characteristics of metal ions in this study to obtain the multiplex detection of cancer biomarkers on a single platform with high sensitivity. The findings were also confirmed using a conventional ELISA, indicating that they might be used in clinical settings [153].

In the realm of quantitative real-time evaluation and early cancer detection, circulating tumor markers (CTMs) such as extracellular vesicles (microvesicles and exosomes), circulating tumor cells (CTC), and circulating nucleic acids in the blood emerged as valuable indicators. These CTMs offer a range of advantages, including their potential as cost-effective, reproducible, dynamic, and non-invasive diagnostic tools for both cancer diagnosis and the monitoring of disease progression during the early stages. Several detection techniques were employed to identify CTMs, including quartz crystal measurement (QCM), microcantilevers, colorimetric assays, the enzyme-linked immunosorbent assay (ELISA), surface-enhanced Raman scattering (SERS), surface plasmon resonance, the polymerase chain reaction (PCR), and electrochemical methods [154].

Moscovici et al. developed a microfabricated glass chip with gold apertures for cell counting using DPV, enabling the specific detection of 125 prostate cancer cells in 15 min, even in complex cell populations [155]. Yang et al. demonstrated a microelectrode-based electrochemical biosensor for micro-RNA detection using nanostructured palladium electrodes, achieving detection as low as 10 aM of a target with enhanced signals using the DPV approach and Fe(III) regeneration of Ru(III) for amplification [156].

Zhang et al. utilized DPV to detect capecitabine in serum specimens without the need for labels, using an electrochemical biosensor based on stacked graphene nanofibers (SGNF) and gold nanoparticles (AuNPs). The biosensor showed a wide linear detection range of 0.05–80.00 M and an exceptional detection limit of 0.017 M for capecitabine electrochemical reduction [157]. Venu et al. published an electrochemical biosensor based on a ZrO_2/rGO nanocomposite for the detection of an anticancer medication (regorafenib; REG). The fabricated biosensor had a wider linear detection range of 11–343 nM, with a remarkable lower detection limit of 59 and a remarkable limit of quantifications of 59 and 17 nM. For the accurate assessment of REG, the biosensor's effectiveness was also good in both blood samples and pharmaceutical formulations. The biosensor was also useful for detecting REG, uric acid, and ascorbic acid all at the same time [158].

3.5.3. Linear Sweep Voltammetry (LSV)

Throughout the scan, the electrode potential is adjusted at a constant rate, and the resultant current is recorded in LSV. Yan et al. developed an immunosensor of carbon combining screen printing with an excellent material, vegetable parchment. The proposed immunosensor, involving linear sweep voltammetry as the electrochemical method and prostate specific antigen (PSA) as a model analyte, showed a limit of detection of 2 pg/mL [159].

In their study, Bo et al. presented an electrochemical immunosensor that employed a double signal amplification strategy utilizing enzyme-encapsulated liposomes and biocatalytic metal deposition. This innovative approach was specifically developed for the detection of human prostate-specific antigen (PSA). Linear sweep voltammetry (LSV) was employed to measure the quantity of deposited silver, which served as an indicator of the target analyte. The experimental findings demonstrated a linear relationship between the anodic stripping peak current and the concentration of PSA within the range of 0.01–100 ng/mL. Impressively, the detection limit achieved by the sensor was as low as 0.007 ng/mL, thereby illustrating its high sensitivity for PSA detection.

3.5.4. Square Wave Voltammetry (SWV)

The excitation signal in SWV consists of a symmetrical square-wave pulse with an amplitude, Esw, superimposed over a staircase waveform with step height E, where the waveform's forward pulse corresponds to the staircase step. The difference between the forward and reverse currents is used to calculate the net current, which is centered on the redox potential. SWV provides various benefits, including high sensitivity, speed, and non-faradic current discrimination.

The combination of autocatalytic deposition and square-wave stripping voltammetry with enlarged gold nanoparticles labeled on goat anti-rabbit immunoglobulin G enabled the detection of the rabbit immunoglobulin G (RIgG) analyte with a remarkably low limit of 1.6 fM, highlighting the sensitivity enhancement of the electrochemical immunoassay

(GaRIgG-Au) method [160]. In a subsequent study, Liu et al. conducted a comparative analysis between square-wave voltammetry (SWV) and electrochemical impedance spectroscopy (EIS) for the development of a label-free electrochemical immunosensor targeting the hormone estradiol. The researchers aimed to assess the performance of these two techniques in terms of sensitivity. Notably, the results revealed that SWV outperformed EIS, with a lower detection limit of 18 pg/mL for estradiol, compared to 26 pg/mL achieved by EIS. This comparative evaluation highlighted the superior sensitivity of SWV in the context of estradiol detection, suggesting its potential for enhanced analytical applications in hormone analysis [161]. Zhang et al. detected CTCs using SWV based on catalytic amplification, and the limit of detection of was 1 cell/mL.

3.5.5. Stripping Voltammetry (SV)

This technique has the lowest detection limits with respect to the commonly used electrochemical techniques. SV is also known as a pre-concentration technique, and the most commonly used stripping voltametric techniques involve anodic stripping voltammetry, cathodic stripping voltammetry, and adsorptive stripping voltammetry.

Despite the fact that each approach has its own distinct characteristics, they all follow the same two procedures. The target analyte is first concentrated onto the working electrode in the sample solution. In the second phase, the potential is used to remove the preconcentrated analyte off the electrode surface, which is then measured. In the stripping process, potential waveforms such as the linear sweep, differential pulse, and square wave can be employed. Due to their ability to distinguish against charging current, differential pulses and square waves are the most prevalent. In addition, compared to the differential pulse, the square wave offers the advantages of a faster scan rate and greater sensitivity.

The Joseph Wang group introduced a novel electrically heated carbon paste electrode specifically designed for conducting adsorptive stripping measurements of trace amounts of nucleic acids. This groundbreaking approach combines the principles of electrochemistry with electrically heated electrodes and adsorptive constant current stripping chronopotentiometry. The integration of these techniques brings forth notable advantages when it comes to the precise detection and quantification of nucleic acids at trace levels. This innovative coupling of electrochemistry with electrically heated electrodes presents a promising avenue for achieving enhanced accuracy and sensitivity in nucleic acid analysis [162].

A recapitulation of the working principle of a biosensor with electrochemical methods available for biosensors is reported in Figure 8.

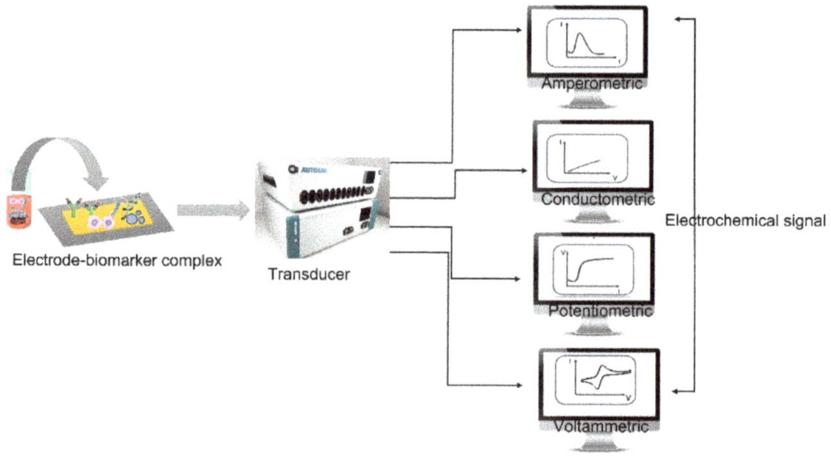

Figure 8. A general presentation illustrating diverse electrochemical techniques employed in biosensing.

4. Sensor Integration

In the advancement of point-of-care and wearable biosensors, miniaturization plays a pivotal role in achieving portability, user-friendliness, and cost-effectiveness as well as maintains high sensitivity with rapid response times. The downsizing of sensing electrodes or their constituents to the nanoscale, as seen in nanoelectrode, nanotextured, nanogap, and field-effect transistor-based biosensors, can significantly enhance the signal-to-noise ratio. It is worth noting, however, that this enhancement can potentially introduce an undesired increase in the biosensor's response time. Therefore, while nanoscale miniaturization offers various advantages, careful consideration must be given to strike a balance between improved sensitivity and response time to ensure the optimal performance of the biosensor.

The above-mentioned electrochemical methods, discussed in the related paragraphs, are well-suited for implementation in miniaturized systems. Achieving miniaturization, along with the features discussed in Section 2, is essential for creating compact, user-friendly, and cost-effective point-of-care, portable, and wearable biosensors. These biosensors offer competitive limits of detection and rapid response times while maintaining their convenient and accessible nature.

Lab-on-a-Chip Platforms: Wearable and Portable Devices for PoC

Lab-on-a-chip (LOC) devices, notably multitasking devices, show the most attractive advantages for performing several lab procedures on a single chip with small volumes of chemicals and great efficiency. One of the most challenging features of LOCs, also known as a micro total analysis system (μTAS), is to fully integrate a microelectromechanical system (MEMS) with automated microfluidic tools to allow their widespread use in medical applications (liquid biopsy, therapeutic follow-ups, and health and disease monitoring). These user-friendly, sensitive, and portable LOC sensors for real-time analysis have various benefits over traditional analytical techniques. In this review, we list several examples of sensors suitable for integration into LOC devices. Nevertheless, the step over technological gap to overcome limitations in the widespread diffusion of such devices is still linked to the scalability of systems and the poor affordability of highly integrated platforms. A few examples, indeed, are the commercially available platforms including electrochemical systems, and their use for monitoring health parameters is going in the direction of self-diagnostics and fitness applications but not toward becoming a gold standard in clinical practices.

Recently, Atkcakoca and coworkers reported on the realization of an electrochemical biosensor with an integrated microheater improving the performance of the nucleic acid hybridization assay based on electrochemical impedance spectroscopy, paving the way for the development of highly sensitive and specific integrated label-free biosensors [163]. Kasturi and collaborators developed a microfluidic channel with integrated valves and an electrochemical biosensor for the detection of beta-amyloid, a biomarker for Alzheimer's disease. Gold microelectrodes are the transducer holding the antibody–antigen complexes inside the fluidic chambers governed by pneumatic valves. A linear response of the sensor, through DPV measurements, was reported for beta-amyloid antigen concentrations from 2.2 pM to 22 μM [164]. One of the most promising applications of an integrated LOC for diagnostic purposes is the realization of smartphone-based and wearable devices, which can really realize all the features of portability, low-cost, and the rapid response needed to fully embody the point-of-care concept.

Flexible electronic devices, making use of specific methods of fabrication and detection, have properties such as flexibility, wearability, conductivity, stretchability, mechanical resistance, and biocompatibility. In this scenario, a plethora of materials suitable as substrates for electrochemical transducers were considered [165]: cellulose-based, polyaniline (PANI), polyimide (PI), and specific fabrication methods were demonstrated. Most of them were able to electrochemically detect ion concentrations from skin and sweat as well as monitor fitness parameters like heartbeat, blood pressure, body temperature, and oxygen concentration [166,167].

Sempionatto and collaborators recently developed a non-invasive device for the simultaneous monitoring of blood pressure and heart rate via ultrasonic transducers and multiple biomarkers via electrochemical sensors. They conformally optimized the integrated device to curved skin, thus ensuring mechanical resistance and the reliable sensing of several compounds: glucose in interstitial fluid; lactate, caffeine, and alcohol in sweat [168].

The integration of small microfluidic channels into wearable devices is also a challenging topic for the development of such devices, since biological fluids, easily exploitable for the liquid biopsy (sweat and interstitial fluid), could be conveyed toward the sensing areas of the devices.

Electrochemical sensors measure the reaction caused by the interaction between the sensing surface and the analytes, and the corresponding response is then converted into electric signals that can be monitored by potentiometry, amperometry, and conductometry measurements [169]. The integration of electrochemical biosensors into point-of-care (PoC) platforms, especially when combined with smartphones, emerged as a powerful tool for personalized health monitoring, thus enhancing the practicality of diagnosis compared to the traditional laboratory-based diagnostic methods [170]. In the last years, smartphone-based biosensors have been widely employed for human health PoC testing to improve the diagnosis and treatment of several diseases, thanks to their cost-effectiveness, ease of use, and portability. Figure 9 illustrates the diagnosis steps in smartphone-based electrochemical biosensors.

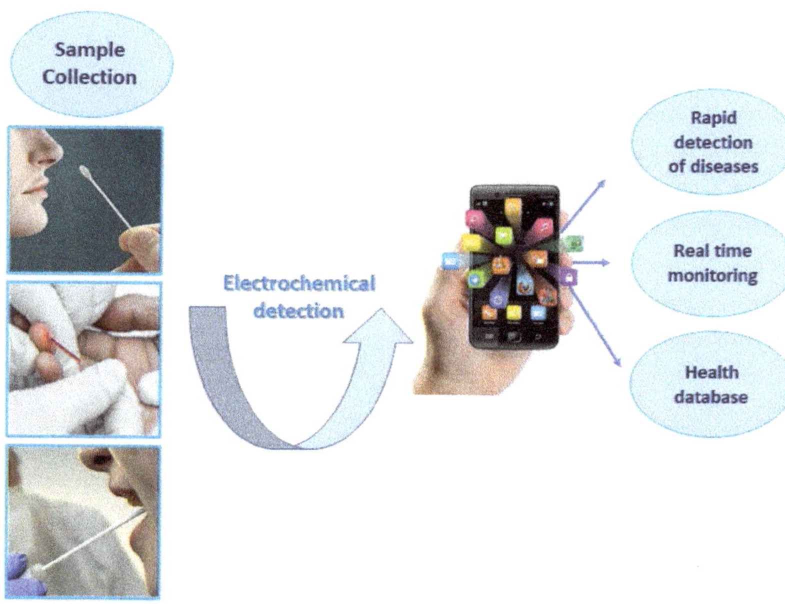

Figure 9. Smartphone-based electrochemical biosensors for health monitoring.

One of the first applications of a smartphone-based electrochemical biosensor system was described as amperometric sensing [171]. Amperometry is a type of voltametric technique in which a constant voltage is applied to the working electrode, and the current provided by the oxidation/reduction of an electroactive analyte is then measured as a function of time. Liu et al. reported an amperometric aptasensor integrated with a smartphone-assisted portable wireless biochip for the simultaneous real-time monitoring of insulin and glucose in saliva with the lowest detectable concentration of 0.85 nM and 0.8 nM, respectively. The sensing platform combined with a Bluetooth transmission system to generate the digital diagnosis by a smartphone signal readout, providing a PoC testing tool for the diagnosis of diabetes and other insulin-resistance-associated diseases [172].

Voltammetry is another electrochemical technique that includes different type of measurements such as differential pulse voltammetry (DPV), cyclic voltammetry (CV), square wave voltammetry (SWV), and linear sweep voltammetry (LSV) and is widely used in smartphone-integrated electrochemical sensors [173]. In 2020, Low et al. developed a DPV-based electrochemical sensor combined with a smartphone for the detection of the circulating miRNA-21 biomarker in saliva. They demonstrated that the smartphone-based biosensing system, equipped with a specific Android application, displayed comparable performances with commercial workstations for the detection of miRNA-21 [174]. Combining a screen-printed immunosensor with a smartphone-based electrochemical system, Fan et al. reported the detection of cancer antigen 125 (CA125) by DPV measurements. Data were transmitted to a smartphone by Bluetooth, acting as the interface for communication with a remote medical center via the Internet. The developed system provided a sensitive detection of CA 125 with an LOD of 2 mU mL^{-1} [96].

Electrochemical impedance spectroscopy (EIS), in which the impedance of the system is measured as a function of frequency, is also widely used in the fabrication of EIS-based electrochemical sensors combined with smartphones. For example, Talukder et al. reported a portable system for the personalized monitoring of the blood cell count, consisting of a smartphone-based microfluidic impedance cytometer [175].

The recent advances in this field highlight the advantages of electrochemical sensing systems integrated with a smartphone, thanks to their high simplicity of fabrication [176]. Moreover, the development of multiplexing smartphone-based electrochemical systems can be used for the simultaneous detection of various biomarkers, thus enabling the remote control of diseases by doctors and accelerating the diagnosis and treatment of a pathology.

5. Comparison of Liquid Biopsy Electrochemical Methods: Advantages and Limitations

Nowadays, cancer is the major disease affecting human health and life, and its large diffusion requires the development of simple, practical, and facile diagnosis methods for simplifying its treatment and improving its cure rate. Compared with medical imaging and a pathological examination, which are the most common cancer diagnosis methods, the liquid biopsy represents a promising strategy for cancer biomarker detection, opening the way to direct and rapid diagnostic methods with high efficacy [177,178]. Among several biomarkers, circulating tumor cells (CTCs) are well-established as promising targets for the detection of tumors via the liquid biopsy. Since tumor cells can be shed into the blood before the formation of visible solid tumor lesions, detecting CTCs before the imaging findings or clinical manifestations is an efficient method for the early diagnosis and monitoring of cancer [179]. However, the content of CTCs in peripheral blood circulation is very small, and the techniques used for its detection (fluorescence imaging, magnetic resonance imaging, and cytological detection [180–182]) have several shortcomings, such as high cost, a large amount of time, low sensitivity, and a lack of specificity, thus limiting their use in clinical applications [183].

In the last years, electrochemical sensing technology has been widely investigated as a good alternative method for the detection of CTCs because of its advantages of high sensitivity, good selectivity, low cost, easy portability, and rapid detection. Compared with traditional detection techniques, electrochemical methods demonstrate competitive results in terms of the LOD and selectivity. CTCs can be electrochemically detected by using two common types of approaches. The first one is often related to impedimetric sensors and exploits the change in the electron transfer produced by CTCs captured on the electrode, usually conjugated to various recognitive materials including antibodies [184], aptamers [185], and receptors [186]. However, this type of sensor usually needs just one electrode to work, thus resulting in a lack of capture efficiency. To overcome this limit, modifying electrodes with nanomaterials can be a good strategy to enhance the capture efficiency for CTCs. For example, Wang et al. conjugated gold nanostars with a high surface area, with CTCs' specific aptamer. Owing to this design, the sensor showed a sensitive detection limit of 5 cells/mL [187]. Cai et al. developed a dual-recognition electrochemical cytosensor for the detection of CTCs. The sensor, based on Cabot carbon black (BP2000)/AuNPs anchoring anti-epithelial cell adhesion molecule (anti-EpCAM) antibodies as capture probes and novel branched PtAuRh trimetallic nanospheres (b-PtAuRh TNS), linked with aptamers targeting mucin1 (MUC1) as signal probes, exhibited a wide linear range of 5–1 × 10^6 cells per mL^{-1} and a low detection limit of 1 cell per mL^{-1} [188]. In another work, Zhang et al. exploited the reaction of $LiFePO_4$ with sodium molybdate to generate an electrochemical signal for detecting CTCs. In particular, they captured CTCs from sample by using Fe_3O_4 magnetic nanospheres (MNs) modified with the EpCAM antibody, while gold-nanoparticle-modified $LiFePO_4$ ($LiFePO_4$/Au) was used as an electrochemical probe. The assay presented a detection range from 3 to 10,000 cells per mL^{-1}, with a detection limit of 1 cell per mL^{-1} [189]. Although several studies demonstrated the advantages of electrochemical sensing technology for the detection of CTCs and the significant advances in the biosensing research area thanks to immunotechnology, microfluidics, and nanotechnology, the clinical use of such biosensors is still limited. In fact, the number of CTCs in the peripheral blood circulation is very little, and detection can be very difficult. Moreover, the existing detection techniques use nucleic acids and antibodies as target molecules, which, lacking specificity for the classification of captured CTCs, cannot be utilized to give precise information about patient-specific tumor biology. Thus, combining advanced technologies such as microfluidics and the DNA walker and exploring more cell-specific targets could be a significative strategy to improve the sensitivity and specificity of such biosensors. Recently, Ming et al. developed a new on-skin optoelectronic biosensor based on cyclic voltammetry that can measure various vital signs related to blood flow and oxygenation in a non-invasive and continuous way [190].

As voltammetry is the most commonly used electrochemical technique for liquid biopsies, this review primarily focuses on the DPV technique, a subtype of voltammetry, because DPV is considered to be an important electrochemical method for liquid biopsy applications because it offers high sensitivity, selectivity, a wide dynamic range, rapid analysis, and minimal interference from other components in body fluids.

Scientists can determine which electrochemical methods are best-suited for particular applications by comparing the various electrochemical techniques used in liquid biopsy biosensing, depending on elements like the type of biomarker being analyzed, the concentration range, and the complexity of the sample matrix.

Such a comparison can help to guide the development of new biosensors for liquid biopsy analysis as well as to optimize existing methods for improved performance and sensitivity. Additionally, understanding the advantages and limitations of different electrochemical methods can aid in the interpretation of experimental results and can inform the selection of appropriate analytical methods for a given research question. These aspects have been summarized in Table 2.

Each electrochemical technique has advantages and disadvantages of its own. The choice of method depends on the specific analyte of interest and the requirements of the analysis.

The desire to manufacture micro total analysis systems, low-cost point-of-care diagnostics, and environmental monitoring devices sparked the creation of tiny and portable biosensor devices. So, for the development of such a biosensor, it is essential to understand the electrochemical method on which this biosensor operates.

The efficient transducer surface or immobilization matrix is the most significant step in the fabrication of a miniaturized electrochemical biosensor. For optimal biosensor performance, we need to carefully select the materials, electrochemical methodology, and manufacturing process. The PoC devices used for the liquid biopsy might benefit from a wise device design and efficient detection procedures. Research needs to be conducted to create combinatorial electrochemical biosensors with a high throughput and low cost for cancer diagnosis, therapy, and monitoring, utilizing the liquid biopsy. The commercialization of biosensors will increase when an electrochemical-based biosensing platform effectively works in a real-world sample environment with excellent selectivity, sensitivity, and stability.

Table 2. A comparison of various electrochemical methods employed in biosensing applications for liquid biopsy analysis.

EC Method	Advantages	Limitations
Potentiometric	▪ High selectivity for specific analytes through the use of ion-selective electrodes (ISEs) ▪ Wide range of analytes that can be detected using ISEs, including ions, gases, and molecules ▪ Simple instrumentation, with ISEs often consisting of a single electrode and a reference electrode ▪ Non-destructive, as the sample is not consumed during the analysis	▪ Limited sensitivity compared to other electrochemical methods ▪ Limited dynamic range, as ISEs typically have a limited linear response range ▪ Interference from other ions or molecules in the sample can affect the accuracy of the analysis ▪ Slow response time compared to other electrochemical methods
Impedimetric	▪ High sensitivity for certain analytes ▪ Can measure non-faradaic processes such as adsorption and desorption ▪ Can provide information on both the electron-transfer kinetics and the charge-transfer resistance of the system	▪ Requires complex instrumentation and data analysis ▪ Limited dynamic range compared to other electrochemical methods ▪ Sensitive to electrode fouling and surface defects
Conductometric	▪ High sensitivity, as changes in conductivity can be highly sensitive to analyte concentration ▪ Wide range of analytes that can be detected, including ions, gases, and molecules ▪ Simple instrumentation, with conductometric biosensors often consisting of a pair of electrodes and a transducer ▪ Non-destructive, as the sample is not consumed during the analysis	▪ Limited selectivity compared to other electrochemical methods ▪ Interference from other ions or molecules in the sample can affect the accuracy of the analysis ▪ May be affected by changes in temperature, humidity, and other environmental factors ▪ May require the optimization of electrode and transducer properties to achieve the desired sensitivity and selectivity
Cyclic Voltammetry (CV)	▪ High sensitivity for certain analytes ▪ Simple instrumentation and low cost ▪ Can measure both oxidation and reduction reactions	▪ Limited selectivity; it can be affected by interfering species ▪ Low resolution; the current signal can be difficult to interpret ▪ Slow scan rate that can limit the speed of analysis
Differential Pulse Voltammetry (DPV)	▪ High sensitivity and selectivity for certain analytes ▪ Wide dynamic range ▪ Rapid analysis ▪ Minimal interference from other components in the sample	▪ Limited applicability to certain types of analytes (e.g., those with weak redox activity) ▪ Requires the careful optimization of parameters such as pulse width and amplitude ▪ High background noise can be a problem in complex samples
Stripping Voltammetry (SV)	▪ High sensitivity and selectivity for analytes, like heavy metals and trace elements ▪ Wide range of analytes that can be detected, including ions, gases, and molecules ▪ Non-destructive, as the sample is not consumed during the analysis ▪ Can be used for both qualitative and quantitative analysis	▪ Requires pre-concentration of the analyte before measurement, which can be time-consuming and may limit the speed of analysis ▪ Can be affected by interference from other species in the sample ▪ Limited dynamic range, particularly for quantitative analysis ▪ May require specialized instrumentation, such as a mercury electrode

6. Future Perspectives and Concluding Remarks

This review paper highlights the significance of screening and early diagnosis in disease management, particularly in the context of cancer. It emphasizes the importance of non-invasive analytical methods capable of detecting biomarkers to facilitate successful treatments and improve patient survival rates. The focus of the study is on the electrochemical methods used for the development of biosensors in the liquid biopsy, owing to their ability to offer a rapid response, precise detection, and low detection limits. This review discusses the advancements in electrochemical biosensors, which hold the potential to enhance the specificity and sensitivity of conventional analytical techniques. Electrochemical biosensors demonstrate the ability to detect minute quantities of analytes, including proteins, nucleic acids, and circulating tumor cells, even in complex bodily fluids such as urine, serum, and blood. Among the various detection techniques explored for cancer biomarker detection, voltammetric sensors are extensively discussed due to their advantages and technical characteristics, which led to their widespread use in the quantitative detection of ions and molecules.

This review also provides a comprehensive comparison of different electrochemical techniques to aid in the selection of the appropriate analytical methods based on the specific requirements. This comparative analysis helps researchers and clinicians identify the most suitable approach for their intended applications. Looking toward the future, the development and refinement of electrochemical biosensors hold tremendous potential for advancing diagnostic capabilities in the field of the liquid biopsy. Further advancements in sensor design and surface modification techniques and integration with emerging technologies like nanomaterials and microfluidics are expected to enhance the performance, reliability, and multiplexing capabilities of electrochemical biosensors. These developments will likely contribute to improved disease detection and monitoring, enabling personalized and targeted treatment strategies.

However, it is important to acknowledge that while electrochemical biosensors offer great promise, there are still challenges to overcome. Some of these challenges include improving the selectivity and stability of sensors, standardizing protocols for clinical use, and addressing the complexities associated with analyzing biomarkers in diverse biological matrices. The continuous innovation and optimization of electrochemical biosensing technologies are expected to play a vital role in improving disease diagnosis, monitoring treatment efficacy, and, ultimately, enhancing patient outcomes.

The potential and ambition of the liquid biopsy pose significant challenges and opportunities for personalized medicine and point-of-care diagnostics and follow-ups. Within this context, the utilization of miniaturized and rapid detection tools, such as electrochemical sensors, holds great promise for advancing the field. These devices can be seamlessly integrated into everyday objects like smartphones, smartwatches, and more. Furthermore, as individuals become increasingly health conscious and emphasize early disease screening, the introduction and widespread use of user-friendly sensors, lab-on-a-chip devices, and similar tools for manipulating and analyzing biofluids are poised to greatly benefit self-awareness and disease detection.

Despite the significant advances in electrochemical biosensors for the liquid biopsy, there are still some challenges and limitations that need to be addressed. These include the optimization of the biosensor design, the selection of the most suitable biomarkers and detection techniques, the validation of biosensor performance in clinical samples, and the standardization of biosensor fabrication and operation. Moreover, there is a need for more interdisciplinary collaboration among researchers from different fields, such as chemistry, biology, engineering, and medicine, to develop innovative and effective solutions for the liquid biopsy. Furthermore, there is the potential for combining electrochemical biosensors with other analytical methods, such as optical or magnetic sensors, to achieve complementary and synergistic results. Electrochemical biosensors for the liquid biopsy have a bright future ahead, as they can revolutionize the approach to diseases and improve the quality of life of patients.

Author Contributions: Conceptualization: M.S.C., F.F. and S.U.; methodology: E.P., A.T., S.U. and G.S.; investigation: G.S., I.T. and S.U.; original draft preparation: S.U., G.S., A.T., E.P., M.S.C. and F.F.; supervision: M.S.C.; funding acquisition: M.S.C., F.F. and A.T. All authors have read and agreed to the published version of the manuscript.

Funding: This research was funded by Italian Ministry of University and Research (MUR), Progetto PON ARS01_00906 "TITAN: Nanotecnologie per l'ImmunoTerapia dei Tumori", funded within FESR Programme PON "Ricerca e Innovazione" 2014–2020 Azione II-OS 1.b); PRIN 2022 Projects-"RESOLVE—innovative platfoRm based on fiEld-flow-fractionation and Sample On-chip detection to unravel extraceLlular Vesicles hEterogeneity" (grant number: 202233FTW8), "MAGELLANO—MAnufacturing systems for Graphene nanocompositEs and their appLications as embedded eLectrochemical seNsors in Lab-On-chip" (grant number: 2022NRMS4X), and Regione Puglia within "Tecnopolo per la medicina di precisione" (TecnoMed Puglia): DGR n.2117 del 21/11/2018, CUP: B84I180 0 0540 0 02 and by Italian Ministry of Enterprises and Made in Italy: "MSP4WATER" Project funded within the frame of the Programme "Fondo per la Crescita Sostenibile—Accordi per l'innovazione DM 31/12/2021" (grant number: F/310200/01-04/X56).

Institutional Review Board Statement: Not applicable.

Informed Consent Statement: Not applicable.

Data Availability Statement: Not applicable.

Acknowledgments: The authors acknowledge the Italian Ministry of Research (MUR), Italian Ministry of Enterprises and Made in Italy, and Regione Puglia for the financial support.

Conflicts of Interest: The authors declare no conflict of interest.

References

1. Ferrara, F.; Zoupanou, S.; Primiceri, E.; Ali, Z.; Chiriaco, M.S. Beyond liquid biopsy: Toward non-invasive assays for distanced cancer diagnostics in pandemics. *Biosens. Bioelectron.* **2022**, *196*, 113698. [CrossRef] [PubMed]
2. Nič, M.; Jirát, J.; Košata, B.; Jenkins, A.; McNaught, A. (Eds.) *Compendium of Chemical Terminology*; Blackwell Scientific Publications: Oxford, UK, 1997.
3. Ragavan, K.V.; Kumar, S.; Swaraj, S.; Neethirajan, S. Advances in biosensors and optical assays for diagnosis and detection of malaria. *Biosens. Bioelectron.* **2018**, *105*, 188–210. [CrossRef] [PubMed]
4. Sadasivuni, K.K.; Ponnamma, D.; Kim, J.; Cabibihan, J.J.; AlMaadeed, M.A. (Eds.) *Biopolymer Composites in Electronics*; Elsevier: Amsterdam, The Netherlands, 2017; p. 544.
5. Kelley, S.O. Disease Detector. *Sci. Am.* **2015**, *313*, 48–51. [CrossRef]
6. Thevenot, D.R.; Toth, K.; Durst, R.A.; Wilson, G.S. Electrochemical biosensors: Recommended definitions and classification. *Biosens. Bioelectron.* **2001**, *16*, 121–131. [CrossRef] [PubMed]
7. Brazaca, L.; Ribovski, L.; Janegitz, B.; Zucolotto, V. *Medical Biosensors for Point of Care (POC) Applications*; Elsevier: Amsterdam, The Netherlands, 2017; pp. 229–254.
8. Naresh, V.; Lee, N. A Review on Biosensors and Recent Development of Nanostructured Materials-Enabled Biosensors. *Sensors* **2021**, *21*, 1109. [CrossRef] [PubMed]
9. Garzarelli, V.; Ferrara, F.; Primiceri, E.; Chiriacò, M.S. Biofluids manipulation methods for liquid biopsy in minimally-invasive assays. *MethodsX* **2022**, *9*, 101759. [CrossRef]
10. de Andrade, A.C.; Resende, R.R.; Marques, A.P.A. Lab-on-chip technologies for point-of-care diagnostics: Advances and challenges for clinical applications. *Biosens. Bioelectron.* **2020**, *1*, 63–73.
11. Primiceri, E.; Chiriacò, M.S.; Ionescu, R.E.; D'Amone, E.; Cingolani, R.; Rinaldi, R.; Maruccio, G. Development of EIS cell chips and their application for cell analysis. *Microelectron. Eng.* **2009**, *86*, 1477–1480. [CrossRef]
12. Gamal, W.; Wu, H.; Underwood, I.; Jia, J.; Smith, S.; Bagnaninchi, P.O. Impedance-based cellular assays for regenerative medicine. *Phil. Trans. R. Soc. B* **2018**, *373*, 20170226. [CrossRef]
13. Evangelos, S.; Evangelos, A.; George, K.; Emmanouil, A.V.K.; Angeliki, T.; Dimitris, T. Flow determination via nanoparticle strain sensors for easy Lab on Chip integration. *Sens. Actuators A. Phys.* **2022**, *344*, 113765.
14. Lucile, A.; Amel, B.; Iago, P.; Madad, A.; Simon, D.; Laurent, M.; Thanh, D.M.; Stéphanie, D. Modular microfluidic system for on-chip extraction. preconcentration and detection of the cytokine biomarker IL-6 in biofluid. *Sci. Rep.* **2022**, *12*, 9468.
15. Shen, C.; Liu, S.; Li, X.; Yang, M. Electrochemical Detection of Circulating Tumor Cells Based on DNA Generated Electrochemical Current and Rolling Circle Amplification. *Anal. Chem.* **2019**, *91*, 11614–11619. [CrossRef] [PubMed]
16. Daulton, E.; Wicaksono, A.N.; Tiele, A.; Kocher, H.M.; Debernardi, S.; Crnogorac-Jurcevic, T.; Covington, J.A. Volatile organic compounds (VOCs) for the non-invasive detection of pancreatic cancer from urine. *Talanta* **2021**, *221*, 121604. [CrossRef]
17. Gilany, K.; Minai-Tehrani, A.; Savadi-Shiraz, E.; Rezadoost, H.; Lakpour, N. Exploring the human seminal plasma proteome: An unexplored gold mine of biomarker for male Infertility and male reproduction disorder. *J. Reprod. Infertil.* **2015**, *16*, 61–71. [PubMed]
18. Park, J.; Kim, N.E.; Yoon, H.; Shin, C.M.; Kim, N.; Lee, D.H.; Park, J.Y.; Choi, C.H.; Kim, J.G.; Kim, Y.K. Fecal Microbiota and Gut Microbe-Derived Extracellular Vesicles in Colorectal Cancer. *Front. Oncol.* **2021**, *11*, 650026. [CrossRef]

19. Sun, Y.; Haglund, T.A.; Rogers, A.J.; Ghanim, A.; Sethu, P. Review: Microfluidics technologies for blood-based cancer liquid biopsies. *Anal. Chim. Acta* **2018**, *1012*, 10–29. [CrossRef]
20. Liu, J.; Bao, N.; Liu, X.; Ding, S.-N. Nonenzymatic amperometric Aptamer cytosensor for ultrasensitive detection of circulating tumor cells and dynamic evaluation of cell surface N-Glycan expression. *ACS Omega* **2018**, *3*, 8595–8604. [CrossRef] [PubMed]
21. Cimmino, I.; Bravaccini, S.; Cerchione, C. Urinary Biomarkers in Tumors: An Overview. In *Springer EBooks*; Springer: Berlin/Heidelberg, Germany, 2021; pp. 3–15.
22. Paul, A.; Chiriaco, M.; Primiceri, E.; Srivastava, D.N.; Maruccio, G. Picomolar detection of retinol binding protein 4 for early management of type II diabetes. *Biosens. Bioelectron.* **2019**, *128*, 122–128. [CrossRef]
23. Siciliano, G.; Chiriaco, M.; Ferrara, F.; Turco, A.; Velardi, L.; Signore, M.A.; Esposito, M.; Gigli, G.; Primiceri, E. Development of a MIP based electrochemical sensor for TGF-β1 detection and its application in liquid biopsy. *Analyst* **2023**, *148*, 4447–4455. [CrossRef]
24. Radi, A.-E.; Abd-Ellatief, M.R. Electrochemical APTasensors: Current status and future Perspectives. *Diagnostics* **2021**, *11*, 104. [CrossRef]
25. Gao, Y.; Qin, Y.; Wan, C.; Sun, Y.; Meng, J.; Huang, J.; Hu, Y.; Jin, H.; Yang, K. Small extracellular vesicles: A novel avenue for cancer management. *Front. Oncol.* **2021**, *11*, 638357. [CrossRef]
26. Chiriaco, M.; Bianco, M.; Nigro, A.; Primiceri, E.; Ferrara, F.; Romano, A.; Quattrini, A.; Furlan, R.; Arima, V.; Maruccio, G. Lab-on-Chip for exosomes and microvesicles detection and characterization. *Sensors* **2018**, *18*, 3175. [CrossRef]
27. Garzarelli, V.; Chiriacò, M.S.; Cereda, M.; Gigli, G.; Ferrara, F. Ultrasensitive qPCR platform for rapid detection of bacterial contamination of raw biological samples at the point of care. *Heliyon* **2023**, *9*, e16229. [CrossRef] [PubMed]
28. Goda, T.; Masuno, K.; Nishida, J.; Kosaka, N.; Ochiya, T.; Matsumoto, A.; Miyahara, Y. A label-free electrical detection of exosomal microRNAs using microelectrode array. *Chem. Commun.* **2012**, *48*, 11942–11944. [CrossRef]
29. Bari, S.M.I.; Hossain, F.B.; Nestorova, G.G. Advances in Biosensors Technology for Detection and Characterization of Extracellular Vesicles. *Sensors* **2021**, *21*, 7645. [CrossRef] [PubMed]
30. Zhuang, L.; You, Q.; Su, X.; Chang, Z.; Ge, M.; Mei, Q.; Yang, L.; Dong, W.; Li, L. High-Performance Detection of Exosomes Based on Synergistic Amplification of Amino-Functionalized Fe_3O_4 Nanoparticles and Two-Dimensional MXene Nanosheets. *Sensors* **2023**, *23*, 3508. [CrossRef]
31. Jeong, S.; Park, J.; Pathania, D.; Castro, C.M.; Weissleder, R.; Lee, H. Integrated Magneto−Electrochemical Sensor for Exosome Analysis. *ACS Nano* **2016**, *10*, 1802–1809. [CrossRef]
32. Ruiyi, L.; Ling, L.; Hongxia, B.; Zaijun, L. Nitrogen-doped multiple graphene aerogel/gold nanostar as the electrochemical sensing platform for ultrasensitive detection of circulating free DNA in human serum. *Biosens. Bioelectron.* **2016**, *79*, 457. [CrossRef] [PubMed]
33. Wang, K.; Peng, Z.; Lin, X.; Nian, W.; Zheng, X.; Wu, J. Electrochemical Biosensors for Circulating Tumor DNA Detection. *Biosensors* **2022**, *12*, 649. [CrossRef] [PubMed]
34. Su, S.; Cao, W.; Liu, W.; Lu, Z.; Zhu, D.; Chao, J.; Weng, L.; Wang, L.; Fan, C. Dual-mode electrochemical analysis of microRNA-21 using gold nanoparticle-decorated MoS_2 nanosheet. *Biosens. Bioelectron.* **2017**, *94*, 552. [CrossRef]
35. Gurudatt, N.G.; Chung, S.; Kim, J.M.; Kim, M.H.; Jung, D.K.; Han, J.Y.; Shim, Y.B. Separation detection of different circulating tumor cells in the blood using an electrochemical microfluidic channel modified with a lipid-bonded conducting polymer. *Biosens. Bioelectron.* **2019**, *146*, 111746. [CrossRef]
36. Tian, L.; Qi, J.; Qian, K.; Oderinde, O.; Liu, Q.; Yao, C.; Song, W.; Wang, Y. Copper (II) oxide nanozyme based electrochemical cytosensor for high sensitive detection of circulating tumor cells in breast cancer. *J. Electroanal. Chem* **2018**, *812*, 1–9. [CrossRef]
37. Tang, S.; Shen, H.; Hao, Y.; Huang, Z.; Tao, Y.; Peng, Y.; Guo, Y.; Xie, G.; Feng, W. A Novel cytosensor based on Pt@Ag nanoflowers and AuNPs/acetylene black for ultrasensitive and highly specific detection of circulating tumor cells. *Biosens. Bioelectron.* **2018**, *104*, 72–78. [CrossRef]
38. Aydın, M.; Aydın, E.B.; Sezgintürk, M.K. A highly selective electrochemical immunosensor based on conductive carbon black and star PGMA polymer composite material for IL-8 biomarker detection in human serum and saliva. *Biosens. Bioelectron.* **2018**, *117*, 720–728. [CrossRef]
39. Soleymani, L.; Li, F. Mechanistic challenges and advantages of biosensor miniaturization into the nanoscale. *ACS Sens.* **2017**, *2*, 458–467. [CrossRef]
40. Martinsson, H.; Sandstrom, T.; Bleeker, A.J.; Hintersteiner, J.D. Current status of optical maskless lithography. *J. Micro-Nanolithogr. Mems Moems* **2005**, *4*, 011003. [CrossRef]
41. Chen, C.; Liu, Y.J.; Jiang, Z.-Y.; Shen, C.; Zhang, Y.; Zhong, F.; Chen, L.; Zhu, S.; Liu, H. Large-area long-wave infrared broadband all-dielectric metasurface absorber based on maskless laser direct writing lithography. *Opt. Express* **2022**, *30*, 13391. [CrossRef] [PubMed]
42. Heiskanen, S.; Maasilta, I. Superconducting tunnel junction fabrication on three-dimensional topography based on direct laser writing. *Appl. Phys. Lett.* **2020**, *117*, 232601. [CrossRef]
43. Ye, J.; Tan, H.; Wu, S.; Ni, K.; Pan, F.; Liu, J.; Tao, Z.; Qu, Y.; Ji, H.; Simon, P.; et al. Direct Laser Writing of Graphene Made from Chemical Vapor Deposition for Flexible, Integratable Micro-Supercapacitors with Ultrahigh Power Output. *Adv. Mater.* **2018**, *30*, 1801384. [CrossRef] [PubMed]
44. Dotan, T.; Berg, Y.; Migliorini, L.; Villa, S.; Santaniello, T.; Milani, P.; Shacham-Diamand, Y. Soft and flexible gold microelectrodes by supersonic cluster beam deposition and femtosecond laser processing. *Microelectron. Eng.* **2021**, *237*, 111478. [CrossRef]

45. Van Der Velden, G.; Fan, D.; Staufer, U. Fabrication of a microfluidic device by using two-photon lithography on a positive photoresist. *Micro Nano Eng.* **2020**, *7*, 100054. [CrossRef]
46. Viehrig, M.; Thilsted, A.H.; Matteucci, M.; Wu, K.; Catak, D.; Schmidt, M.S.; Zór, K.; Boisen, A. Injection-Molded Microfluidic Device for SERS Sensing Using Embedded Au-Capped Polymer Nanocones. *ACS Appl. Mater. Interfaces* **2018**, *10*, 37417–37425. [CrossRef] [PubMed]
47. Perrone, E.; Cesaria, M.; Zizzari, A.; Bianco, M.; Ferrara, F.; Raia, L.; Guarino, V.; Cuscunà, M.; Mazzeo, M.; Gigli, G.; et al. Potential of CO_2-laser processing of quartz for fast prototyping of microfluidic reactors and templates for 3D cell assembly over large scale. *Mater. Today Bio* **2021**, *12*, 100163. [CrossRef] [PubMed]
48. Timilsina, S.S.; Durr, N.; Yafia, M.; Sallum, H.M.; Jolly, P.; Ingber, D.E. Ultrarapid Method for Coating Electrochemical Sensors with Antifouling Conductive Nanomaterials Enables Highly Sensitive Multiplexed Detection in Whole Blood. *Adv. Healthc. Mater.* **2021**, *11*, 2102244. [CrossRef]
49. Li, X.-L.; Zhan, C.; Qiqi, H.; He, M.; Yang, C.; Yang, C.; Huang, X.; Chen, M.; Xie, X.; Chen, H.-J. Smart Diaper Based on Integrated Multiplex Carbon Nanotube-Coated Electrode Array Sensors for In Situ Urine Monitoring. *ACS Appl. Nano Mater.* **2022**, *5*, 4767–4778. [CrossRef]
50. Tseng, C.C.; Kung, C.-T.; Chen, R.; Tsai, M.-H.; Chao, H.-R.; Wang, Y.; Fu, L.-M. Recent advances in microfluidic paper-based assay devices for diagnosis of human diseases using saliva, tears and sweat samples. *Sens. Actuators B-Chem.* **2021**, *342*, 130078. [CrossRef]
51. Dong, T.; Pires, N.; Yang, Z.; Jiang, Z. Advances in Electrochemical Biosensors Based on Nanomaterials for Protein Biomarker Detection in Saliva. *Adv. Sci.* **2022**, *10*, 2205429. [CrossRef]
52. Destgeer, G.; Lee, K.H.; Jung, J.H.; Alazzam, A.; Sung, H.J. Continuous separation of particles in a PDMS microfluidic channel via travelling surface acoustic waves (TSAW). *Lab Chip* **2013**, *13*, 4210. [CrossRef]
53. Go, D.B.; Atashbar, M.Z.; Ramshani, Z.; Chang, H.-C. Surface acoustic wave devices for chemical sensing and microfluidics: A review and perspective. *Anal. Methods* **2017**, *9*, 4112–4134. [CrossRef]
54. Cardoso, S.; Leitao, D.C.; Dias, T.; Valadeiro, J.; Silva, M.; Chícharo, A.; Silverio, V.; Gaspar, J.; Freitas, P.P. Challenges and trends in magnetic sensor integration with microfluidics for biomedical applications. *J. Phys. D* **2017**, *50*, 213001. [CrossRef]
55. Gonzalez-Gallardo, C.L.; Díaz, A.; Casanova-Moreno, J.R. Improving plasma bonding of PDMS to gold-patterned glass for electrochemical microfluidic applications. *Microfluid. Nanofluidics* **2021**, *25*, 20. [CrossRef]
56. Zoupanou, S.; Chiriaco, M.; Tarantini, I.; Ferrara, F. Innovative 3D Microfluidic Tools for On-Chip fluids and Particles manipulation: From design to experimental validation. *Micromachines* **2021**, *12*, 104. [CrossRef]
57. Guascito, M.R.; Filippo, E.; Mattei, G.; Manno, D.; Serra, A.; Turco, A. A new amperometric nanostructured sensor for the analytical determination of hydrogen peroxide. *Biosens. Bioelectron.* **2008**, *24*, 1057–1063. [CrossRef] [PubMed]
58. De Benedetto, G.E.; Corvaglia, S.; Pompa, P.P.; Mattei, G. An innovative and simple all electrochemical approach to functionalize electrodes with a carbon nanotubes/polypyrrole molecularly imprinted nanocomposite and its application for sulfamethoxazole analysis. *J. Colloid Interface Sci.* **2021**, *599*, 676–685.
59. Guascito, M.R.; Chirizzi, D.; Mattei, G.; Mazzotta, E.; Siciliano, M.; Siciliano, T.; Tepore, A.; Turco, A. Low-potential sensitive H_2O_2 detection based on composite micro tubular Te adsorbed on platinum electrode. *Biosens. Bioelectron.* **2011**, *26*, 3562–3569. [CrossRef]
60. Zhang, Q.; Xu, J.-J.; Liu, Y.; Chen, H.-Y. In-situ synthesis of poly(dimethylsiloxane)–gold nanoparticles composite films and its application in microfluidic systems. *Lab Chip* **2008**, *8*, 352–357. [CrossRef]
61. Anshori, I.; Harimurti, S.; Rizalputri, L.N.; Hartono, M.S.; Althof, R.R.; Handayani, M.; Mengko, T.L.R.; Yuliarto, B. Modified screen-printed electrode using graphene ink for electrochemical sensor application. *J. Phys. Conf. Ser.* **2021**, *1912*, 012022. [CrossRef]
62. Pachauri, N.; Lakshmi, G.B.V.S.; Sri, S.; Gupta, P.K.; Solanki, P.R. Silver molybdate nanoparticles based immunosensor for the non-invasive detection of Interleukin-8 biomarker. *Mater. Sci. Eng. C* **2020**, *113*, 110911. [CrossRef]
63. Ko, E.; Tran, V.-K.; Geng, Y.; Chung, W.C.; Park, C.S.; Kim, M.; Jin, G.H.; Seong, G.H. Continuous electrochemical detection of hydrogen peroxide by Au-Ag bimetallic nanoparticles in microfluidic devices. *J. Electroanal. Chem.* **2017**, *792*, 72–78. [CrossRef]
64. Xuecui, M.; Jiao, Y.; Xinge, Y.; Zhengchun, P.; Guanghui, Z.; Yingchun, L. Wearable molecularly imprinted electrochemical sensor with integrated nanofiber-based microfluidic chip for in situ monitoring of cortisol in sweat. *Sens. Actuators B Chem.* **2023**, *381*, 133451.
65. Zheng, F.; Pu, Z.; He, E.; Huang, J.; Yu, B.; Li, D.; Li, Z. From functional structure to packaging: Full-printing fabrication of a microfluidic chip. *Lab Chip* **2018**, *18*, 1859–1866. [CrossRef]
66. Cinti, S.; Arduini, F.; Moscone, D.; Palleschi, G.; Gonzalez-Macia, L.; Killard, A.J. Cholesterol biosensor based on inkjet-printed Prussian blue nanoparticle-modified screen-printed electrodes. *Sens. Actuators B-Chem.* **2015**, *221*, 187–190. [CrossRef]
67. Lee, C.-W.; Chang, H.-Y.; Wu, J.-K.; Tseng, F.-G. Ultra-sensitive electrochemical detection of bacteremia enabled by redox-active gold nanoparticles (raGNPs) in a nano-sieving microfluidic system (NS-MFS). *Biosens. Bioelectron.* **2019**, *133*, 215–222. [CrossRef] [PubMed]
68. Beck, F.; Horn, C.; Baeumner, A.J. Dry-reagent microfluidic biosensor for simple detection of NT-proBNP via Ag nanoparticles. *Anal. Chim. Acta* **2021**, *1191*, 339375. [CrossRef] [PubMed]
69. Gu, S.; Lu, Y.; Ding, Y.; Li, L.; Song, H.; Wang, J.; Wu, Q. A droplet-based microfluidic electrochemical sensor using platinum-black microelectrode and its application in high sensitive glucose sensing. *Biosens. Bioelectron.* **2014**, *55*, 106–112. [CrossRef]

70. Sun, J.; Xianyua, Y.; Jiang, X. Point-of-care biochemical assays using gold nanoparticle-implemented microfluidics. *Chem. Soc. Rev.* **2014**, *43*, 6239–6253. [CrossRef] [PubMed]
71. Aravamudhan, S.; Kumar, A.; Mohapatra, S.S.; Bhansali, S. Sensitive estimation of total cholesterol in blood using Au nanowires based micro-fluidic platform. *Biosens. Bioelectron.* **2007**, *22*, 2289–2294. [CrossRef] [PubMed]
72. Wang, X.; He, X.; He, Z.; Liwei, H.; Ge, C.; Wang, L.; Li, S.; Xu, Y. Detection of prostate specific antigen in whole blood by microfluidic chip integrated with dielectrophoretic separation and electrochemical sensing. *Biosens. Bioelectron.* **2022**, *204*, 114057. [CrossRef]
73. Vural, T.; Yaman, Y.T.; Ozturk, S.; Abaci, S.; Denkbaş, E.B. Electrochemical immunoassay for detection of prostate specific antigen based on peptide nanotube-gold nanoparticle-polyaniline immobilized pencil graphite electrode. *J. Colloid Interface Sci.* **2018**, *510*, 318–326. [CrossRef]
74. Sharma, S.; Bhatia, V. Magnetic nanoparticles in microfluidics-based diagnostics: An appraisal. *Nanomedicine* **2021**, *16*, 1329–1342. [CrossRef]
75. Kurbanoglu, S.; Mayorga-Martinez, C.C.; Medina-Sánchez, M.; Rivas, L.; Ozkan, S.A.; Merkoçi, A. Antithyroid drug detection using an enzyme cascade blocking in a nanoparticle-based lab-on-a-chip system. *Biosens. Bioelectron.* **2015**, *67*, 670–676. [CrossRef]
76. Liu, Z.; Jin, M.; Cao, J.-P.; Ruiwen, N.; Li, P.; Zhou, G.; Yu, Y.; Van Den Berg, A.; Shui, L. Electrochemical sensor integrated microfluidic device for sensitive and simultaneous quantification of dopamine and 5-hydroxytryptamine. *Sens. Actuators B-Chem.* **2018**, *273*, 873–883. [CrossRef]
77. Yang, J.; Yu, J.H.; Strickler, J.R.; Chang, W.-J.; Gunasekaran, S. Nickel nanoparticle–chitosan-reduced graphene oxide-modified screen-printed electrodes for enzyme-free glucose sensing in portable microfluidic devices. *Biosens. Bioelectron.* **2013**, *47*, 530–538. [CrossRef] [PubMed]
78. Cincotto, F.H.; Fava, E.L.; Moraes, F.; Fatibello-Filho, O.; Faria, R.C. A new disposable microfluidic electrochemical paper-based device for the simultaneous determination of clinical biomarkers. *Talanta* **2019**, *195*, 62–68. [CrossRef]
79. Chunhua, W.; Zhang, Y.; Tang, W.; Wang, C.; Han, Y.; Qiang, L.; Gao, J.; Liu, H.; Heinegård, D. Ultrasensitive, high-throughput and multiple cancer biomarkers simultaneous detection in serum based on graphene oxide quantum dots integrated microfluidic biosensing platform. *Anal. Chim. Acta* **2021**, *1178*, 338791.
80. Wisitsoraat, A.; Sritongkham, P.; Karuwan, C.; Phokharatkul, D.; Maturos, T.; Tuantranont, A. Fast cholesterol detection using flow injection microfluidic device with functionalized carbon nanotubes based electrochemical sensor. *Biosens. Bioelectron.* **2010**, *26*, 1514–1520. [CrossRef] [PubMed]
81. Materon, E.M.; Lima, R.A.; Joshi, N.; Shimizu, F.M.; Oliveira, O.N. Graphene-Containing Microfluidic and Chip-Based Sensor Devices for Biomolecules. In *Elsevier EBooks*; Elsevier: Amsterdam, The Netherlands, 2019; pp. 321–336.
82. Narang, J.; Malhotra, N.; Singhal, C.; Mathur, A.; Chakraborty, D.; Anil, A.; Ingle, A.; Pundir, C.S. Point of care with micro fluidic paper based device integrated with nano zeolite–graphene oxide nanoflakes for electrochemical sensing of ketamine. *Biosens. Bioelectron.* **2017**, *88*, 249–257. [CrossRef]
83. Dolati, A.; AbdelFatah, T.; Sanati, A.; Jalali, M.; Flynn, S.J.; Mahshid, S.; Mahshid, S. A Nanostructured Gold/Graphene Microfluidic Device for Direct and Plasmonic-Assisted Impedimetric Detection of Bacteria. *ACS Appl. Mater. Interfaces* **2020**, *12*, 23298–23310.
84. Zhang, S.; Zahed, A.; Sharifuzzaman, M.; Yoon, S.; Hui, X.; Barman, S.C.; Sharma, S.; Yoon, H.J.; Park, C.; Park, J.Y. A wearable battery-free wireless and skin-interfaced microfluidics integrated electrochemical sensing patch for on-site biomarkers monitoring in human perspiration. *Biosens. Bioelectron.* **2021**, *175*, 112844. [CrossRef]
85. Ciftci, S.; Cánovas, R.; Neumann, F.; Paulraj, T.; Nilsson, M.; Crespo, G.A.; Madaboosi, N. The sweet detection of rolling circle amplification: Glucose-based electrochemical genosensor for the detection of viral nucleic acid. *Biosens. Bioelectron.* **2020**, *151*, 112002. [CrossRef]
86. Liu, J.; Zhang, Y.; Xie, H.; Li, H.; Zheng, L.; Ye, H. Applications of Catalytic Hairpin Assembly Reaction in Biosensing. *Small* **2019**, *15*, 1902989. [CrossRef]
87. He, R.; Niu, Y.; Li, Z.; Li, A.; Yang, H.-Y.; Xu, F.; Li, F. A Hydrogel Microneedle Patch for Point-of-Care Testing Based on Skin Interstitial Fluid. *Adv. Healthc. Mater.* **2020**, *9*, 1901201. [CrossRef] [PubMed]
88. Lin, Y.; Bariya, M.; Nyein, H.Y.Y.; Kivimäki, L.; Uusitalo, S.; Jansson, E.; Ji, W.; Yuan, Z.; Happonen, T.; Liedert, C.; et al. Porous Enzymatic Membrane for Nanotextured Glucose Sweat Sensors with High Stability toward Reliable Noninvasive Health Monitoring. *Adv. Funct. Mater.* **2019**, *29*, 1902521. [CrossRef]
89. Zhong, R.; Tang, Q.; Wang, S.; Zhang, H.; Zhang, F.; Xiao, M.; Man, T.; Qu, X.; Li, L.; Zhang, W.; et al. Self-Assembly of Enzyme-Like Nanofibrous G-Molecular Hydrogel for Printed Flexible Electrochemical Sensors. *Adv. Mater.* **2018**, *30*, 1706887. [CrossRef] [PubMed]
90. Zhao, F.; Shi, Y.; Pan, L.; Yu, G. Multifunctional Nanostructured Conductive Polymer Gels: Synthesis, Properties, and Applications. *Acc. Chem. Res.* **2017**, *50*, 1734–1743. [CrossRef] [PubMed]
91. Li, L.; Wang, Y.; Pan, L.; Shi, Y.; Cheng, W.; Shi, Y.; Yu, G. A Nanostructured Conductive Hydrogels-Based Biosensor Platform for Human Metabolite Detection. *Nano Lett.* **2015**, *15*, 1146–1151. [CrossRef] [PubMed]
92. Kayser, L.V.; Lipomi, D.J. Stretchable Conductive Polymers and Composites Based on PEDOT and PEDOT:PSS. *Adv. Mater.* **2019**, *31*, 1806133. [CrossRef]
93. Xu, J.; Xu, K.; Han, Y.; Wang, D.; Li, X.; Hu, T.; Yi, H.; Ni, Z. A 3D porous graphene aerogel@GOx based microfluidic biosensor for electrochemical glucose detection. *Analyst* **2020**, *145*, 5141–5147. [CrossRef]

94. Zhang, Q.; Ma, S.; Zhang, K.; Zhang, L.; Liu, C.; Shi, H.; Wang, C.; Wang, N.; Zhu, A. A facile integrated microfluidic chip based on Chitosan-Gold Nanoparticles-Anchored Three-Dimensional graphene fiber film for monitoring prostate specific antigen. *Microchem. J.* **2023**, *184*, 108171. [CrossRef]
95. Chand, R.; Neethirajan, S. Microfluidic platform integrated with graphene-gold nano-composite aptasensor for one-step detection of norovirus. *Biosens. Bioelectron.* **2017**, *98*, 47–53. [CrossRef]
96. Fan, Y.; Shi, S.; Ma, J.; Guo, Y. Smartphone-based electrochemical system with multi-walled carbon nanotubes/thionine/gold nanoparticles modified screen-printed immunosensor for cancer antigen 125 detection. *Microchem. J.* **2021**, *174*, 107044. [CrossRef]
97. Di Giulio, T.; Barca, A.; Verri, T.; De Gennaro, M.; Giancane, G.; Mazzotta, E.; Malitesta, C. Molecular imprinting based on metal-ion mediated recognition: Electrosynthesis of artificial receptors for the selective detection of peptides. *Sens. Actuators B-Chem.* **2023**, *383*, 133589. [CrossRef]
98. Mazouz, Z.; Rahali, S.; Fourati, N.; Zerrouki, C.; Aloui, N.; Zerrouki, C.; Kalfat, R.; Chehimi, M.M.; Othmane, A.; Kalfat, R. Highly Selective Polypyrrole MIP-Based Gravimetric and Electrochemical Sensors for Picomolar Detection of Glyphosate. *Sensors* **2017**, *17*, 2586. [CrossRef] [PubMed]
99. Yarman, A.; Kurbanoglu, S.; Jetzschmann, K.J.; Ozkan, S.A.; Wollenberger, U.; Scheller, F.W. Electrochemical MIP-Sensors for Drugs. *Curr. Med. Chem.* **2017**, *25*, 4007–4019. [CrossRef]
100. Mazouz, Z.; Mokni, M.; Fourati, N.; Zerrouki, C.; Barbault, F.; Zerrouki, C.; Kalfat, R.; Kalfat, R.; Omezzine, A.; Bouslema, A.; et al. Computational approach and electrochemical measurements for protein detection with MIP-based sensor. *Biosens. Bioelectron.* **2020**, *151*, 111978. [CrossRef]
101. Liustrovaite, V.; Pogorielov, M.; Boguzaite, R.; Ratautaite, V.; Ramanavicius, A.; Pilvenyte, G.; Holubnycha, V.; Korniienko, V.; Diedkova, K.; Viter, R.; et al. Towards Electrochemical Sensor Based on Molecularly Imprinted Polypyrrole for the Detection of Bacteria—Listeria monocytogenes. *Polymers* **2023**, *15*, 1597. [CrossRef]
102. Wang, L.; Pagett, M.; Zhang, W. Molecularly imprinted polymer (MIP) based electrochemical sensors and their recent advances in health applications. *Sens. Actuators Rep.* **2023**, *5*, 100153. [CrossRef]
103. Li, M.; Wang, X.; Zhang, L.; Wei, X.-P. A high sensitive epitope imprinted electrochemical sensor for bovine serum albumin based on enzyme amplifying. *Anal. Biochem.* **2017**, *530*, 68–74. [CrossRef]
104. Majd, S.M.; Mirzapour, F.; Shamsipur, M.; Manouchehri, I.; Babaee, E.; Pashabadi, A.; Moradian, R. Design of a novel aptamer/molecularly imprinted polymer hybrid modified Ag–Au@Insulin nanoclusters/Au-gate-based MoS2 nanosheet field-effect transistor for attomolar detection of BRCA1 gene. *Talanta* **2023**, *257*, 124394. [CrossRef]
105. Choi, D.J.; Yang, J.; Hong, S.W.; Park, J.Y. Molecularly imprinted polymer-based electrochemical impedimetric sensors on screen-printed carbon electrodes for the detection of trace cytokine IL-1β. *Biosens. Bioelectron.* **2022**, *204*, 114073. [CrossRef]
106. Ratautaite, V.; Boguzaite, R.; Brazys, E.; Plausinaitis, D.; Ramanavicius, S.; Samukaite-Bubniene, U.; Bechelany, M.; Ramanavicius, A. Evaluation of the interaction between SARS-CoV-2 spike glycoproteins and the molecularly imprinted polypyrrole. *Talanta* **2023**, *253*, 123981. [CrossRef]
107. Shumyantseva, V.V.; Bulko, T.V.; Sigolaeva, L.V.; Kuzikov, A.V.; Pogodin, P.V.; Archakov, A.I. Molecular imprinting coupled with electrochemical analysis for plasma samples classification in acute myocardial infarction diagnostic. *Biosens. Bioelectron.* **2018**, *99*, 216–222. [CrossRef]
108. Hussein, H.; Kandeil, A.; Gomaa, M.R.; Nashar, R.M.E.; El-Sherbiny, I.M.; Hassan, R.Y.A. SARS-CoV-2-Impedimetric Biosensor: Virus-Imprinted Chips for Early and Rapid Diagnosis. *ACS Sens.* **2021**, *6*, 4098–4107. [CrossRef] [PubMed]
109. Liu, W.; Ma, Y.-K.; Sun, G.; Wang, S.; Deng, J.-Y.; Wei, H. Molecularly imprinted polymers on graphene oxide surface for EIS sensing of testosterone. *Biosens. Bioelectron.* **2017**, *92*, 305–312. [CrossRef] [PubMed]
110. Motia, S.; Bouchikhi, B.; Bari, N.E. An electrochemical molecularly imprinted sensor based on chitosan capped with gold nanoparticles and its application for highly sensitive butylated hydroxyanisole analysis in foodstuff products. *Talanta* **2021**, *223*, 121689. [CrossRef] [PubMed]
111. Shoaie, N.; Daneshpour, M.; Azimzadeh, M.; Mahshid, S.; Khoshfetrat, S.M.; Jahanpeyma, F. Electrochemical sensors and biosensors based on the use of polyaniline and its nanocomposites: A review on recent advances. *Microchim. Acta* **2019**, *186*, 465. [CrossRef]
112. Shoaie, N.; Forouzandeh, M.; Omidfar, K. Voltammetric determination of the Escherichia coli DNA using a screen-printed carbon electrode modified with polyaniline and gold nanoparticles. *Microchim. Acta* **2018**, *185*, 217. [CrossRef]
113. Eggins, B.R. *Chemical Sensors and Biosensors*; John Wiles & Sons: New York, NY, USA, 2002.
114. Zhang, X.; Li, F.; Wei, Q.; Du, B.; Wu, D.; Li, H. Ultrasensitive nonenzymatic immunosensor based on bimetallic gold–silver nanoclusters synthesized by simple mortar grinding route. *Sens. Actuator B-Chem* **2014**, *194*, 64–70. [CrossRef]
115. Zouari, M.; Campuzano, S.; Pingarrón, J.M.; Raouafi, N. Competitive RNA-RNA hybridization-based integrated nanostructured disposable electrode for highly sensitive determination of miRNAs in cancer cells. *Biosens. Bioelectron.* **2017**, *91*, 40–45. [CrossRef]
116. Francisco, G.O.; German, E.G.; Chiara, B.; Ines, C.G.; Carmen, G.N.; Richard, F.D.; María, P.M.V.; Maria, J.S.; Germ, A.M.; Jose, E.H.; et al. Microfluidic amperometric immunosensor based on porous nanomaterial towards claudin7 determination for colorectal cancer diagnosis. *Talanta* **2022**, *251*, 123766.
117. Grieshaber, D.; MacKenzie, R.; Vörös, J.; Reimhult, E. Electrochemical Biosensors—Sensor Principles and Architectures. *Sensors* **2008**, *8*, 1400–1458. [CrossRef] [PubMed]
118. Bard, A.J.; Faulkner, L.R. *Electrochemical Methods: Fundamentals and Applications*, 2nd ed.; Wiley: New York: NY, USA, 2008.

119. Jia, Y.; Qin, M.; Zhang, H.; Niu, W.; Li, X.; Wang, L.; Li, X.; Bai, Y.; Cao, Y.; Feng, X. Label-free biosensor: A novel phage modified Light Addressable Potentiometric Sensor system for cancer cell monitoring. *Biosens. Bioelectron.* **2007**, *22*, 3261. [CrossRef] [PubMed]
120. Wang, Y.; Zhang, Z.; Jain, V.; Yi, J.; Mueller, S.; Sokolov, J.; Liu, Z.; Levon, K.; Rigas, B.; Rafailovich, M.H. Potentiometric sensors based on surface molecular imprinting: Detection of cancer biomarkers and viruses. *Sens. Actuators B Chem.* **2010**, *146*, 381. [CrossRef]
121. Mathur, A.; Blais, S.; Goparaju, C.M.V.; Neubert, T.; Pass, H.; Levon, K. Development of a Biosensor for Detection of Pleural Mesothelioma Cancer Biomarker Using Surface Imprinting. *PLoS ONE* **2013**, *8*, 1. [CrossRef]
122. Shaibani, P.M.; Etayash, H.; Naicker, S.; Kaur, K.; Thundat, T. Metabolic Study of Cancer Cells Using a pH Sensitive Hydrogel Nanofiber Light Addressable Potentiometric Sensor. *ACS Sens.* **2017**, *2*, 151. [CrossRef]
123. Gu, Y.; Ju, C.; Li, Y.; Shang, Z.; Wu, Y.; Jia, Y.; Niu, Y. Detection of circulating tumor cells in prostate cancer based on carboxylated graphene oxide modified light addressable potentiometric sensor. *Biosens. Bioelectron.* **2015**, *66*, 24. [CrossRef]
124. Mani, G.K.; Morohoshi, M.; Yasoda, Y.; Yokoyama, S.; Kimura, H.; Tsuchiya, K. ZnO-Based Microfluidic pH Sensor: A Versatile Approach for Quick Recognition of Circulating Tumor Cells in Blood. *ACS Appl. Mater Interfaces* **2017**, *9*, 5193–5203. [CrossRef]
125. Elshafey, R.; Tavares, A.C.; Siaj, M.; Zourob, M. Electrochemical impedance immunosensor based on gold nanoparticles–protein G for the detection of cancer marker epidermal growth factor receptor in human plasma and brain tissue. *Biosens. Bioelectron.* **2013**, *50*, 143149. [CrossRef]
126. Han, L.; Liu, P.; Petrenko, V.A.; Liu, A. A label-free electrochemical impedance cytosensor based on specific peptide fused phage selected from landscape phage library. *Sci. Rep.* **2016**, *6*, 22199. [CrossRef]
127. Hu, Y.; Zuo, P.; Ye, B.C. Label-free electrochemical impedance spectroscopy biosensor for direct detection of cancer cells based on the interaction between carbohydrate and lectin. *Biosens. Bioelectron.* **2013**, *43*, 7983. [CrossRef]
128. Azzouzi, S.; Mak, W.C.; Kor, K.; Turner, A.; Ali, M.; Beni, V. An integrated dual functional recognition/amplification bio-label for the one-step impedimetric detection of Micro-RNA-21. *Biosens. Bioelectron.* **2017**, *92*, 154–161. [CrossRef]
129. Kilic, T.; De Sousa Valinhas, A.T.; Wall, I.; Renaud, P.; Carrara, S. Label-free detection of hypoxiainduced extracellular vesicle secretion from MCF-7 cells. *Sci. Rep.* **2018**, *9*, 27203.
130. Ronkainen, N.J.; Halsall, H.B.; Heineman, W.R. Electrochemical biosensors. *Chem. Soc. Rev.* **2010**, *39*, 1747–1763. [CrossRef]
131. Labib, M.; Hedström, M.; Amin, M.; Mattiasson, B. A capacitive immunosensor for detection of cholera toxin. *Anal. Chim. Acta* **2009**, *634*, 255–261. [CrossRef]
132. Liang, J.; Wang, J.; Zhang, L.; Wang, S.; Yao, C.; Zhang, Z. Conductometric immunoassay of alpha-fetoprotein in sera of liver cancer patients using bienzymefunctionalized nanometer-sized silica beads. *Analyst* **2019**, *144*, 265. [CrossRef]
133. Bhardwaj, S.K.; Sharma, A.L.; Bhardwaj, N.; Kukkar, M.; Gill, A.A.S.; Kim, K.H.; Deep, A. TCNQ-doped Cu-metal organic framework as a novel conductometric immunosensing platform for the quantification of prostate cancer antigen. *Sens. Actuators B Chem.* **2017**, *240*, 10. [CrossRef]
134. Lin, Y.H.; Lin, W.S.; Wong, J.C.; Hsu, W.C.; Peng, Y.S.; Chen, C.L. Bottom-up assembly of silicon nanowire conductometric sensors for the detection of apolipoprotein A1, a biomarker for bladder cancer. *Microchim. Acta* **2017**, *184*, 2419. [CrossRef]
135. Pourali, A.; Rashidi, M.R.; Barar, J.; Pavon-Djavid, G.; Omidi, Y. Voltammetric biosensors for analytical detection of cardiac troponin biomarkers in acute myocardial infarction. *TrAC Trends Anal. Chem.* **2021**, *2134*, 116123. [CrossRef]
136. Malhotra, B.D. *Biosensors: Fundamentals Applications*; Smithers Rapra Technology Ltd. Shawbury: Shrewsbury, UK, 2017.
137. Kumar, S.; Kumar, S.; Tiwari, S.; Srivastava, S.; Srivastava, M.; Yadav, B.K.; Kumar, S.; Tran, T.T.; Dewan, A.K.; Mulchandani, A.; et al. Biofunctionalized nanostructured zirconia for biomedical application: A smart approach for oral cancer detection. *Adv. Sci.* **2015**, *2*, 1500048. [CrossRef]
138. Wang, H.; Zhang, Y.; Yu, H.; Wu, D.; Ma, H.; Li, H.; Du, B.; Wei, Q. Label-free electrochemical immunosensor for prostate-specific antigen based on silver hybridized mesoporous silica nanoparticles. *Anal. Biochem.* **2013**, *434*, 123127. [CrossRef]
139. Kumar, S.; Lei, Y.; Alshareef, N.H.; Quevedo-Lopez, M.; Salama, K.N. Biofunctionalized two-dimensional Ti3C2 MXenes for ultrasensitive detection of cancer biomarker. *Biosens. Bioelectron.* **2018**, *121*, 243249. [CrossRef] [PubMed]
140. Taleat, Z.; Cristea, C.; Marrazza, G.; Mazloum-Ardakani, M.; Săndulescu, R. Electrochemical immunoassay based on aptamer–protein interaction and functionalized polymer for cancer biomarker detection. *J. Electroanal. Chem.* **2014**, *717*, 119124. [CrossRef]
141. Feng, X.; Gan, N.; Zhang, H.; Yan, Q.; Li, T.; Cao, Y.; Hu, F.; Yu, H.; Jiang, Q. A novel strategy for multiplexed immunoassay of tumor markers based on electrochemiluminescence coupled with cyclic voltammetry using graphene-polymer nanotags. *Electrochim. Acta* **2015**, *170*, 292299. [CrossRef]
142. Lin, C.W.; Wei, K.C.; Liao, S.S.; Huang, C.Y.; Sun, C.L.; Wu, P.J.; Lu, Y.J.; Yang, H.W.; Ma, C.C. A reusable magnetic graphene oxide-modified biosensor for vascular endothelial growth factor detection in cancer diagnosis. *Biosens. Bioelectron.* **2015**, *67*, 431437. [CrossRef]
143. Amjadi, M.; Khoshraj, J.M.; Majidi, M.R.; Baradaran, B.; de la Guardia, M. Evaluation of Flavonoid Derivative and Doxorubicin Effects in Lung Cancer Cells (A549) Using Differential Pulse Voltammetry Method. *Adv. Pharm. Bull.* **2018**, *8*, 637–642. [CrossRef]
144. Wang, K.; He, M.Q.; Zhai, F.H.; He, R.H.; Yu, Y.L. A novel electrochemical biosensor based on polyadenine modified aptamer for label-free and ultrasensitive detection of human breast cancer cells. *Talanta* **2017**, *166*, 8792. [CrossRef]
145. Pacheco, J.G.; Silva, M.S.; Freitas, M.; Nouws, H.P.; Delerue-Matos, C. Molecularly imprinted electrochemical sensor for the point-of-care detection of a breast cancer biomarker (CA 15-3), Sens. *Actuators B Chem.* **2018**, *256*, 905912. [CrossRef]

146. Fakunle, E.S.; Fritsch, I. Low-temperature co-fired ceramic microchannels with individually addressable screen-printed gold electrodes on four walls for self-contained electrochemical immunoassays. *Anal. Bioanal. Chem.* **2010**, *398*, 26052615. [CrossRef] [PubMed]
147. Zani, A.; Laschi, S.; Mascini, M.; Marrazza, G. A New Electrochemical Multiplexed Assay for PSA Cancer Marker Detection. *Electroanalysis* **2011**, *23*, 9199. [CrossRef]
148. Erdem, A.; Congur, G. Label-free voltammetric detection of microRNAs at multi-channel screen printed array of electrodes comparison to graphite sensors. *Talanta* **2014**, *118*, 713. [CrossRef]
149. Ahn, J.; Cho, J.Y. Current serum lung cancer biomarkers. *J. Mol. Biomark. Diagn* **2013**, *4*, 2.
150. Alatas, F.; Alatas, Ö.; Metintas, M.; Colak, O.; Harmanci, E.; Demir, S. Diagnostic value of CEA, CA 15-3, CA 19-9, CYFRA 21-1, NSE and TSA assay in pleural effusions. *Lung Cancer* **2001**, *31*, 916. [CrossRef]
151. Wu, J.; Zhang, Z.; Fu, Z.; Ju, H. A disposable two-throughput electrochemical immunosensor chip for simultaneous multianalyte determination of tumor markers. *Biosens. Bioelectron.* **2007**, *23*, 114120. [CrossRef] [PubMed]
152. Wu, D.; Guo, A.; Guo, Z.; Xie, L.; Wei, Q.; Du, B. Simultaneous electrochemical detection of cervical cancer markers using reduced graphene oxide-tetraethylene pentamine as electrode materials and distinguishable redox probes as labels. *Biosens. Bioelectron.* **2014**, *54*, 634639. [CrossRef]
153. Xu, T.; Jia, X.; Chen, X.; Ma, Z. Simultaneous electrochemical detection of multiple tumor markers using metal ions tagged immunocolloidal gold. *Biosens. Bioelectron.* **2014**, *56*, 174179. [CrossRef] [PubMed]
154. Zhou, Y.G.; Kermansha, L.; Zhang, L.; Mohamadi, R.M. Miniaturized electrochemical sensors to facilitate liquid biopsy for detection of circulating tumor markers. In *Applications of Microfluidic Systems in Biology and Medicine*; Tokeshi, M., Ed.; Springer: Singapore, 2019; p. 7198.
155. Moscovici, M.; Bhimji, A.; Kelley, S.O. Rapid and specific electrochemical detection of prostate cancer cells using an aperture sensor array. *Lab Chip* **2013**, *13*, 940946. [CrossRef]
156. Yang, H.; Hui, A.; Pampalakis, G.; Soleymani, L.; Liu, F.F.; Sargent, E.H.; Kelley, S.O. Direct, electronic microRNA detection for the rapid determination of differential expression profiles. *Angew. Chem. Int. Ed.* **2009**, *48*, 84618464. [CrossRef]
157. Zhang, Q.; Shan, X.; Fu, Y.; Liu, P.; Li, X.; Liu, B.; Zhang, L.; Li, D. Electrochemical determination of the anticancer drug capecitabine based on a graphene-gold nanocomposite-modified glassy carbon electrode. *Int. J. Electrochem. Sci.* **2017**, *12*, 1077310782. [CrossRef]
158. Venu, M.; Venkateswarlu, S.; Reddy, Y.V.M.; Seshadri Reddy, A.; Gupta, V.K.; Yoon, M.; Madhavi, G. Highly sensitive electrochemical sensor for anticancer drug by a zirconia nanoparticle-decorated reduced graphene oxide nanocomposite. *ACS Omega* **2018**, *3*, 1459714605. [CrossRef] [PubMed]
159. Yan, M.; Zang, D.; Ge, S.; Ge, L.; Yu, J. A disposable electrochemical immunosensor based on carbon screen-printed electrodes for the detection of prostate specific antigen. *Biosens. Bioelectron.* **2012**, *38*, 355–361. [CrossRef]
160. Liao, K.; Huang, H. Femtomolar immunoassay based on coupling gold nanoparticle enlargement with square wave stripping voltammetry. *Anal. Chim. Acta* **2005**, *538*, 159–164. [CrossRef]
161. Liu, X.; Duckworth, P.A.; Wong, D.K.Y. Square wave voltammetry versus electrochemical impedance spectroscopy as a rapid detection technique at electrochemical immunosensors. *Biosens. Bioelectron.* **2010**, *25*, 1467–1473. [CrossRef]
162. Anderson, J.L.; Coury, L.A.; Leddy, J. Dynamic electrochemistry: Methodology and application. *Anal. Chem.* **1998**, *70*, 519–589. [CrossRef]
163. Akcakoca, I.; Ghorbanpoor, H.; Blair, E.O.; Öztürk, Y.; Dizaji, A.N.; Kocagöz, T.; Avci, H.; Corrigan, D.K.; Guzel, F.B. An electrochemical biosensor with integrated microheater to improve the sensitivity of electrochemical nucleic acid biosensors. *J. Micromech. Microeng.* **2022**, *32*, 045008. [CrossRef]
164. Kasturi, S.; Torati, S.R.; Eom, Y.; Kim, C. Microvalve-controlled miniaturized electrochemical lab-on-a-chip based biosensor for the detection of β-amyloid biomarker. *J. Ind. Eng. Chem.* **2021**, *97*, 349–355. [CrossRef]
165. Liu, E.; Cai, Z.; Ye, Y.; Zhou, M.; Liao, H.; Yi, Y. An Overview of Flexible Sensors: Development, Application, and Challenges. *Sensors* **2023**, *23*, 817. [CrossRef] [PubMed]
166. Chen, S.-W.W.; Qi, J.; Fan, S.; Qiao, Z.; Yeo, J.G.; Lim, C.T. Flexible Wearable Sensors for Cardiovascular Health Monitoring. *Adv. Healthc. Mater.* **2021**, *10*, 2100116. [CrossRef] [PubMed]
167. Vaghasiya, J.V.; Mayorga-Martinez, C.C.; Pumera, M. Wearable sensors for telehealth based on emerging materials and nanoarchitectonics. *Npj Flex. Electron.* **2023**, *7*, 26. [CrossRef]
168. Sempionatto, J.R.; Lin, M.; Yin, L.; De La Paz, E.; Pei, K.; Sonsa-Ard, T.; De Loyola E Silva, A.N.; Khorshed, A.A.; Zhang, F.; Tostado, N.; et al. An epidermal patch for the simultaneous monitoring of haemodynamic and metabolic biomarkers. *Nat. Biomed. Eng.* **2021**, *5*, 737–748. [CrossRef]
169. Subramaniam, S. The smartphone biosensors for point-of-care detection of human infectious diseases: Overview and perspectives—A systematic review. *Curr. Opin. Electrochem.* **2021**, *32*, 100925.
170. Rauf, S.; Lahcen, A.A.; Aljedaibi, A.; Beduk, T.; De Oliveira Filho, J.P.; Salama, K.N. Gold nanostructured laser-scribed graphene: A new electrochemical biosensing platform for potential point-of-care testing of disease biomarkers. *Biosens. Bioelectron.* **2021**, *180*, 113116. [CrossRef] [PubMed]
171. Chugh, B.; Thakur, S.; Singh, A.K.; Joany, R.M.; Rajendran, S.; Nguyen, T.V. Electrochemical sensors for agricultural application. In *Elsevier EBooks*; Elsevier: Amsterdam, The Netherlands, 2022; pp. 147–164.
172. Liu, S.; Shen, Z.; Deng, L.; Liu, G. Smartphone assisted portable biochip for non-invasive simultaneous monitoring of glucose and insulin towards precise diagnosis of prediabetes/diabetes. *Biosens. Bioelectron.* **2022**, *209*, 114251. [CrossRef] [PubMed]

173. Umapathi, R.; Ghoreishian, S.M.; Sonwal, S.; Rani, G.M.; Yu, J.S. Portable electrochemical sensing methodologies for on-site detection of pesticide residues in fruits and vegetables. *Coord. Chem. Rev.* **2022**, *453*, 214305. [CrossRef]
174. Low, S.S.; Pan, Y.; Ji, D.; Li, Y.; Lu, Y.; He, Y.; Chen, Q.; Liu, Q. Smartphone-based portable electrochemical biosensing system for detection of circulating microRNA-21 in saliva as a proof-of-concept. *Sens. Actuators B-Chem.* **2020**, *308*, 127718.
175. Talukder, N.; Furniturewalla, A.; Le, T.A.; Chan, M.T.V.; Hirday, S.; Cao, X.; Xie, P.; Lin, Z.; Gholizadeh, A.; Orbine, S.; et al. A portable battery powered microfluidic impedance cytometer with smartphone readout: Towards personal health monitoring. *Biomed. Microdevices* **2017**, *19*, 36. [CrossRef]
176. Rosati, G.; Urban, M.; Zhao, L.; Yang, Q.; De Carvalho Castro E Silva, C.; Bonaldo, S.; Parolo, C.; Nguyen, E.P.; Ortega, G.; Fornasiero, P.; et al. A plug, print & play inkjet printing and impedance-based biosensing technology operating through a smartphone for clinical diagnostics. *Biosens. Bioelectron.* **2022**, *196*, 113737.
177. Tang, X.-Y.; Li, Y.; Ma, J.; Wang, X.; Zhao, W.; Hossain, A.; Yang, Y. Adenovirus-mediated specific tumor tagging facilitates CAR-T therapy against antigen-mismatched solid tumors. *Cancer Lett.* **2020**, *487*, 1–9. [CrossRef]
178. Kumar, A.; Deep, G. Exosomes in hypoxia-induced remodeling of the tumor microenvironment. *Cancer Lett.* **2020**, *488*, 1–8. [CrossRef]
179. Gai, P.; Ji, Y.; Wang, W.; Song, R.; Zhu, C.; Chen, Y.; Zhang, J.-R.; Zhu, J.-J. Ultrasensitive self-powered cytosensor. *Nano Energy* **2016**, *19*, 541–549. [CrossRef]
180. Wang, M.; Liu, Y.-X.; Shao, B.; Liu, X.; Hu, Z.; Wang, C.; Li, H.; Zhu, L.; Li, P.; Yang, Y. HER2 status of CTCs by peptide-functionalized nanoparticles as the diagnostic biomarker of breast cancer and predicting the efficacy of anti-HER2 treatment. *Front. Bioeng. Biotechnol.* **2022**, *10*, 1015295. [CrossRef]
181. Zoupanou, S.; Volpe, A.; Primiceri, E.; Gaudiuso, C.; Ancona, A.; Ferrara, F.; Chiriacò, M.S. SMILE Platform: An Innovative Microfluidic Approach for On-Chip Sample Manipulation and Analysis in Oral Cancer Diagnosis. *Micromachines* **2021**, *12*, 885. [CrossRef] [PubMed]
182. Tran, H.V.; Ngo, N.; Medhi, R.; Srinoi, P.; Liu, T.; Rittikulsittichai, S.; Lee, T.R. Multifunctional Iron Oxide Magnetic Nanoparticles for Biomedical Applications: A Review. *Materials* **2022**, *15*, 503. [CrossRef] [PubMed]
183. Khoshfetrat, S.M.; Mehrgardi, M.A. Amplified detection of leukemia cancer cells using an aptamer-conjugated gold-coated magnetic nanoparticles on a nitrogen-doped graphene modified electrode. *Bioelectrochemistry* **2017**, *114*, 24–32. [CrossRef]
184. Salahandish, R.; Ghaffarinejad, A.; Naghib, S.M.; Majidzadeh-A, K.; Zargartalebi, H.; Sanati-Nezhad, A. Nano-biosensor for highly sensitive detection of HER2 positive breast cancer. *Biosens. Bioelectron.* **2018**, *117*, 104–111. [CrossRef]
185. Shen, C.; Zhong, L.; Xiong, L.; Liu, C.; Yu, L.; Chu, X.; Luo, X.; Zhao, M.; Liu, B. A novel sandwich-like cytosensor based on aptamers-modified magnetic beads and carbon dots/cobalt oxyhydroxide nanosheets for circulating tumor cells detection. *Sens. Actuators B-Chem.* **2021**, *331*, 129399. [CrossRef]
186. Chen, Y.; Peng, J.; Lai, Y.; Wu, B.; Sun, L.; Weng, J. Ultrasensitive label-free detection of circulating tumor cells using conductivity matching of two-dimensional semiconductor with cancer cell. *Biosens. Bioelectron.* **2019**, *142*, 111520. [CrossRef]
187. Wang, S.; Zhao, X.; Liu, F.-F.; Younis, M.; Xia, X.-H.; Wang, C. Direct Plasmon-Enhanced Electrochemistry for Enabling Ultrasensitive and Label-Free Detection of Circulating Tumor Cells in Blood. *Anal. Chem.* **2019**, *91*, 4413–4420. [CrossRef] [PubMed]
188. Cai, J.; Shen, H.; Wang, Y.; Peng, Y.; Tang, S.; Zhu, Y.; Liu, Q.; Li, B.; Xie, G.; Feng, W. A dual recognition strategy for accurate detection of CTCs based on novel branched PtAuRh trimetallic nanospheres. *Biosens. Bioelectron.* **2021**, *176*, 112893. [CrossRef]
189. Zhang, W.; Chen, H.; Yang, M.; Liao, L. Electrochemical assay for detection of circulating tumor cells based on $LiFePO_4$ as electrochemical probe. *Mater. Lett.* **2020**, *276*, 128219. [CrossRef]
190. Ming, X.; Yu, T.; Jiang, Y.; Tao, R.; Li, J.; Ran, F.; Zhu, T.; Huang, J.; Zhang, J.; Zhang, J.-H.; et al. Multi-vital on-skin optoelectronic biosensor for assessing regional tissue hemodynamics. *SmartMat* **2022**, *4*, e1157.

chemosensors

MDPI

Review

Surface Plasmon Electrochemistry: Tutorial and Review

Zohreh Hirbodvash [1,2] and Pierre Berini [1,2,3,*]

1 Department of Physics, University of Ottawa, 150 Louis Pasteur, Ottawa, ON K1N 6N5, Canada
2 NEXQT Institute, University of Ottawa, 25 Templeton St., Ottawa, ON K1N 6N5, Canada
3 School of Electrical Engineering and Computer Science, University of Ottawa, 25 Templeton St., Ottawa, ON K1N 6N5, Canada
* Correspondence: berini@eecs.uottawa.ca

Abstract: Surface plasmon polaritons (SPPs) are optical surface waves propagating along a metal surface. They exhibit attributes such as field enhancement and sub-wavelength localization, which make them attractive for surface sensing, as they are heavily exploited in surface plasmon biosensors. Electrochemistry also occurs on metal surfaces, and electrochemical techniques are also commonly applied in biosensors. As metal surfaces are integral in both, it is natural to combine these techniques into a single platform. Motivations include: (i) realising a multimodal biosensor (electrochemical and optical), (ii) using SPPs to probe the electrochemical double layer or to probe electrochemical activity, thus revealing complementary information on redox reactions, or (iii) using SPPs to pump electrochemical reactions by creating non-equilibrium energetic electrons and holes in a working electrode through the absorption of SPPs thereon. The latter is of interest as it may yield novel redox reaction pathways (i.e., plasmonic electrocatalysis).

Keywords: surface plasmon; electrochemistry; waveguide electrode; cyclic voltammetry; chronoamperometry; convolutional voltammetry

Citation: Hirbodvash, Z.; Berini, P. Surface Plasmon Electrochemistry: Tutorial and Review. *Chemosensors* **2023**, *11*, 196. https://doi.org/10.3390/chemosensors11030196

Academic Editors: Iulia Gabriela David and Dana Elena Popa

Received: 9 February 2023
Revised: 13 March 2023
Accepted: 16 March 2023
Published: 19 March 2023

1. Introduction

Analyzing biomolecules is required in many fields, ranging from food safety inspection to medical diagnostics [1–3]. Many labelled detection methods have been used to detect biomolecules. These labeled detection techniques include enzyme-linked immunosorbent assay (ELISA) [4,5], colorimetric and fluorescence detection [6–8], polymerase chain reaction (PCR) [9,10], radioactive isotopes (radio-immunoassays) [11], vibrational spectroscopy (e.g., infrared and Raman spectroscopy) [12], and some other techniques [13,14]. Labelled techniques generally create problems due to the use of labels, e.g., the interference of fluorophores with binding kinetics, or changes in fluorescence dynamics due to dye–protein interactions [15]. Problems such as these do not exist in label-free methods to detect biomolecules, such as mass spectrometry (MS) [16], quartz crystal microbalance (QCM) techniques [17], surface plasmon resonance (SPR) [18] or localized surface plasmon resonance (LSPR) [19] techniques, and methods based on long-range surface plasmon polaritons (LRSPPs) [20] or the anomalous reflection from gold [21].

A biosensor is a transduction device that is used to measure biological or biochemical reactions by generating quantitative signals (electrical, thermal, optical) proportional to the concentration of an analyte [2,22,23]. A biosensor has a recognition element (e.g., enzymes, nucleic acids, cells, and micro-organisms or antibodies) to selectively capture the analyte of interest in a sample. Depending on the underlying transducer technology, one can identify several types of biosensors: electrochemical [24,25], piezoelectric [26,27], thermal [28], and optical [29].

The most common type of biosensor is electrochemical [30–33]. Electrodes have an essential role as solid support for the immobilization of biomolecules and electron transfer to/from the redox species. As a result of certain electroactive species undergoing a redox reaction in the system, a voltage or a current is generated. One of the critical aspects of an electrochemical biosensor is the use of enzymes as predominant recognition elements due to their specific binding capabilities and biocatalytic activity [31,32]. In this sensing method, different electrochemical detection techniques such as amperometric, potentiometric, and conductometric techniques are used [33].

QCM biosensors exploit the piezoelectric effect and are the most widespread example of a piezoelectric sensor [34]. In such biosensors, changes in accumulated surface mass density change the resonant frequency, which is monitored in real time, while sensing in gaseous or fluidic environments [35].

Thermometric biosensors use thermometry (measurement of temperature) to monitor biochemical interactions, using, for instance, a thermometer [28] or sensitive thermistors [36], as heat-producing biochemical absorption takes place. Thermometry has also been integrated with ELISA in a new method known as thermometric ELISA (TELISA) [37,38].

Optical biosensors measure an optical parameter such as absorption or refractive index associated with surface mass loading resulting from biomolecular interaction, optical emission such as fluorescence emitted by a label, or Raman scattering specific to a molecular species [39–41]. Optical biosensors can provide highly sensitive, direct, real-time, and label-free detection of many biological and chemical substances [42]. Optical biosensors operate in label-based or label-free mode [42]. In label-based sensors, a fluorescent label is used, and the optical emission produced (or quenched) by a biochemical reaction is monitored. In label-free devices, the detected signal is generated directly by the interaction of the analyte with recognition elements immobilized on the transducer surface [42]. There are different varieties of optical biosensors, such as optrode fibre optics, evanescent-wave waveguides, flow immunosensors, and SPR.

Surface plasmon polaritons (SPPs) are optical surface waves propagating at the interface between a metal and a dielectric. Biosensors based on SPPs are among the most studied and used because of the large overlap of SPPs with an adlayer forming on the metal, and their large surface fields [43]. These features make SPPs very sensitive to changes in the refractive index near the metallic surface. However, propagation losses associated with the absorption of light in the metal is a limitation. This problem can be addressed using symmetric structures based on a thin metal layer to support long-range SPPs (LRSPPs) [44]. LRSPPs require that the top and bottom claddings have a similar index of refraction. Biosensing occurs in an aqueous environment, which limits the choice of materials for the bottom cladding to, for example, low-index polymers such as Teflon or Cytop [20]. The use of such polymers imposes limitations during the fabrication of devices, particularly with regards to thermal processing. An alternative approach consists of using a one-dimensional photonic crystal (1DPC) [43]. A 1DPC can be used on one of the sides of a thin metal layer and can mimic the optical properties of the medium on the other side over a limited wavelength range. The modes supported in such a structure are called Bloch LRSPPs.

A trend in (bio)chemical sensor research is to combine electrochemical techniques with surface plasmons. This combination is natural because both techniques involve reactions occurring on a metal surface. In such systems, surface plasmons can be used to probe the electrochemical double layer or to influence electrochemical reactions by involving energetic carriers that are naturally generated in the working electrode as surface plasmons are absorbed therein [30]. In this paper, we introduce concepts related to optical sensing using SPPs and concepts related to electrochemical sensing in a tutorial-like fashion. We then review some representative literature on the integration of both techniques, aiming to realize multimodal sensors, or to investigate fundamentals by optically probing electrochemical activity or uncovering electrochemical effects due to optical absorption (pumping) of working electrodes. Section 2 introduces concepts of importance to SPPs on planar structures. Section 3 introduces concepts related to electrochemistry, and it non-exhaustively but representatively reviews some of the literature on the integration of electrochemical techniques with surface plasmons. Section 4 then provides a brief conclusion.

2. Surface Plasmons

2.1. Optics of Metals

Metals are traditionally used for circuitry and guiding microwaves or far-infrared EM waves. Metals are often modelled as perfect electric conductors (PECs) in these low-frequency applications because the conduction electrons cancel out the external field, effectively preventing wave propagation through the metal. Field penetration increases as the frequency increases into the near-infrared and visible regions of the electromagnetic spectrum, which increases absorption and dissipation. Metals exhibit complex, frequency-dependent dielectric functions in this region of the spectrum. The metal permittivity can be described by various models. In the following, two well-known models, the Drude-Sommerfeld and Drude-Lorentz models, are briefly explained.

The permittivity of metals at optical frequencies can be modelled simply as a plasma (free electron gas). Optical waves induce electron oscillations relative to the lattice of positive ions. An oscillation can be described by its frequency and interaction with the lattice by the relaxation time, which is inversely proportional to the collision frequency of the free electrons. The expression for the dielectric function of a (non-magnetized) plasma is given by the Drude-Sommerfeld equation [45]:

$$\varepsilon(\omega) = 1 - \frac{\omega_p^2}{\omega^2 + i\gamma\omega} \tag{1}$$

where ω_p is the plasma frequency, γ is the collision frequency that is proportional to the inverse of the relaxation time of the electrons ($\tau = 1/\gamma$) [46], and $\varepsilon(\omega)$ is the relative permittivity. The plasma frequency is given by:

$$\omega_p = \sqrt{\frac{n_e e^2}{m^* \varepsilon_0}} \tag{2}$$

In the above, e is the electronic charge, n_e is electron density, m^* is the effective electron mass in the metal, and ε_0 is the permittivity of free space. The plasma frequency occurs in the visible range for most metals.

The frequency dependent dielectric function of plasma has real and imaginary components:

$$Re\,\{\varepsilon(\omega)\} = 1 - \frac{\omega_p^2}{\omega^2 + \gamma^2} \tag{3}$$

$$Im\,\{\varepsilon(\omega)\} = \frac{\gamma\omega_p^2}{\gamma(\omega^2 + \gamma^2)} \tag{4}$$

The plasma (free electron gas) is metallic if $\omega_p > \omega$ and $\omega \gg \gamma$, as the permittivity is mostly real and negative.

For $\omega_p \leq \omega$, corresponding to short visible wavelengths and the ultraviolet region for most metals, the plasma is dielectric as the real part of permittivity becomes positive. In this region, too, the free electron gas model typically does not match the measured permittivity due to the onset of vertical (inter-band) electron transitions in the metal, which dramatically increase the imaginary part of the permittivity [45]. In this case, the Drude-Sommerfeld model is improved by adding terms [47,48]:

$$\varepsilon(\omega) = \varepsilon_\infty - \frac{\omega_p^2}{\omega^2 + i\gamma\omega} + \varepsilon_{int}(\omega) \tag{5}$$

where ε_∞ is the high-frequency relative permittivity, and $\varepsilon_{int}(\omega)$ is a fitted relative permittivity that models inter-band transitions:

$$\varepsilon_{int}(\omega) = 1 - \frac{\widetilde{\omega}_p{}^2}{(\omega_\circ{}^2 - \omega^2) - i\gamma\omega} \tag{6}$$

In the above, $\widetilde{\omega}_p$ is similar to ω_p of the Drude-Sommerfeld model. The corrected model (5) and (6) is termed the Drude-Lorentz model.

The optical permittivity is usually modelled from experimental data [48,49]. Figure 1 shows the relative permittivity of Au given by the Drude-Sommerfeld and Drude-Lorentz formulas, along with experimental data [50]. The Drude-Sommerfeld equation fits the experimental data only at long wavelengths, $\lambda_0 > 700$ nm, whereas the Drude-Lorentz model captures the measurements accurately for $\lambda_0 > 500$ nm on the long-wavelength side of the absorption peak caused by inter-band transitions. Thus, Au behaves as a good optical metal at wavelengths longer than the absorption edge, $\lambda_0 > 700$ nm, into the infra-red, approaching a perfect electric conductor at long wavelengths beyond about 10 μm.

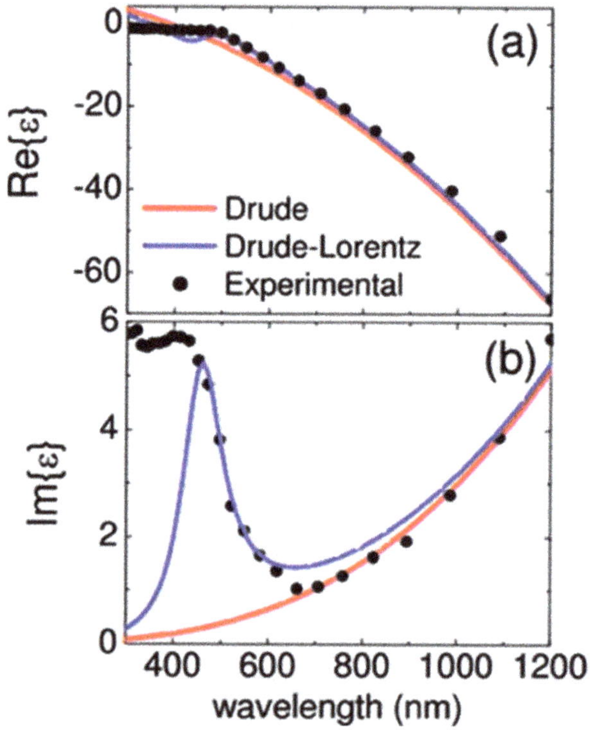

Figure 1. (**a**) Real and (**b**) imaginary parts of the permittivity at optical wavelengths. Black dots correspond to experimental data. The red and blue curves are fitted to the Drude-Sommerfeld and Drude-Lorentz formulas. Reprinted with permission from [50]. Copyright 2011 American Chemical Society.

2.2. *Surface Plasmon Polaritons (SPPs)*

Electromagnetic (EM) waves in the visible and near-infrared ranges can couple, under the right circumstances, to electron oscillations on the surface of a metal. The resulting excitation is termed a surface plasmon polariton (SPP), and it propagates as a surface wave along the metal-dielectric interface with fields that are maximum at the interface and decay exponentially (evanescently) away [45,50–53]. The two main types of SPPs include localized SPPs, which exist as resonant modes on nanostructures, such as metal nanoparticles, and propagating SPPs, which exist on a plane metal surface bounded by a dielectric, as sketched in Figure 2.

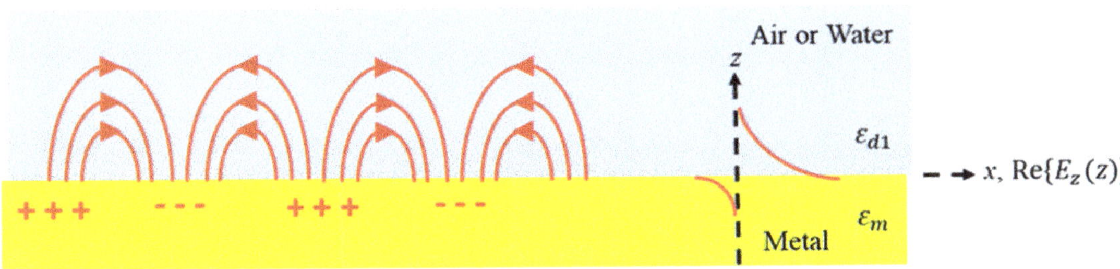

Figure 2. Coupled excitation involving EM fields (curves) and an electron surface charge density wave (+ and −) forming a propagating SPP along a metal-dielectric interface.

As depicted in Figure 2, the coupled excitation involves EM fields that induce a coherent oscillation of surface charges indicated by the "+" and "−" signs. The curved arrows indicate the associated electric fields. The SPP will propagate along the interface in the x-direction with a complex propagation constant K_{spp}, implying an exponential decay with propagation distance due to absorption.

The electric displacement (D_z) must be continuous across the boundary, i.e., $\varepsilon_m E_{z,m} = \varepsilon_d E_{z,d}$, where ε_m is the relative permittivity of the metal, ε_d is the relative permittivity of the dielectric, and E_z is the normal electrical field in both regions. Since $\varepsilon_m \neq \varepsilon_d$, there is a discontinuity at the surface in the electric field, resulting in surface charges at the interface. The real permittivity of a metal at optical frequencies is negative, while the permittivity of dielectrics is positive, resulting in a change in direction of the normal electric field across the interface, which enables the existence of an optical surface wave, in this case the SPP. TE (transverse-electric) polarized waves (s-polarized) have no electric field component perpendicular to the surface, so they cannot induce a polarization. Only TM (transverse-magnetic) polarized waves (p-polarized) can propagate.

Solving the vector wave equations for a planar interface between optically semi-infinite metal and dielectric regions (i.e., a single interface) yields the dispersion equation for propagating SPPs as follows [54]:

$$K_{spp} = \frac{\omega}{c}\sqrt{\frac{\varepsilon_d \varepsilon_m}{\varepsilon_d + \varepsilon_m}} \tag{7}$$

in which ω is the optical angular frequency, and c is the speed of light in free space. Based on the Drude-Sommerfeld model, ε_m can be expressed as Equation (3). For visible and near-infrared light, $\gamma \ll \omega_p$, Equation (3) is often simplified to:

$$\varepsilon_m = 1 - \frac{\omega_p^2}{\omega^2} \tag{8}$$

The dispersion curve of the propagating SPP can be obtained by inserting Equation (8) in Equation (7), yielding Figure 3. SPPs behave like free-space photons at low frequencies, but their dispersion curves move increasingly to the right of the light line as frequency increases. At $\omega_L = \omega_p/\sqrt{2}$, K_{spp} reaches an asymptotic limit. SPPs have greater momentum than incident light at the same frequency, so the dispersion curve bends to the right of the light line, which means that SPPs cannot be excited directly by light. A grating or a prism coupler can be used to increase the momentum of the incident light. Such approaches compensate the momentum mismatch, allowing light to couple into propagating SPPs travelling along the metal-dielectric interface.

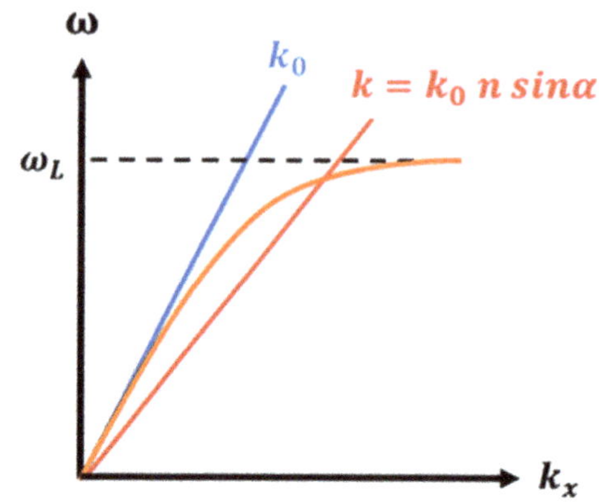

Figure 3. Dispersion plot of a propagating SPP showing the problem of momentum mismatch between illuminating light (blue line) and the SPP (orange curve). It is necessary to overcome momentum mismatch in order to couple the incident light to the SPP, because the SPP is on the right side of the light line, having higher momentum (K_{spp}) than a free space photon (k_0) of the same frequency. In order to provide additional momentum, one can use evanescent wave coupling in total internal reflection from a prism, $k = k_0 n \sin\alpha$ (red line).

Figure 4a sketches the real part of the normal electric field profile of the single-interface SPP on a metal of relative permittivity ε_m bound by a dielectric of relative permittivity ε_{d1} [47]. The field is observed to peak at the interface and decay exponentially on each side. A single-interface SPP propagates along the surface until its energy is dissipated via absorption in the metal or scattered. This leads to a propagation length typically of the order of 10 to 100 μm at visible and near-infrared wavelengths. The metals most useful to support SPPs over this wavelength range are Al and Ag, with Au and Cu being limited primarily to the near-infrared [55,56].

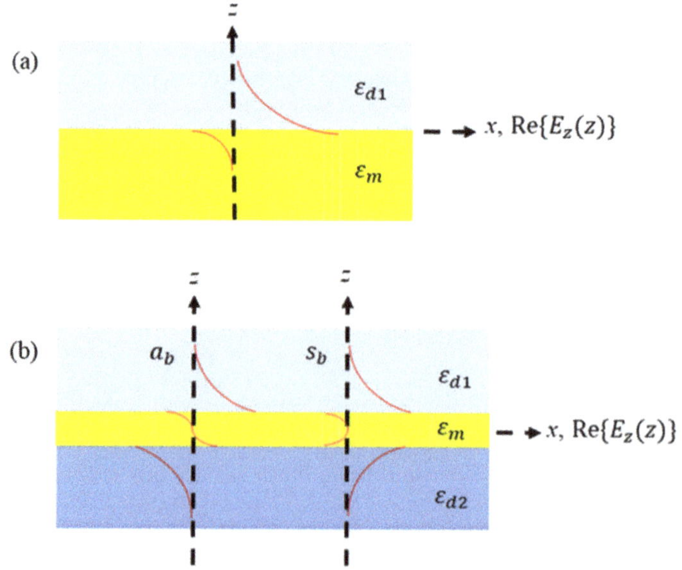

Figure 4. Sketch of 1D normal electric field profiles (real part), $Re\{E_z(z)\}$, for (**a**) the single-interface SPP, and (**b**) two coupled SPP modes on a thin metal film, a_b and s_b. The propagation direction may be taken along the y axis.

Figure 4b illustrates a metal slab of finite thickness bound by dielectrics on both sides. The SPPs supported by the top and bottom interfaces couple through the metal to become coupled modes (supermodes). They are described by symmetric (s_b) or asymmetric (a_b) transverse electric field profiles, as sketched in Figure 4b.

An interesting condition that occurs for the modes on the metal slab if the permittivity of the bounding dielectrics is lossless (lower-cladding and upper-cladding) is that they are equal ($\varepsilon_{d1} = \varepsilon_{d2}$) [44]. The attenuation of the symmetric mode drops as the metal thickness decreases and as the fields therein are expelled. Conversely, the attenuation of the asymmetric mode increases as the metal thickness decreases because confinement to the metal increases. On a thin metal slab, the symmetric mode is termed the long-range surface plasmon polariton (LRSPP) as they have a lower attenuation (10× to 100× lower) than the asymmetric mode and the corresponding single-interface SPP. A lower attenuation allows for longer optical interaction lengths with the analyte in sensor applications.

The metal slab sketched in Figure 4b can be limited to a width w, forming a metal stripe, as depicted in Figure 5, which introduces lateral (horizontal) confinement. This limitation dramatically changes the modal solutions, adding modes to the system that are symmetric or asymmetric about the z-axis. This leads to four fundamental modes, including one that is symmetric about both the x and z axes, referred to as the ss_b° mode [57]. If the permittivity of the lower cladding equals that of the upper cladding, the ss_b° mode exhibits vanishing attenuation as the metal stripe vanishes ($w, t \to 0$) and is referred to as the LRSPP mode of the metal stripe waveguide [57]. Furthermore, the symmetric LRSPP profile is amenable to efficient end-fire excitation by an incident Gaussian beam or by butt-coupling to an optical fibre [44]. The metal stripe waveguide enables integrated optical structures such as straight waveguides, bends, splitters, and Mach-Zehnder interferometers, useful for sensing applications [58–60].

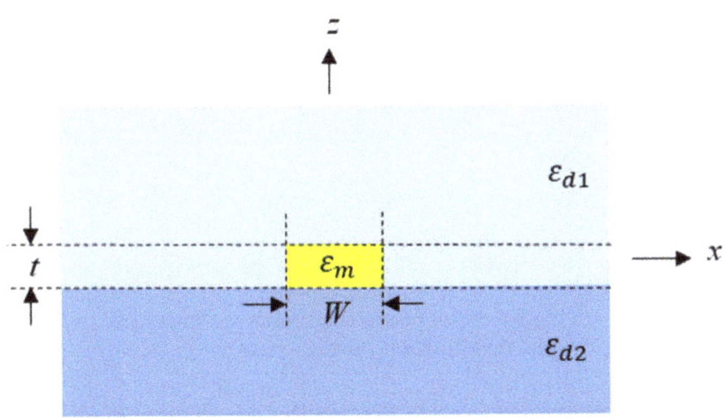

Figure 5. Metal stripe of thickness t and width w bound by dielectrics. The substrate and cladding are generally not the same, but $\varepsilon_{d1} = \varepsilon_{d2}$ is required for LRSPP propagation. The propagation direction is along the y axis.

The lateral confinement changes the modal solutions dramatically, and modal analysis requires the use of numerical methods. Numerical methods for modal calculations include the method of lines (MoL) [61,62], the finite element method (FEM) [62], or the finite difference method (FDM) [63].

For the waveguide structures shown in Figures 4 and 5, the SPP mode propagates along y following an $e^{-\gamma y}$ dependence, where $\gamma = \alpha + jK_{spp}$ is the complex propagation constant, K_{spp} is the phase constant, and α is the attenuation constant ($e^{+j\omega t}$ time harmonic form implied). The mode power attenuation (MPA) [dB/m], computed from α, is [46]:

$$MPA = 20\alpha \log_{10} e \quad (9)$$

The mode power is reduced by a factor of $1/e$ at a distance from the launch point defined as the propagation length of the mode (L_e) [46]:

$$L_e = \frac{1}{2\alpha} \quad (10)$$

2.3. Excitation of Surface Plasmon Polaritons

To excite SPPs, both energy and momentum (phase matching) conservation must be fulfilled. Phase matching requires that the propagation constant of the input light ($k = \omega/c$) be equal to the phase constant of the SPP (K_{spp}). Figure 3 shows that K_{spp} is always larger than k for single-interface SPPs, so the direct excitation of SPPs by light is not possible.

The most common technique to excite single-interface SPPs is via prism coupling, where the attenuated total reflection (ATR) condition is indicative of SPP excitation at the metal dielectric interface. The Otto configuration, sketched in Figure 6a, is one prism-coupling method used to excite SPPs [64,65], where a dielectric gap exists between the prism and the metal surface. The Kretschmann–Raether configuration, sketched in Figure 6b, is another technique for exciting SPPs, and is more convenient as the metal film is deposited directly onto the base of the prism [66,67]. A prism-coupled system based on the Kretschmann–Raether configuration is the most common method of exciting SPPs in SPR sensors [68–70].

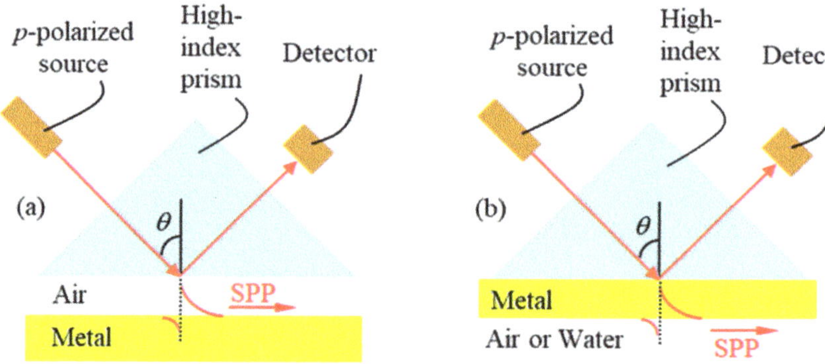

Figure 6. Excitation of SPPs using (**a**) Otto and (**b**) Kretschmann–Raether configurations.

In the Kretschmann geometry, the momentum matching condition for the excitation of SPPs on the metal film is described as:

$$k_x = \left(\frac{2\pi}{\lambda}\right) n_p \sin\theta = Re\{K_{spp}\} \quad (11)$$

where k_x is the in-plane wavenumber of the incident light at incident angle θ that can couple into SPPs, n_p is the refractive index of the prism, and λ is the free-space wavelength.

Prism coupling has been employed extensively to excite SPPs at a metal liquid interface, where SPPs were used to address a large number of biodetection problems, e.g., bulk refractive index sensing [71], biosensing of pesticides in environmental applications [72], sensing the effects of toxins on cells [73], sensing involving two parameters [74], biosensing proteins [75], monitoring cellular motion within fibroblast cells [76], detecting *E. coli* bacteria [77], detecting pathogens in food [78], and for sensing Bovine Serum Albumin (BSA) [79], cardiac troponin [80], Benzo[a]pyrene (BaP) [81], and pituitary hormones [82].

LRSPP excitation can be performed either by a prism coupled system based on attenuated total reflection (ATR), or, as discussed in the previous sub-section, in an end-fire coupling scheme using a TM-polarized Gaussian beam or a polarization-maintaining single mode fibre (PM-SMF) (butt-coupling) [83,84]. In a prism-coupled system, the incident angle can be changed beyond the critical angle and the reflected power is monitored using a photodetector as shown in Figure 7.

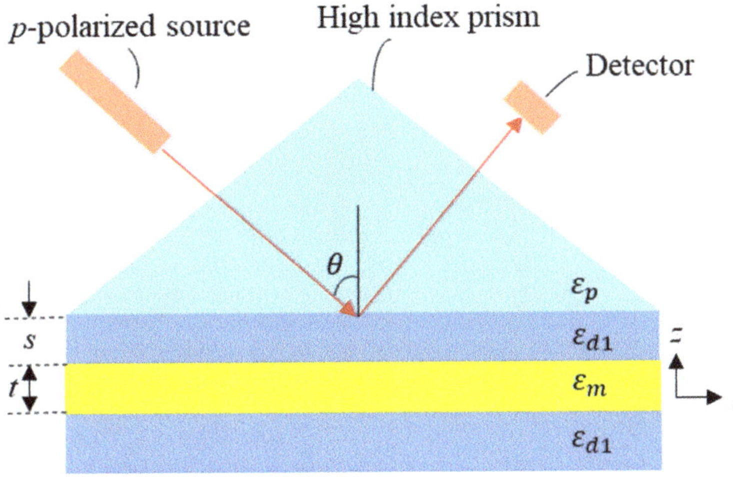

Figure 7. Prism coupling technique for exciting LRSPPs; ε_p is the relative permittivity of the prism ($=n_p^2$) and ε_{d1} is the relative permittivity of the dielectrics bounding the metal, the outer one being the sensing solution.

The in-plane wavenumber of the incident beam is equal to the wavenumber of the propagating LRSPP mode at a certain angle, and a drop in the reflected power is observed (similar to the Kretschmann–Raether configuration for single-interface SPPs). In this configuration, a thin metal slab is bound by two dielectric (insulator) layers, forming an insulator-metal-insulator (IMI) structure that can support LRSPPs. For sensing applications, the dielectric layer separating the metal film from the prism should have a relative permittivity of ε_{d1} matching that of the aqueous sensing fluid on the other side of the film.

Another method of exciting SPPs and LRSPPs is with grating couplers. Figure 8 shows a grating, formed of periodic bumps of period Λ. Grating couplers are designed and realized to satisfy momentum conservation:

$$\Lambda = m(\lambda n_{eff} - n_c \sin\theta) \tag{12}$$

where n_c is the refractive index of the medium of incidence, λ is the free-space wavelength, m (integer) is the order of the grating, θ is the angle of incidence, and n_{eff} is the average effective index of the SPP propagating along the waveguide bearing the grating.

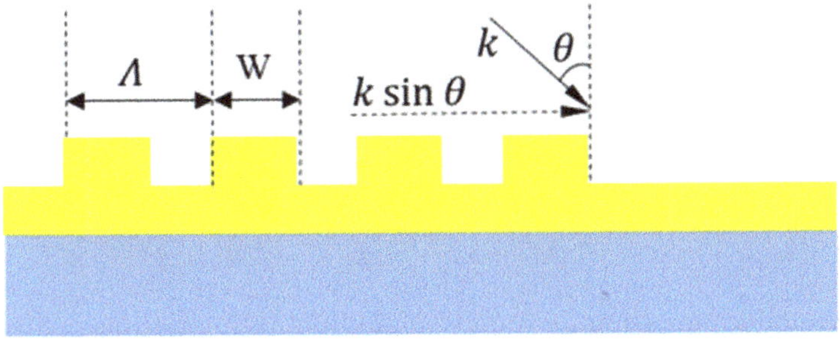

Figure 8. Geometry of a grating coupler for SPPs as a step-in-height metal pattern.

Figure 8 illustrates a metal grating, but the grating can be patterned in another material such as a dielectric. Grating couplers operate by scattering incoming light of wavenumber k at the angle of incidence θ into SPPs of phase constant K_{spp}. Using Equation (12), the phase matching condition (momentum conservation) is written:

$$K_{spp} = k \sin\theta \pm m\frac{2\pi}{\Lambda} \quad (13)$$

Grating couplers are easy to align and excite but the coupling efficiency is limited due to the nature of the structure. This approach has been applied to metal stripes embedded in Cytop [85], and to metal stripes on a truncated photonic crystal [43,86]. As an alternative to raised metal bumps, nanoholes can be milled into the metal film and used for similar purposes, or support resonances therein [87,88].

As mentioned previously, end-fire coupling is a particularly apt scheme for exciting LRSPPs on the metal stripe waveguide (*cf.* Figure 5) because the mode fields of the LRSPP, Figure 5, overlap very well with those of a PM-SMF or with an incident Gaussian beam [44]. In such an excitation scheme, the coupling efficiency can be obtained by estimating the overlap factor C of the LRSPP mode with the source field:

$$C = \frac{\iint E_{z1} \cdot E_{z2}^{*}\, dA}{\sqrt{\left(\iint E_{z1} \cdot E_{z1}^{*}\, dA\right)\left(\iint E_{z2} \cdot E_{z2}^{*}\, dA\right)}} \quad (14)$$

which is computed from the spatial distribution of the main transverse electric field component of the LRSPP, E_{z1}, and of the source field, E_{z2}. The mode power coupling efficiency is given by $|C|^2$ if there is no discontinuity in the materials at the coupling plane. The mode power coupling loss (C_{pl}) is expressed in dB:

$$C_{pl} = -20\ \log_{10}|C| \quad (15)$$

Butt coupling to a PM-SMF can produce coupling efficiencies to LRSPPs on a metal stripe greater than 90%, and has been heavily used to excite biosensors based on such waveguides [20].

2.4. Bloch Long-Range Surface Plasmon Polaritons (Bloch LRSPPs)

In biosensing applications based on metal stripe waveguides (Figure 5), the lower cladding material must be selected in such a way that its refractive index is close to that of the sensing fluid, which also acts as the upper cladding. This is needed to ensure that the LRSPP is supported, because a symmetric dielectric environment is required near the metal stripe, and to ensure that any microfluidic channels etched into the cladding to define the sensing region are non-invasive optically once they are filled with sensing fluid. Sensing fluids in biosensors are generally aqueous, e.g., high-purity buffer-carrying analyte, or a patient sample such as urine or blood diluted in the buffer. The material used for the cladding should have a refractive index close to that of water, e.g., Cytop [20,89] or Teflon [90].

An alternative approach consists of using a truncated one-dimensional photonic crystal (1DPC) as a substrate [43]. This approach has the advantage of material flexibility because several inorganic materials can be used to implement the stack. A 1DPC can be used as a lower cladding because it can be designed to mimic the optical properties of the medium on the other side of the metal stripe or slab over a range of wavenumber and wavelength. A mode supported in this structure is termed a Bloch LRSPP.

Figure 9a shows such a structure. The 1DPC is formed on a Si substrate and covered by an optically semi-infinite Cytop layer or by the sensing medium. The 1DPC depicted was formed as a SiO_2/Ta_2O_5 multilayer stack, designed to tailor the field decay into the stack and minimize losses from light tunnelling into the Si substrate [43].

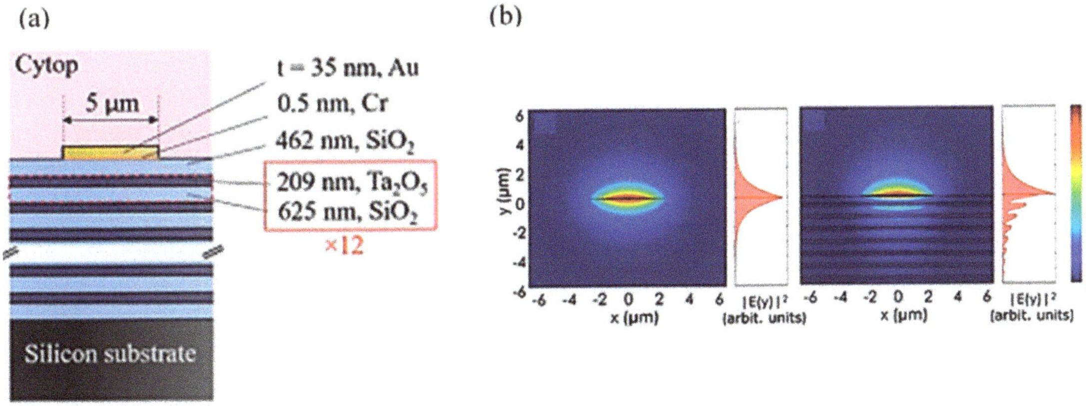

Figure 9. (**a**) Au stripe on a truncated 1DPC covered by Cytop, supporting Bloch LRSPPs. (**b**) Profiles of the squared magnitude of the perpendicular electric field of the LRSPP on a Au stripe buried in Cytop (**left panels**), and of the Bloch LRSPP on a Au stripe on a truncated 1DPC covered in Cytop (**right panels**). Reprinted with permission from [43]. Copyright 2017 American Chemical Society.

The distribution of the squared magnitude of the perpendicular electric field component of the LRSPP in the corresponding Cytop embedded structure is shown on the left panels of Figure 9b, and of the Bloch LRSPP on the truncated 1DPC on the right panels of Figure 9b. The field distributions extend similarly into the claddings, with the exception that the Bloch LRSPP has an oscillatory character as it decays into the IDPC. Both modes present essentially the same surface sensitivities as they overlap similarly with a thin adlayer on the metal stripe [91,92].

Figure 10 illustrates schematically an excitation scheme for Bloch LRSPPs consisting of exciting an Au grating coupler on a Au stripe with a *p*-polarized Gaussian beam launched using an aligned PM-SMF. Such an excitation scheme has the advantage of not requiring a high-quality input facet as well as simpler optical alignments in comparison to butt-coupling. However, typically, the coupling efficiency is lower than end-fire coupling in such a scheme. The grating consists of rectangular ridges of width W and height H defined over the period Λ. The grating was modeled with the 2D FEM in the frequency domain using commercial software, and the coupling efficiency was calculated by computing the power carried by the Bloch LRSPP and normalising it to the incident power, as described in [93]. A coupling efficiency of 16% was deduced at $\lambda_0 = 1310$ nm over about 70 nm of optical bandwidth [43].

2.5. Optical Interrogation of Surface Plasmon Biosensors

Four primary optical interrogation methods can be applied to a surface plasmon sensor [94], depending in part on the excitation scheme used—prism, grating, or end-fire coupling. Firstly, phase interrogation, where a phase shift in the collected light is measured indicating a change in refractive index, with the wavelength and excitation conditions remaining constant—this scheme must be incorporated into an interferometer to convert phase changes to changes in intensity. Secondly, angular interrogation, where a single wavelength (monochromatic) laser is used and the shift in angle of SPP coupling is measured indicating refractive index changes—this scheme, commonly termed SPR, requires a prism or grating, and, fundamentally, also interrogates phase, albeit by varying the coupling conditions. Thirdly, spectral interrogation uses a broadband source with a spectrograph or a tunable laser with a power sensor, measuring a shift in the resonant or coupling wavelength in reflection or transmission to examine the change in refractive index—this scheme is broadly applicable using prism, grating, or end-fire coupling. Fourthly, intensity interrogation measures the change in transmitted or reflected light intensity, while keeping other excitation conditions constant.

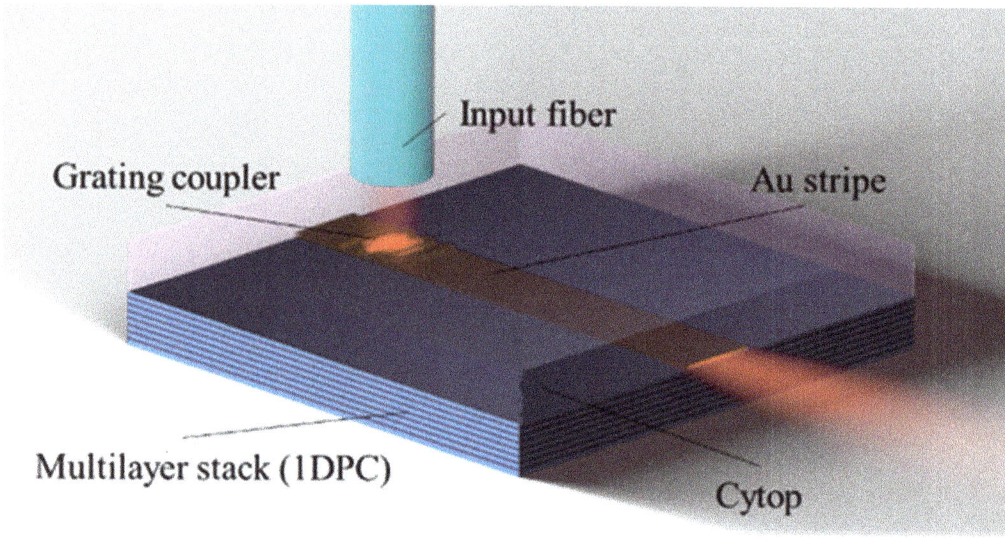

Figure 10. Isometric sketch of a metal stripe waveguide with a grating coupler on a 1DPC, illustrating an excitation arrangement involving a polarisation-maintaining single-mode fibre. Reprinted with permission from [43]. Copyright 2017 American Chemical Society.

Structures for sensing applications can be characterized by the surface sensitivity, the bulk sensitivity, and the figure of merit (FOM). The bulk sensitivity is defined as a change in measurand, e.g., a coupling or resonant wavelength λ_{res} vs. the change in the refractive of the bounding medium (e.g., $\partial\lambda_{res}/\partial n_c$). By changing the refractive index of the bounding medium from a nominal value to another, step by step, one can determine the bulk sensitivity.

The metal surface in a biosensor is functionalized chemically with a receptor chemistry to react selectively with the target analyte, which is in contact with the sensing solution, also operating as the optical bounding medium (cladding). As analyte binds to the receptors, an adlayer of refractive index higher than the sensing solution forms, causing the measurand to change. The surface sensitivity is defined as the change in measurand, e.g., a resonance wavelength λ_{res}, as a function of adlayer thickness a forming at the metallic/solution interface (e.g., $\partial\lambda_{res}/\partial a$).

Conventional SPR biosensors based on the Kretschmann–Raether configuration use a glass prism to excite SPPs. An immobilized bio-recognition element is coated on the metal surface as shown in Figure 11. Plasmon excitation occurs at a specific angle, termed the SPR angle. At this angle, the incident light is coupled to SPPs, which leads to a decrease in the intensity of the reflected beam. The refractive index of the sensing solution and the presence of an adlayer at the metal/solution interface determine the SPR angle. Biochemical reactions change the thickness of the adlayer, which changes the SPR angle required to maintain excitation of SPPs. Changing the refractive index of the dielectric medium on the other side of the metal film, or forming an adlayer thereon, will result in significant changes in the SPP coupling angle (Equation (11), as K_{spp} is altered). Plotting the intensity vs. the incidence angle over time produces a sensorgram from which binding kinetics can be extracted by tracking features in the angular response, such as the coupling angle (minimum reflected intensity), as shown on the left panel of Figure 11 [68,69].

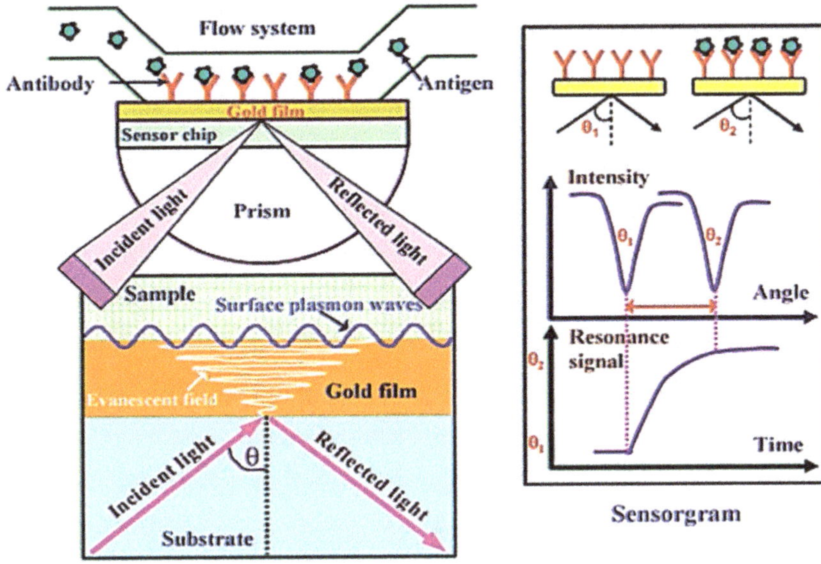

Figure 11. Schematic illustrating the SPR immunoassay technique. Adapted with permission from [69]. Copyright 2007 Elsevier.

Wavelength interrogation is also possible, where the excitation of SPPs is manifested by the appearance of a dip in the reflectance spectrum at a fixed angle of incidence. The full width at half-maximum (FWHM) of such a dip is typically ~50 nm [70]. The reflected field near the dip also undergoes a phase change. SPR biosensors conventionally exploit single-interface SPPs, which have a high attenuation and broad resonance conditions.

3. Electrochemical Surface Plasmon Sensors

A recent trend in (bio)chemical sensors is to combine electrochemical techniques with surface plasmon sensors. Such a merger is natural because both techniques exploit reactions occurring on a metal surface, often Au due to its inertness, in both types of system. In such systems, surface plasmons can be used to probe the electrochemical double layer or to influence electrochemical reactions by involving energetic carriers that are naturally generated in the working electrode as surface plasmons are absorbed therein. For instance, electrochemical surface plasmon resonance (EC-SPR) probes faradaic processes by monitoring the change in refractive index that happens with the change in redox state at the electrode surface in electrochemical Kretschmann SPR systems [95]. EC-SPR is an important application of SPR to study local electrochemical reactions on the surface of the electrode. We briefly discuss cyclic voltammetry before reviewing the literature on such systems.

3.1. Cyclic Voltammetry

Cyclic voltammetry (CV) is an electrochemical technique that probes the reduction and oxidation reactions of a redox species [96]. Figure 12 shows cyclic voltammograms resulting from the application of this technique [97]. Here, the x-axis corresponds to the potential applied to the system (V), and the y-axis represents the response (measurand), which is the resulting current (C). CV data are commonly reported using two conventions, but the sign convention used to obtain and plot the data is rarely stated. As shown in Figure 12, the potential axis clarifies the convention.

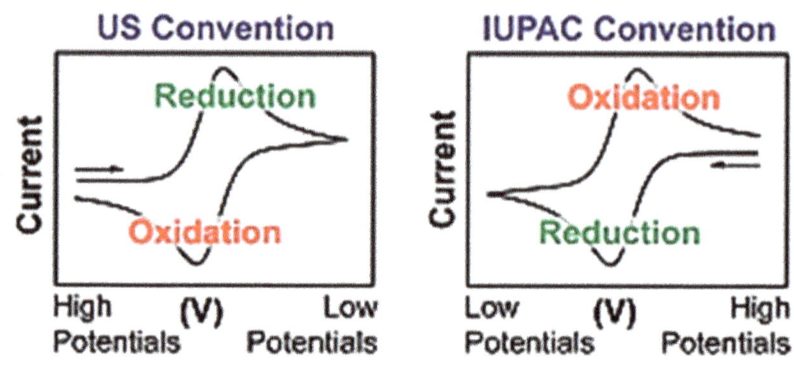

Figure 12. Schematic voltammograms. To report CV data, US and IUPAC conventions are commonly used. The data reported in the two conventions appear rotated by 180°. Reprinted with permission from [97]. Copyright 2018 American Chemical Society.

During a CV experiment, the potential is varied linearly at a rate of a few millivolts per second. This parameter is called the scan rate and it is one of the most important parameters in cyclic voltammetry. Figure 13a shows the relation between time and applied potential, with the potential plotted on the *x*-axis, commensurate with the corresponding voltage voltammogram of Figure 13b. In Figure 13a, the potential is swept in the positive direction during the forward scan, from the initial potential E_1 to the switching potential E_2, forming the anodic portion of the trace. The scan direction is then reversed in such a way that the potential is swept in the negative direction to E_1, forming the cathodic portion of the trace.

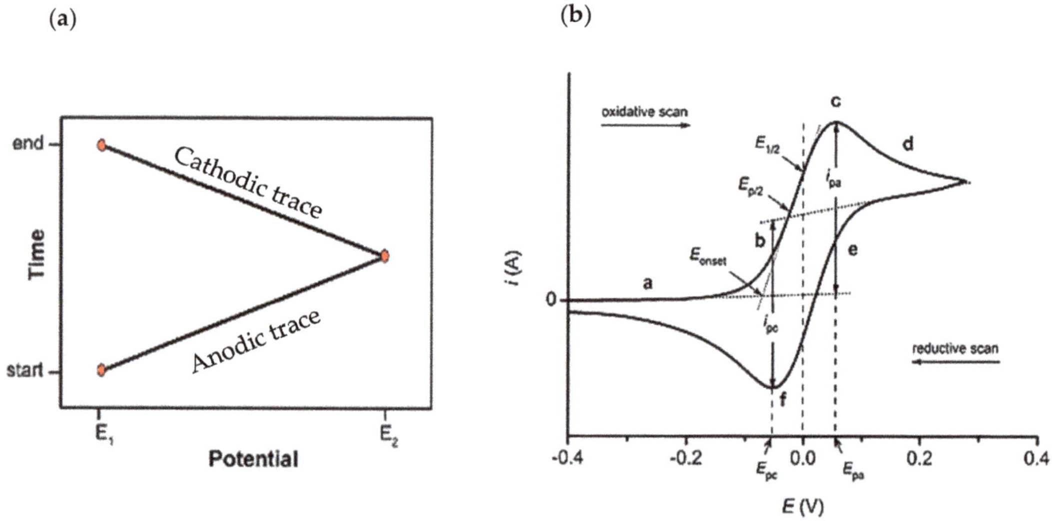

Figure 13. (**a**) Applied potential vs. time in a cyclic voltammetry experiment, with the starting, switching, and ending potentials identified. (**b**) Voltammogram of a reversible reaction (IUPAC convention). Reprinted with permission from [97]. Copyright 2018 American Chemical Society.

The Nernst equation holds at equilibrium between the oxidized and reduced species [97]:

$$E = E^{\circ} + \frac{RT}{nF} \ln \frac{([OX])}{([Red])} = E^{\circ} + 2.3026 \log_{10} \frac{([OX])}{([Red])} \quad (16)$$

The Nernst equation relates the relative concentrations of the oxidized and reduced species in the system at equilibrium to the standard potential of the redox species (E°) and the potential of the electrochemical cell (E). [Red] and [Ox] are the concentration of the reduced and oxidized species, T is the temperature, R is the universal gas constant, F is Faraday's constant, and n is the number of electrons transferred in the reaction.

The formal potential is specific to the experimental conditions and is often determined by $E_{1/2}$, the average of potentials f and c on Figure 13b. The Nernst equation predicts how a system will respond to a change in concentration of the redox species or a change in electrode potential. For $E = E^{\circ} = E_{1/2}$, the Nernst equation predicts that the oxidized species will be reduced until the concentration of the oxidized and reduced species are equal at equilibrium. The concentration of the species near the electrode changes following the Nernst equation as the potential is varied during a CV experiment.

Reduction locally at the electrode occurs when a solution of oxidized species is scanned to negative potentials (cathodic direction) from the rightmost point in Figure 13b, resulting in a current and in depletion of the oxidized species at the electrode surface. At potential f shown on Figure 13b, where the maximum cathodic current (i_{pc}) is observed, the current is limited by the diffusion of additional oxidized species from the bulk solution. During the scan, the diffusion layer continues to grow at the surface of the electrode containing the reduced species. As a result, mass transport of the oxidized species slows. Therefore, upon scanning to more negative potentials, the diffusion of oxidized species from the bulk to the electrode surface slows, thereby decreasing the current.

When the minimum potential is reached, the scanning direction is reversed to scan the potential in the positive (anodic) direction. At the electrode surface, the concentration of reduced species then decreases while the concentration of oxidized species increases, satisfying the Nernst equation. Following the Nernst equation, at $E = E_{1/2}$, the concentrations of reduced and oxidized species at the electrode's surface are equal. These concentrations occur at the two points corresponding to the potential between the two peaks, oxidation—c and reduction—f, from which E° is estimated for a reversible 1-electron transfer reaction. The two peaks are separated (not coincident) due to diffusion of the redox species to and from the electrode.

When the reaction on the electrode is limited by diffusion, a plot of the peak currents (anodic or cathodic), i_p, vs. the square root of the scan rate (ϑ), satisfies the Randle–Sevcik equation [83]:

$$i_p = 0.4463\, nFAC\left(\frac{n\, F\vartheta D}{RT}\right)^{1/2} \tag{17}$$

where A is the electrode area, D is the diffusion coefficient, i_p is the peak current, ϑ is the scan rate, and C is the concentration of the redox species.

3.2. Electrochemical Surface Plasmon Resonance (EC-SPR)

An interesting study of SPR to probe electrochemical reactions was reported by Wang et al. [98]. In this work, EC-SPR led to a new way to measure convolution voltammetry directly without the need for numerical integration of the electrochemical current response. With convolutional voltammetry, it is possible to determine diffusion constants, bulk concentrations, and the number of electrons transferred between electroactive species.

In EC-SPR, the output optical signal is proportional to the time convolution of the electrochemical current density [98]:

$$\Delta\theta(t) = B(\alpha_R D_R^{-1/2} - \alpha_0 D_0^{-1/2})\left(nF\pi^{\frac{1}{2}}\right)^{-1} \int_0^t i(t')\,(t-t')^{-1/2}\,dt' \tag{18}$$

where $\Delta\theta = \theta(t) - \theta_0$ measures the changes in SPR resonance (coupling) angle. $\theta_0 = B(\alpha_0 C_0^0 + \alpha_R C_R^0)$ is the SPR angle at $t = 0$ and D_0 and D_R are the diffusion coefficients of the redox species. The constant B is the bulk sensitivity given by the change in

SPR angle relative to the change in the bulk refractive index, and can be determined for a particular SPR system and redox species.

The experiments reported used a prism-based SPR imaging setup as illustrated in Figure 14A. This experiment utilized a BK7 prism and a red collimated LED (wavelength 670 nm) as the light source, with a high-speed CCD camera for detection. An Au-coated microscope slide was used as the SPR sensing surface, placed on the prism using intervening index matching fluid. A Pt wire counter electrode and a Ag/AgCl reference electrode were inserted into the electrolyte through an opening in the electrochemical cell.

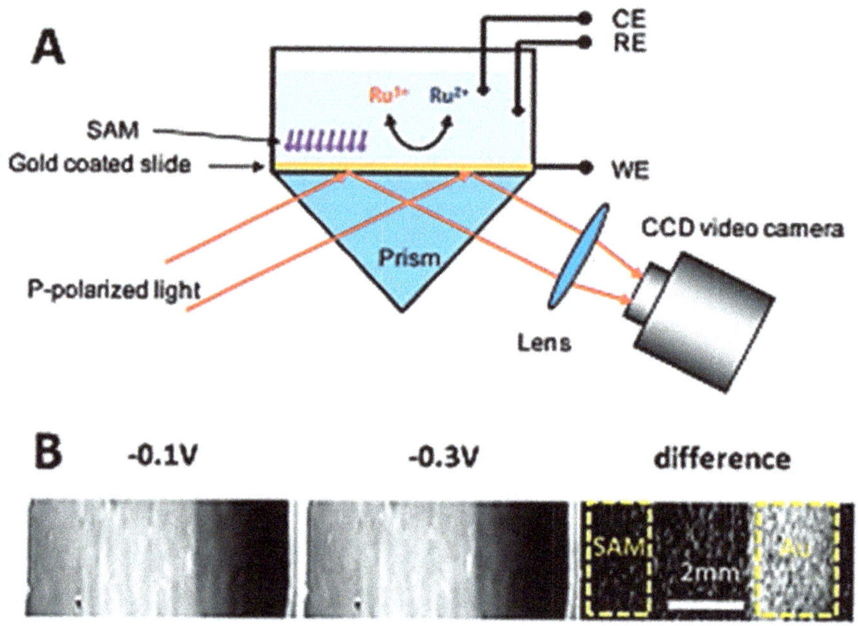

Figure 14. (**A**) Sketch of the SPR setup. (**B**) SPR images in 11.78 mM Ru $(NH_3)_6^{3+}$ dissolved in phosphate buffer (0.5 M pH 7), at potentials of −0.1 and −0.3 V (vs. Ag/AgCl), and their difference. The Au (bare) and SAM-coated regions used for data extraction and processing are highlighted by yellow squares. Reprinted with permission from [98]. Copyright 2010 American Chemical Society.

In Figure 15A, the cyclic voltammograms are plotted for three scan rates, 0.01, 0.1, and 1.0 V/s, indicating that the peak current increases by 10× when the scan rate is increased from 0.01 to 1.0 V/s. The SPR voltammograms recorded simultaneously at different scan rates, plotted in Figure 15B, show that the SPR response is not strongly dependent on the scan rate, as expected from the theory (*cf.* Equation (18)). Figure 15C plots the computed SPR voltammograms using the measured current responses of Figure 15A, yielding reasonable agreement with the directly measured SPR voltammograms plotted in Figure 15B.

Another study involving EC-SPR investigated how electrochemical reactions affect the metal/liquid interface to shift the reflection minimum. A prism coated with a Ag film was used in an EC-SPR configuration with halide and perchlorate electrolytes to probe optically the electrochemical reactions [99]. As the electrode potential becomes increasingly positive, the resonance shifts to a smaller wavevector at a fixed wavelength, as shown in Figure 16. Changes in ion adsorption, in the optical properties of the ionic double layer, and in electron density at the metal surface are some of the factors contributing to this shift.

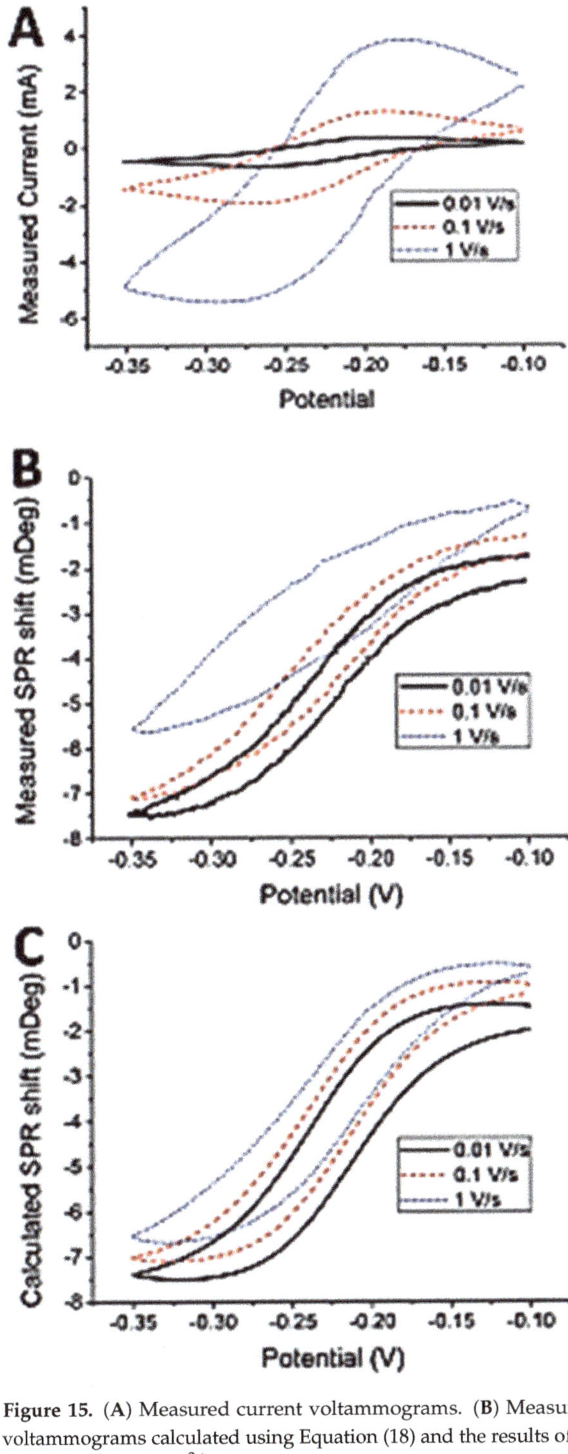

Figure 15. (**A**) Measured current voltammograms. (**B**) Measured SPR voltammograms. (**C**) SPR voltammograms calculated using Equation (18) and the results of panel (**A**) as inputs. The electrolyte was 3 mM Ru $(NH_3)_6^{3+}$ in phosphate buffer and the electrode/SPR surface was bare gold. Reprinted with permission from [98]. Copyright 2010 American Chemical Society.

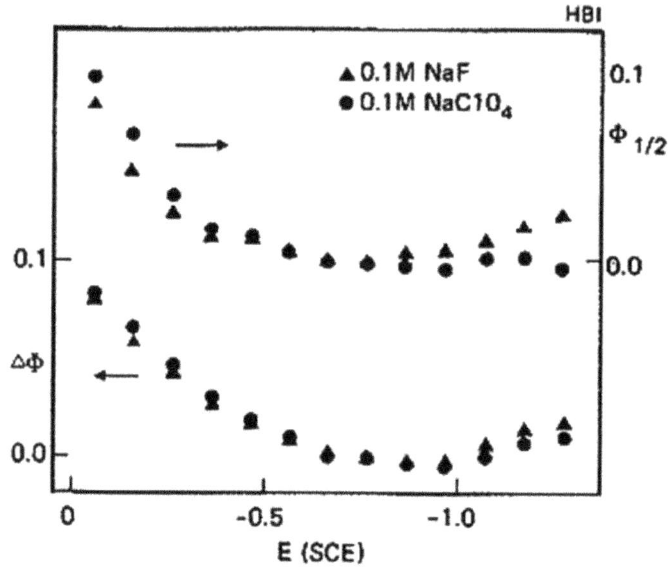

Figure 16. Shift in SPR peak position and width vs. potential for a silver coated prism in 0.1 M NaF and 0.1 M $NaClO_4$. The shifts are relative to the position and width at −0.76 V [99]. Reprinted with permission from [99]. Copyright 1980 Elsevier.

EC-SPR was also demonstrated to be a highly effective tool for detecting and measuring intermediates, and it is being explored as a potential tool for studying reaction kinetics [100–102]. The semiquinone radical anion BQ^- was detected in the hydroquinone-benzoquinone ion system by producing a large negative shift in a flow-through EC-SPR system, as shown in Figure 17 [100]. This study shows that flow-through EC-SPR can be used to monitor chemical reactions in solution in addition to surface interactions.

During potentiostatic oxydoreduction, an EC-SPR biosensor with an absorptive redox mediator film, was used to detect reversible refractive index changes in the film [101]. Using the organic dye Methylene Blue (MB) as an electroactive label, this work examined the theoretical and experimental foundations of EC-SPR sensing. Electrochemical influence on the SPR response is dependent on the local MB concentration and can be used to design highly selective and sensitive biosensing systems [101].

3.3. Energetic (Hot) Charge Carriers in EC-SPR

Energetic (hot) charge carriers refer to either photoexcited holes or electrons that exist in non-equilibrium high-kinetic-energy states in photoactive materials, e.g., metals and semiconductors, after being exposed to photons [103,104]. Photoexcited, non-equilibrium hot carriers in metallic structures could lead to bandgap-free photodetection and selective photocatalysis [105]. However, hot carrier devices must be significantly improved to meet practical application requirements. A promising pathway to increase the efficiency of these systems is to involve the excitation of SPPs [105].

The energy in an SPP is dissipated as free-space radiation (radiative loss) through scattering, or as absorption (nonradiative decay) in the metal. The absorption of SPPs produces energetic carriers—electrons and holes—in the metal that are not in thermal equilibrium with the lattice. These non-equilibrium hot carriers enable energy-harvesting applications in photovoltaics, photodetection, photon up-conversion, and photocatalysis [103–108].

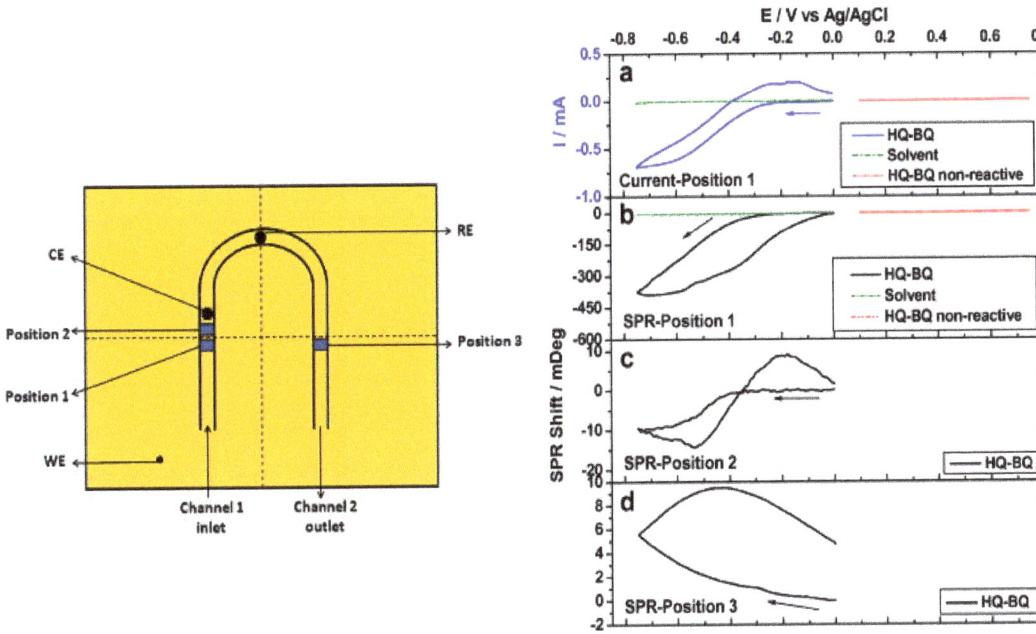

Figure 17. (**Left panel**): Top-view sketch of a Au-coated sensing chip with two flow channels connected in a U-shape and the electrodes. The width, height and length of each channel are, respectively, 2.0, 0.5, and 5 mm. The working electrode (WE) is connected to the sensing chip by a contact pin. The counter electrode (CE) and reference electrode (RE) are in the flow channel at the locations indicated. The solution is injected into the cell via channel 1 and emerges from channel 2. Position 1 is on the WE, and positions 2 and 3 are downstream. It takes 0.1 and 5 s for the solution to flow from position 1 to positions 2 and 3, for a flow rate of 100 μL/min. (**Right panel**): Characterization of $BQ^{\bullet -}$ by CV-SPR at positions 1, 2, and 3. (**a**) Current of the HQ–BQ and of the solvent (0.1 M Bu_4NPF_6 in acetonitrile), and of HQ–BQ over a non-reactive potential window. (**b**) SPR signal of the HQ–BQ and of the solvent (0.1 M Bu_4NPF_6 in acetonitrile), and of HQ–BQ in a non-reactive potential window at position 1. (**c**) SPR signal of the HQ–BQ at position 2. (**d**) SPR signal of the HQ–BQ at position 3. Flow rate: 100 μL/min. Potential scan rate: 0.1 V/s. Reprinted with permission from [100]. Copyright 2010 Elsevier.

Recent attention has focused on studying energetic carriers created from SPP absorption and their role in electrochemical reactions [30]. Plasmonic electrocatalysis could improve the efficiency of chemical processes, or could enable reaction pathways that are not accessible or difficult to access thermally. However, the details of hot carrier transfer in photochemical processes are nebulous, specifically in processes where hot holes are involved. Using photoelectrochemistry, hot holes and hot electrons can be localized on photoanodes and photocathodes, allowing the investigation of hole-transfer and electron-transfer dynamics in oxidation and reduction reactions separately.

Electrodes in electrochemical setups are classified as anodes on which oxidation reactions occur, and as cathodes on which reduction reactions occur, both of which can be used as plasmonic photoelectrodes. The working principles and structures used for photoanodes (metal or metal/n-type semiconductor) and photocathodes (metal or metal/p-type semiconductor) are briefly discussed given their prevalence in the literature [106,107].

Heterostructures as a metal on an n-type semiconductor have been extensively investigated as photoanodes in plasmonic photocatalysis. The Fermi level is near the conduction band in an n-type semiconductor. When semiconductors come into contact with metals, they donate electrons to the metal to equalize Fermi levels. An upward band bending occurs at metal/semiconductor interface (called a "Schottky barrier"). When hot electrons created in the metal have sufficient energy, they can overcome the Schottky barrier to reach the conduction band of the semiconductor, as depicted in Figure 18a [106]. However, due to upward band bending, they are rapidly swept from the interface into the semiconductor. Such heterostructures are useful as a means of extending the lifetime of energetic electrons, which is longer in a semiconductor than in a metal. Electrons are then carried by the external circuit to the counter electrode, as shown in Figure 18a, where they participate in reduction reactions, e.g., H_2O reduction. Hot holes on the metal surface drive oxidation reactions, e.g., H_2O oxidation, as shown in Figure 18a, producing an anodic photocurrent on the working electrode.

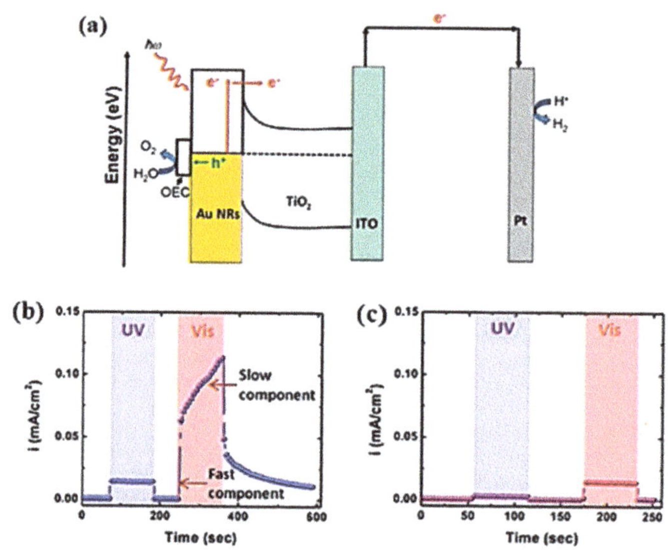

Figure 18. Plasmonic metal/n-type semiconductor photoanode working electrode for redox of water. (**a**) Energy band diagram of a Au/TiO_2 photoanode. Absorption of visible light in Au generates energetic electron-hole pairs therein. Hot electrons emitted from Au into TiO_2 enable water reduction on the counter electrode (acting as cathode), and hot holes in Au are extracted by a Co-based oxygen evolution catalyst (Co-OEC) on Au to enable water oxidation. Current vs. time for Au/TiO_2 (**b**) with Co-OEC, and (**c**) without Co-OEC. Reprinted with permission from [106]. Copyright 2012 American Chemical Society.

Plasmonic photocathodes, which consist of a heterostructures formed of a metal on a p-type semiconductor, as sketched in Figure 19a, have also been reported [107]. Plasmon-generated hot electrons are extracted from the metal to drive reduction reactions thereon, but plasmon-generated hot holes in the metal are captured by the p-type semiconductors. Theoretically, photoexcited holes above the interband edge in Au are considerably hotter than their photoexcited electron counterparts, which suggests a greater collection efficiency for hot holes compared to hot electrons, given similar Schottky barriers. However, due to the comparatively short mean free path of hot holes in a metal and the lack of p-type semiconductors with wide bandgaps, harvesting hot holes from plasmonic metals is a challenge. p-GaN has recently been used as the semiconductor in plasmonic photocathodes, as sketched in Figure 19a. In Au/p-GaN heterostructures, excited holes in the Au nanoparticles transfer into the valence band of p-GaN. During plasmon-driven CO_2 reduction, electrons trapped in the Au nanoparticles contribute to the cathodic photocurrent, as shown in Figure 19b.

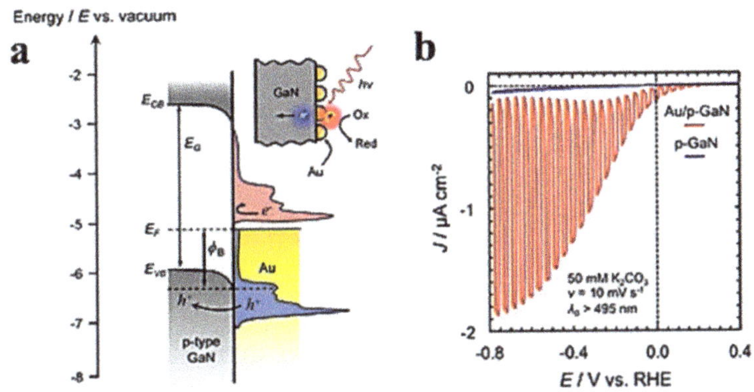

Figure 19. Plasmonic photocathode comprised of a metal on a p-type semiconductor and reduction of CO_2. (**a**) Energy band diagram of a Au/p-GaN photocathode, illustrating the conduction band edge E_{CB}, the valence band edge E_{VB}, the Fermi level energy E_F, the bandgap energy E_G and the Schottky barrier potential ϕ_B. Plasmon absorption creates hot electrons (red) and hot holes (blue) above and below E_F, respectively. Hot holes with energies larger than $e\phi_B$ can overcome the Schottky barrier to populate available valence band levels in p-GaN. (**b**) Linear sweep voltammetry of Au/p-GaN (red) and bare p-GaN (blue) photocathodes. Reprinted with permission from [107]. Copyright 2018 American Chemical Society.

Creating energetic carriers through SPP absorption on metals increases the local temperature and heat diffuses into the adjacent reaction volume. Due to the temperature dependence of electrochemical reactions, isolating the roles of temperature and energetic carriers is essential to provide an understanding of results [108–113].

An analytical and experimental investigation of thermal effects on the electrochemical response of working electrodes bearing plasmonic nanostructures was reported in [113]. The time-dependent temperature profiles near illuminated electrodes was computed by considering heat flow via conduction, away from the interface between a planar electrode and the solution, yielding expressions for the temperature distribution in the system, for the corresponding increase in electrochemical reaction rate due to enhanced diffusion [113].

Figure 20 gives the distribution of velocity (left) and temperature (right) near a cylindrical glassy carbon electrode embedded in an insulating sheath dipped in an electrochemical cell filled with an electrolyte, assuming an input power of 10 Wcm^{-2} dissipated by the electrode over 10 s [113]. Heating the electrode surface produces significant solution flow in the vertical direction along the electrode surface and away from its end. Such convection has the effect of lowering the temperature at the electrode surface by a factor of $\sim$2, serving as a mechanism for transferring heat away from the electrode surface.

Increasing local temperatures affects electrochemical reaction rates through enhancements in mass transfer due to convection and by altering the diffusion coefficients of the redox species. Furthermore, heterogeneous electron transfer rates and redox potentials are temperature dependent. As predicted by the theoretical analyses of heat transfer by conduction and convection, optical absorption on electrodes would produce a temperature increase resulting in solution flow. The enhancement of mass transfer alone would produce significant current increases in any electrochemical reaction in the system concerned.

Models of diffusion, convection, and mass transport predict that redox currents increase approximately linearly with heating power rather than exponentially (as might be expected from the Arrhenius law), and that the current rises due to convection within 10 s of heating (e.g., [113]). Thus, temperature trends are not evident, implying that electrochemical cells should be stabilized and electrode temperatures monitored. Furthermore, independent thermal control experiments should be carried out and optical parameters other than intensity should be varied to isolate the effects of energetic carriers from those due to temperature.

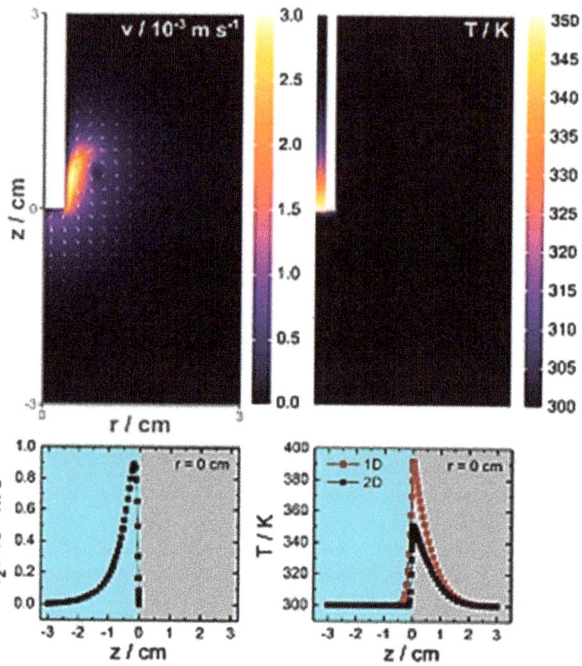

Figure 20. (**Top left**): velocity and (**top right**): temperature distributions of the electrolyte near a glassy carbon electrode (1.5 mm radius) embedded in an insulating sheath (white rectangle) in an electrochemical cell, obtained via FEM modelling of heat transfer including conduction and convection in electrolyte for 10 Wcm^{-2} heat input at the surface. The magnitude of the velocity and the temperature are shown on their respective color bars 10 s after the onset of heating. The (**bottom panels**) give the corresponding vertical distributions of the *z*-component of the velocity, v_z, and of the temperature. The red curve on the bottom right panel gives the temperature profile computed using a simple 1D model neglecting convection. Reprinted with permission from [113]. Copyright 2019 American Chemical Society.

Most plasmonic catalysis research has involved arrangements of colloidal Au nanoparticles excited by visible light. This scenario, however, poses certain challenges. The temperature near nanoparticles can be difficult to measure or predict due to collective effects [110], and carriers excited at visible photon energies above the interband threshold of Au ($h\upsilon$~2 eV) have short lifetimes due to electron-electron scattering at high carrier energies [114]. Nanoparticles arranged lithographically on a substrate can alleviate the former challenge—-for instance, arrangements supporting surface lattice resonances [115], which can also be used in catalysis [116].

Alternatively, electrochemistry can be carried out with lithographically-defined and evaporated Au microelectrodes on a substrate [117–119], provided the electrodes are annealed prior to use [117]. Such electrodes were shaped as a stripe and used simultaneously as a surface plasmon waveguide supporting Bloch LRSPPs at infrared wavelengths excited by grating couplers, as sketched in Figure 21 [43,118,119]. The use of infrared wavelengths ensures that energetic carriers created in the stripe by LRSPP absorption are long lived. Counter electrodes can also be defined on chip, enabling an integrated plasmonic/electrochemical sensing chip.

The chip bearing an Au stripe waveguide also serving as a working electrode and a Pt counter electrode was integrated into a three-electrode electrochemical cell using an external reference electrode, and cyclic voltammograms were obtained while varying the incident optical power and wavelength, as shown in Figure 22A [118]. By studying oxidation and reduction reactions separately, energetic hole processes were separated from energetic electron processes. Excitation of LRSPPs increased the redox current density by 10×, as shown in Figure 22A. The reduction, oxidation, and equilibrium potentials dropped by about 2×, and they split following the photon energy beyond a clearly observed threshold with SPP power, as shown in Figure 22C. Electrochemical impedance spectroscopy measurements showed that under LRSPP excitation, the charge transfer resistance was reduced by a factor of almost 2. During LRSPP excitation, the working electrode temperature was monitored in situ and in real time. Chronoamperometry measurements obtained as the LRSPPs are on-off modulated at 600 Hz to yield a current response modulated at the same frequency, which excludes thermally-enhanced mass transfer. The opening of non-equilibrium redox channels due to the transfer of energetic carriers to the redox species was invoked to explain these observations [118].

The optical power output from a waveguide working electrode that supports propagating Bloch LRSPPs is proportional to the time-convolution of the electrochemical current density, which enables real-time convolutional electrochemistry [119]. The optical response of a waveguide working electrode was formulated theoretically, and then verified experimentally via cyclic voltammetry and chronoamperometry measurements for different concentrations of potassium ferricyanide in potassium nitrate used as the electrolyte [119]. By increasing the optical power, the LRSPP no longer acts solely as a probe of electrochemical activity, but also as a pump creating hot (energetic) electrons and holes, leading to significantly enhanced currents in this regime (*cf.* Figure 22). The output optical power remains proportional to the time-convolution of the current density in this high-power regime, even when energetic carriers are responsible for the redox reactions [119].

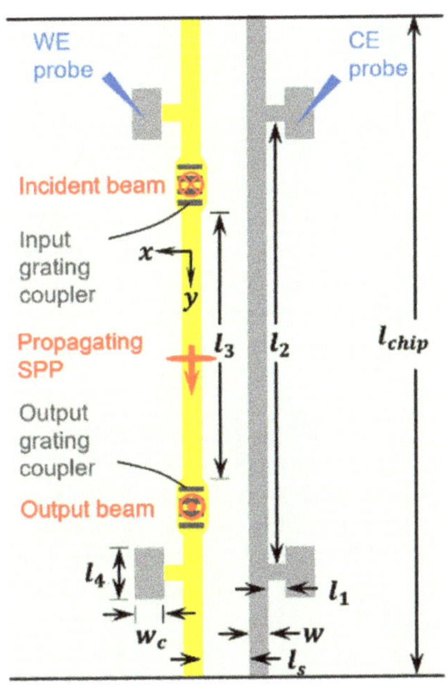

Figure 21. Schematic in top view of the sensing chip. The thickness of the Au working electrode (WE, yellow) and of the Pt counter electrode (CE, gray) is t = 35 nm and their dimensions are: l_1 = 29 μm, l_2 = 2600 μm, l_3 = 1850 μm, l_4 = 250 μm, l_{chip} = 3000 μm, w_c = 100 μm, w = 5 μm, l_s = 40 μm. The Bloch LRSPPs propagating along the WE are excited using an optical beam normally incident on an input grating coupler. The output grating coupler produces a normally emerging output optical beam that can be monitored. Reprinted with permission from [119]. Copyright 2022 American Chemical Society.

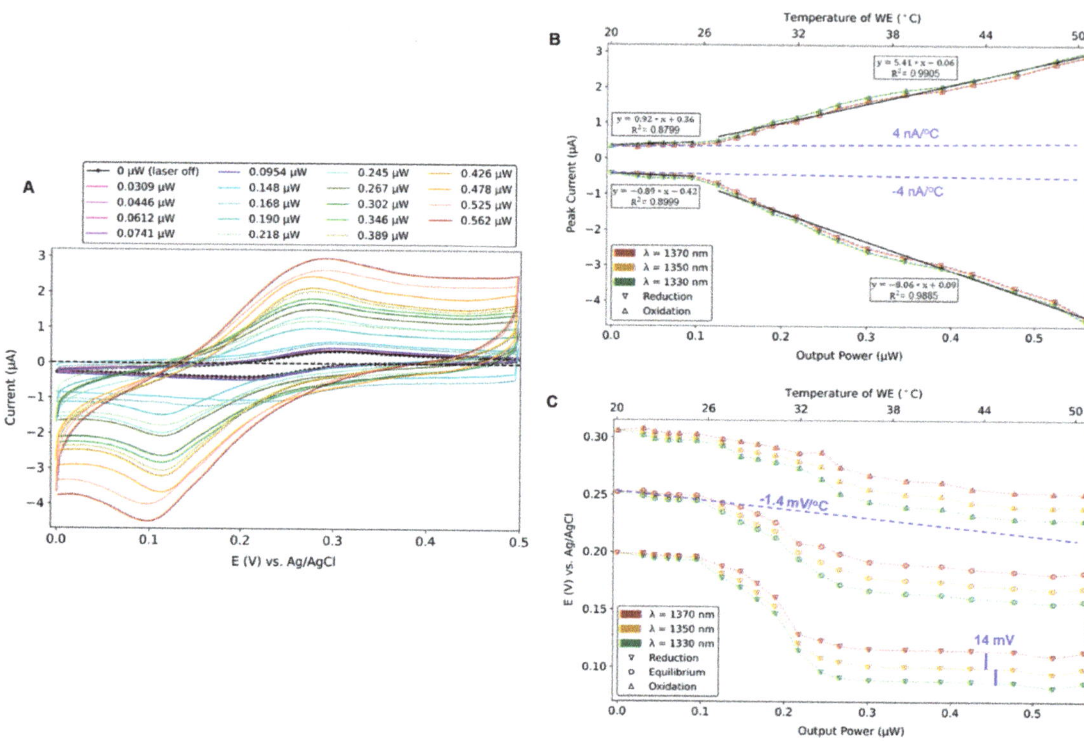

Figure 22. Cyclic voltammetry under optical illumination. (**A**) CV curves obtained at a scan rate of 100 mV/s, on a Au WE in 0.5 mM $K_3[Fe(CN)_6]$ + 100 mM KNO_3 electrolyte, as the optical power at λ_0 = 1350 nm increases (legend—the output power was monitored). The incident optical power ranged from 0 to 6.3 mW. The black dots plot the reference CV curve (laser off). (**B**) Redox current peaks, and (**C**) potentials vs. output optical power, from CV curves measured at λ_0 = 1330, 1350 and 1370 nm. The blue dashed lines on Parts (**B**,**C**) plot the linear thermal trends measured independently, ruling out thermal effects. Reprinted with permission from [118]. Copyright 2022 The Authors, licensed under a Creative Commons Attribution (CC BY) license.

4. Conclusions

We introduced in a tutorial-like fashion basic electrochemical concepts and concepts pertaining to SPPs on planar structures, and we reviewed representative literature on their integration in the same structure. Integration is motivated by the prospect of multimodal biosensors where the strengths of both techniques can be leveraged. Other motivating factors include the use of SPPs to probe electrochemical activity, leading naturally to real-time convolutional voltammetry by monitoring the output optical signal, or the use of SPPs as a "pump" affecting electrochemical reactions. Pumping occurs through SPP absorption in the working electrode, leading to the creation of energetic (hot) electrons and holes that can transfer more readily to the redox species. Enhanced carrier transfer can lead to enhanced electrochemical currents or may open new redox channels (pathways).

Integration challenges remain on several fronts. For instance, clearly differentiating thermal effects from the effects of energetic carriers in plasmonic hot carrier electrochemistry and developing robust fabrication methods that integrate on-chip sealed microfluidic channels with optical (plasmonic) working and counter electrodes. Non-trivially, three types of interface must be simultaneously integrated on-chip: (i) sealed fluidic interfaces to couple microfluidic channels to external tubing, (ii) optical structures to couple incident light to plasmonic working electrodes and extract output light therefrom, and (iii) integrating isolated electrical contacts to connect to working and counter electrodes. These motivating factors and challenges, and the promise of new and exciting applications, are driving a vigorous global expansion of the field of surface plasmon electrochemistry.

Author Contributions: Writing—original draft preparation, Z.H.; writing—review and editing, P.B. All authors have read and agreed to the published version of the manuscript.

Funding: This research received no external funding.

Institutional Review Board Statement: Not applicable.

Informed Consent Statement: Not applicable.

Data Availability Statement: No new data were created or analyzed in this study. Data sharing is not applicable to this article.

Conflicts of Interest: The authors declare no conflict of interest.

References

1. Jia, M.; Li, S.; Zang, L.; Lu, X.; Zhang, H. Analysis of Biomolecules Based on the Surface Enhanced Raman Spectroscopy. *J. Nanomater.* **2018**, *8*, 730. [CrossRef]
2. Bhalla, N.; Jolly, P.; Formisano, N.; Estrela, N.P. Introduction to biosensors. *Essays Biochem.* **2016**, *60*, 1–8. [PubMed]
3. Cooper, M.A. Optical biosensors in drug discovery. *Nat. Rev. Drug Discov.* **2002**, *1*, 515–528. [CrossRef] [PubMed]
4. Lequin, R.M. Enzyme Immunoassay (EIA)/Enzyme-Linked Immunosorbent Assay (ELISA). *Clin. Chem.* **2005**, *51*, 2415–2418. [CrossRef]
5. Sun, Z.; Lv, J.; Liu, X.; Tang, Z.; Wang, X.; Xu, Y.; Hammock, B.D. Development of a Nanobody-Avi Tag Fusion Protein and Its Application in a Streptavidin-Biotin-Amplified Enzyme-Linked Immunosorbent Assay for Ochratoxin A in Cereal. *Anal. Chem.* **2018**, *90*, 10628–10634. [CrossRef] [PubMed]
6. Endo, T.; Okuyama, A.; Matsubara, Y.; Nishi, K.; Kobayashi, M.; Yamamura, S.; Morita, Y.; Takamura, Y.; Mizukami, H.; Tamiya, E. Fluorescence-based assay with enzyme amplification on a micro-flow immunosensor chip for monitoring coplanar polychlorinated biphenyls. *Anal. Chim. Acta* **2005**, *531*, 7–13. [CrossRef]
7. Yuan, F.; Chen, M.; Leng, B.Y.; Wang, B.S. An efficient autofluorescence method for screening limonium bicolor mutants for abnormal salt gland density and salt secretion. *S. Afr. J. Bot.* **2013**, *88*, 110–117. [CrossRef]
8. Deng, Y.; Feng, Z.; Yuan, F.; Guo, J.; Suo, S.; Wang, B. Identification and functional analysis of the autofluorescent substance in limonium bicolor salt glands. *J. Plant Physiol.* **2015**, *97*, 20–27. [CrossRef]
9. Harris, E. *A Low-Cost Approach to PCR: Appropriate Transfer of Biomolecular Techniques*; Oxford University Press: Oxford, UK, 1998.
10. Chang, L.; Li, J.; Wang, L. Immuno-PCR: An ultrasensitive immunoassay for biomolecular Detection. *Anal. Chem. Acta* **2016**, *910*, 12–24. [CrossRef]
11. Chard, T. *Introduction to Radioimmunoassay and Related Techniques*, 5th ed.; Elsevier: Amsterdam, The Netherlands, 1995.
12. Liu, R.; Xiong, Y.; Tang, W.; Guo, Y.; Yan, X.; Si, M. Near-infrared surface-enhanced Raman spectroscopy (NIR-SERS) studies on oxyheamoglobin (OxyHb) of liver cancer based on PVA-Ag nanofilm. *J. Raman Spectrosc.* **2013**, *44*, 362–369. [CrossRef]
13. Munch, M.; Nielsen, L.P.; Handberg, K.J.; Ørgensen, P.H. Detection and subtyping (H5 and H7) of avian type A influenza virus by reverse transcription-PCR and PCR-ELISA. *Arch. Virol.* **2001**, *146*, 87–97. [CrossRef] [PubMed]
14. Daly, P.; Collier, T.; Doyle, S. PCR-ELISA detection of Escherichia coli in milk. *Lett. Appl. Microbiol.* **2002**, *34*, 222–226. [CrossRef] [PubMed]
15. Medintz, I.L.; Clapp, A.R.; Melinger, J.S.; Deschamps, J.R.; Mattoussi, H. A reagentless biosensing assembly based on quantum dot-donor Forster resonance energy transfer. *Adv. Mater.* **2005**, *17*, 2450–2455. [CrossRef]
16. Nedelkov, D.; Nelson, R.W. Practical considerations in BIA/MS: Optimizing the biosensor-mass spectrometry interface. *J. Mol. Recognit.* **2000**, *13*, 140–145. [CrossRef]
17. Yao, C.; Zhu, T.; Qi, Y.; Zhao, Y.; Xia, H.; Fu, W. Development of a quartz crystal microbalance biosensor with aptamers as bio-recognition element. *Sensors* **2010**, *10*, 5859–5871. [CrossRef]
18. Dostálek, J.; Čtyroký, J.; Homola, J.; Brynda, E.; Skalský, M.; Nekvindová, P.; Špirková, J.; Škvor, J.; Schröfel, J. Surface plasmon resonance biosensor based on integrated optical waveguide. *Sens. Actuators B Chem.* **2001**, *76*, 8–12. [CrossRef]
19. Haes, A.J.; Van Duyne, R.P. A unified view of propagating and localized surface plasmon resonance biosensors. *Anal. Bioanal. Chem.* **2004**, *379*, 920–930. [CrossRef]
20. Krupin, O.; Asiri, H.; Wang, C.; Tait, R.N.; Berini, P. Biosensing using straight long-range surface plasmon waveguides. *Opt. Express* **2013**, *21*, 698–709. [CrossRef]
21. Watanabe, M.; Kajikawa, K. An optical fiber biosensor based on anomalous reflection of gold. *Sens. Actuators B Chem.* **2003**, *89*, 126–130. [CrossRef]

22. Yoon, J.; Shin, M.; Lee, T.; Choi, J.W. Highly sensitive biosensors based on biomolecules and functional nanomaterials depending on the types of nanomaterials: A perspective review. *Materials* **2020**, *13*, 299. [CrossRef]
23. Homola, J. Surface plasmon resonance sensors for detection of chemical and biological Species. *Chem. Rev.* **2008**, *108*, 462–493. [CrossRef] [PubMed]
24. Tang, D.; Yuan, R.; Chai, Y.; Liu, Y.; Dai, J.J.; Zhong, X. Novel potentiometric immunosensor for determination of diphtheria antigen based on compound nanoparticles and bilayer two-dimensional sol-gel as matrices. *Anal. Bioanal. Chem.* **2005**, *381*, 674–680. [CrossRef] [PubMed]
25. Darain, F.; Park, S.U.; Shim, Y.B. Disposable amperometric immunosensor system for rabbit IgG using a conducting polymer modified screen-printed electrode. *Biosens. Bioelectr.* **2003**, *18*, 773–780. [CrossRef] [PubMed]
26. Miura, N.; Higobashi, H.; Sakai, G.; Takeyasu, A.; Uda, T.; Yamazoe, N. Piezoelectric crystal immunosensor for sensitive detection of methamphetamine (stimulant drug) in human urine. *Sens. Actuators B Chem.* **1993**, *13*, 188–191. [CrossRef]
27. Zhang, B.; Mao, Q.; Zhang, X.; Jiang, T.; Chen, M.; Yu, F.; Fu, W. A novel piezoelectric quartz micro-array immunosensor based on self-assembled monolayer for determination of human chorionic gonadotropin. *Biosens. Bioelectr.* **2004**, *19*, 711–720. [CrossRef]
28. Ramanathan, K.; Danielsson, B. Principles and applications of thermal biosensors. *Biosens. Bioelectr.* **2001**, *16*, 417–423. [CrossRef]
29. Kumbhat, S.; Sharma, K.; Gehlot, R.; Solanki, A.; Joshi, V. Surface plasmon resonance based immunosensor for serological diagnosis of dengue virus infection. *J. Pharm. Biomed. Anal.* **2010**, *52*, 255–259. [CrossRef]
30. Wang, J. Towards Genoelectronics: Electrochemical Biosensing of DNA Hybridization. *Chem-Eur. J.* **1999**, *5*, 1681–1685. [CrossRef]
31. Cho, I.H.; Kim, D.H.; Park, S. Electrochemical biosensors: Perspective on functional nanomaterials for on-site analysis. *Biomater. Res.* **2020**, *24*, 6. [CrossRef]
32. Zhu, C.; Yang, G.; Li, H.; Du, D.; Lin, Y. Electrochemical sensors and biosensors based on nanomaterials and nanostructures. *Anal. Chem.* **2015**, *87*, 230–249. [CrossRef]
33. Grieshaber, D.; MacKenzie, R.; Vörös, J.; Reimhult, E. Electrochemical Biosensors—Sensor Principles and Architectures. *Sensors* **2008**, *8*, 1400–1458. [CrossRef] [PubMed]
34. Dertien, E.; Regtien, P.P. *Sensors for Mechatronics*; Elsevier: Amsterdam, The Netherlands, 2018.
35. Yao, C.; Qi, Y.; Zhao, Y.; Xiang, Y.; Chen, Q.; Fu, W. Aptamer-based piezoelectric quartz crystal microbalance biosensor array for the quantification of IgE. *Biosens. Bioelectr.* **2009**, *24*, 2499–2503. [CrossRef]
36. Danielsson, B.; Mattiasson, B.; Mosbach, K. Enzyme thermistor devices and their analytical applications. *Appl. Biochem.* **1981**, *3*, 97–143.
37. Mattiasson, B.; Borrebaeck, C.; Sanfridson, B.; Mosbach, K. Thermometric enzyme linked immunosorbent assay: TELISA. *Biochim. Biophys. Acta* **1977**, *483*, 221–227. [CrossRef] [PubMed]
38. Kazura, E.; Mernaugh, R.; Baudenbacher, F. A capillary-perfused, nanocalorimeter platform for thermometric enzyme-linked immunosorbent assay with femtomole sensitivity. *Biosensors* **2020**, *10*, 71. [CrossRef] [PubMed]
39. Sawant, S.N. Development of biosensors from biopolymer composites. In *Biopolymer Composites in Electronics*; Elsevier: Amsterdam, The Netherlands, 2017; pp. 353–383.
40. Long, F.; Zhu, A.; Shi, H. Recent advances in optical biosensors for environmental monitoring and early warning. *Sensors* **2013**, *13*, 13928–13948. [CrossRef]
41. Mol, N.J.; Fischer, M.J.E. *Surface Plasmon Resonance: Methods and Protocols*; Springer: Berlin/Heidelberg, Germany, 2010.
42. Damborský, P.; Švitel, J.; Katrlík, J. Optical biosensors. *Essays Biochem.* **2016**, *60*, 91–100.
43. Fong, N.R.; Menotti, M.; Lisicka-Skrzek, E.; Northfield, H.; Olivieri, A.; Tait, N.; Liscidini, M.; Berini, P. Bloch long range surface plasmon polaritons on metal stripe waveguides on a multilayer substrate. *ACS Photonics* **2017**, *4*, 593–599. [CrossRef]
44. Berini, P. Long-range surface plasmon polaritons. *Adv. Opt. Photonics* **2009**, *1*, 484–588. [CrossRef]
45. Maier, S.A. *Plasmonics: Fundamentals and Applications*; Springer: Berlin/Heidelberg, Germany, 2007.
46. Berini, P. Plasmon polariton waves guided by thin lossy metal films of finite width: Bound modes of symmetric structures. *Phys. Rev. B* **2000**, *61*, 10484–10503. [CrossRef]
47. Vargas, W.E. Optical and electrical properties of hydrided palladium thin films studied by an inversion approach from transmittance measurements. *Thin Solid Films* **2006**, *496*, 189–196. [CrossRef]
48. Rakić, A.D. Optical properties of metallic films for vertical-cavity optoelectronic devices. *Appl. Opt.* **1998**, *37*, 5271–5283. [CrossRef]
49. Johnson, P.B.; Christy, R.W. Optical constants of the noble metals. *Phys. Rev. B* **1972**, *6*, 4370. [CrossRef]
50. Giannini, V.; Fernández-Domínguez, A.I.; Heck, S.C.; Maier, S.A. Plasmonic nanoantennas: Fundamentals and their use in controlling the radiative properties of nanoemitters. *Chem. Rev.* **2011**, *111*, 3888–3912. [CrossRef]
51. Boardman, A.D.; Paranjape, B.V. The optical surface modes of metal spheres. *J. Phys. F* **1977**, *7*, 1935–1945. [CrossRef]
52. Schmidt, F.P.; Ditlbacher, H.; Hohenester, U.; Hohenau, A.; Hofer, F.; Krenn, J.R. Universal dispersion of surface plasmons in flat nanostructures. *Nat. Commun.* **2014**, *5*, 3604. [CrossRef]
53. Benson, O. Assembly of hybrid photonic architectures from nanophotonic Constituents. *Nature* **2011**, *480*, 193–199. [CrossRef] [PubMed]
54. Homola, J. Electromagnetic Theory of Surface Plasmons. *Springer Ser. Chem. Sens. Biosens.* **2006**, *4*, 3–44.
55. McPeak, K.M.; Jayanti, S.V.; Kress, S.J.P.; Meyer, S.; Iotti, S.; Rossinelli, A.; Norris, D.J. Plasmonic Films Can Easily Be Better: Rules and Recipes. *ACS Photonics* **2015**, *2*, 326–333. [CrossRef]

56. West, P.R.; Ishii, S.; Naik, G.V.; Emani, N.K.; Shalaev, V.M.; Boltasseva, A. Searching for better plasmonic materials. *Lasers Phot. Rev.* **2010**, *4*, 795–808. [CrossRef]
57. Berini, P. Plasmon-polariton waves guided by thin lossy metal films of finite width: Bound modes of asymmetric structures. *Phys. Rev. B* **2001**, *63*, 125417. [CrossRef]
58. Boltasseva, A.; Nikolajsen, T.; Leosson, K.; Kjaer, K.; Larsen, M.S.; Bozhevolnyi, S.I. Integrated optical components utilizing long-range surface plasmon polaritons. *J. Light. Technol.* **2005**, *23*, 413–422. [CrossRef]
59. Charbonneau, R.; Scales, C.; Breukelaar, I.; Fafard, S.; Lahoud, N.; Mattiussi, G.; Berini, P. Passive integrated optics elements based on long-range surface plasmon polaritons. *J. Light. Technol.* **2006**, *24*, 477–494. [CrossRef]
60. Ebbesen, T.W.; Genet, C.; Bozhevolnyi, S.I. Surface plasmon circuitry. *Phys. Today* **2008**, *61*, 44. [CrossRef]
61. Minh, T.; Tanaka, K.; Tanaka, M. Complex propagation constants of surface plasmon polariton rectangular waveguide by method of lines. *Opt. Express* **2008**, *16*, 9378–9390. [CrossRef]
62. Berini, P.; Buckley, R. On the convergence and accuracy of numerical mode computations of surface plasmon waveguides. *J. Comput. Theor. Nanosci.* **2009**, *6*, 2040–2053. [CrossRef]
63. Veronis, G.; Kocaba, S.; Miller, D.A.B.; Fan, S. Modeling of Plasmonic Waveguide Components and Networks. *J. Comput. Theor. Nanosci.* **2009**, *6*, 1808–1826. [CrossRef]
64. Otto, A. Excitation of nonradiative surface plasma waves in silver by the method of frustrated total reflection. *Z. Für Phys. A* **1968**, *216*, 398–410. [CrossRef]
65. Otto, A. Excitation by light of ω+ and ω- surface plasma waves in thin metal layers. *Z. Für Phys. A* **1969**, *219*, 227–233. [CrossRef]
66. Kretschmann, E. Die Bestimmung optischer Konstanten von Metallen durch Anregung von Oberfl¨achenplasmaschwingungen. *Z. Für Phys. A* **1971**, *241*, 313–324. [CrossRef]
67. Novotny, L.; Hecht, B. *Principles of Nano-Optics*; Cambridge University Press: Cambridge, UK, 2012.
68. Fischer, M.J.E. *Surface Plasmon Resonance*; Springer: Berlin/Heidelberg, Germany, 2010.
69. Shankaran, D.R.; Gobi, V.; Miura, N. Recent advancements in surface plasmon resonance immunosensors for detection of small molecules of biomedical, food and environmental interest. *Sens. Actuators B Chem.* **2007**, *121*, 158–177. [CrossRef]
70. Cooper, M.A. *Label-Free Biosensors: Techniques and Applications*; Cambridge University Press: Cambridge, UK, 2009.
71. Slavik, R.; Homola, J. Ultrahigh resolution long range surface plasmon-based sensor. *Sens. Actuators B Chem.* **2007**, *123*, 10–12. [CrossRef]
72. Mauriz, E.; Calle, A.; Manclus, J.J.; Montoya, A.; Lechuga, L.M. Multi-analyte SPR immunoassays for environmental biosensing of pesticides. *Anal. Bioanal. Chem.* **2007**, *387*, 1449–1458. [CrossRef]
73. Chabot, V.; Miron, Y.; Grandbois, M.; Charette, P.G. Long range surface plasmon resonance for increased sensitivity in living cell biosensing through greater probing depth. *Sens. Actuators B Chem.* **2012**, *174*, 94–101. [CrossRef]
74. Monzon Hernandez, D.; Velazquez-Gonzalez, J.S.; Luna-Moreno, D.; Torres-Cisneros, M.; Hernandez-Romano, I. Prism-Based Surface Plasmon Resonance for Dual-Parameter Sensing. *IEEE Sens. J.* **2018**, *18*, 4030–4037. [CrossRef]
75. Boruah, R.; Mohanta, D.; Choudhury, A.; Nath, P.; Ahmed, G.A. Surface Plasmon Resonance-Based Protein Bio-Sensing Using a Kretschmann Configured Double Prism Arrangement. *IEEE Sens. J.* **2015**, *15*, 6791–6796. [CrossRef]
76. Yang, C.-T.; Mejard, R.; Griesser, H.J.; Bagnaninchi, P.O.; Thierry, B. Cellular Micromotion Monitored by Long-Range Surface Plasmon Resonance with Optical Fluctuation Analysis. *Anal. Chem.* **2015**, *87*, 1456–1461. [CrossRef]
77. Vala, M.; Etheridge, S.; Roach, J.; Homola, J. Long-range surface plasmons for sensitive detection of bacterial analytes. *Sens. Actuators B Chem.* **2009**, *139*, 59–63. [CrossRef]
78. Koubova, A.V.; Brynda, E.; Karasova, L.; Homola, J.; Dostalek, J.; Tobiska, P.; Rosicky, J. Detection of foodborne pathogens using surface plasmon resonance biosensors. *Sens. Actuators B Chem.* **2001**, *74*, 100–105. [CrossRef]
79. Lee, W.; Lee, D.-B.; Oh, B.-K.; Lee, W.H.; Choi, J.-W. Nanoscale fabrication of protein A on self-assembled monolayer and its application to surface plasmon resonance immunosensor. *Enzyme Microb. Technol.* **2004**, *35*, 678–682. [CrossRef]
80. Wei, J.; Mu, Y.; Song, D.; Fang, X.; Liu, X.; Bu, L.; Zhang, H.; Zhang, G.; Ding, J.; Wang, W.; et al. A novel sandwich immunosensing method for measuring cardiac troponin I in sera. *Anal. Biochem.* **2003**, *321*, 209–216. [CrossRef] [PubMed]
81. Gobi, K.V.; Kataoka, C.; Miura, N. Surface plasmon resonance detection of endocrine disruptors using immunoprobes based on self assembled monolayers. *Sens. Actuators B Chem.* **2005**, *108*, 784–790. [CrossRef]
82. Trevino, J.; Calle, A.; Rodriguez-Frade, J.M.; Mellado, M.; Lechuga, L.M. Surface plasmon resonance immunoassay analysis of pituitary hormones in urine and serum samples. *Clin. Chim. Acta* **2009**, *403*, 56–62. [CrossRef] [PubMed]
83. Dostalek, J.; Kasry, A.; Knoll, W. Long range surface plasmons for observation of biomolecular binding events at metallic surfaces. *Plasmonics* **2007**, *2*, 97–106. [CrossRef]
84. Degiron, A.; Cho, S.Y.; Tyler, T.; Jokerst, N.; Smith, D.R. Directional coupling between dielectric and long-range plasmon waveguides. *New J. Phys.* **2009**, *11*, 015002. [CrossRef]
85. Hirbodvash, Z.; Khodami, M.; Fong, N.R.; Lisicka-Skrzek, E.; Olivieri, A.; Northfield, H.; Tait, R.N.; Berini, P. Grating couplers fabricated by e-beam lithography for long range surface plasmon waveguides embedded in a fluoropolymer. *Appl. Opt.* **2019**, *58*, 2994–3002. [CrossRef]
86. Khodami, M.; Hirbodvash, Z.; Krupin, O.; Wong, W.R.; Lisicka-Skrzek, E.; Northfield, H.; Hahn, C.; Berini, P. Fabrication of Bloch Long Range Surface Plasmon Waveguides Integrating Counter Electrodes and Microfluidic Channels for Multimodal Biosensing. *J. Microelectromech. Syst.* **2021**, *30*, 686–695. [CrossRef]

87. Tellez, G.A.C.; Hassan, S.; Tait, R.N.; Berini, P.; Gordon, R. Atomically flat symmetric elliptical nanohole arrays in a gold film for ultrasensitive refractive index sensing. *Lab A Chip* **2013**, *13*, 2541–2546. [CrossRef]
88. Hajebifard, A.; Berini, P. Fano resonances in plasmonic heptamer nano-hole arrays. *Opt. Express* **2017**, *25*, 18566–18580. [CrossRef]
89. Asahi Glass Company. Cytop Technical Brochure. 2009. Available online: http://www.agc.com (accessed on 8 February 2023).
90. Dupont. "Teflon AF Properties". Available online: www.dupont.com (accessed on 8 February 2023).
91. Berini, P. Bulk and surface sensitivities of surface plasmon waveguides. *New J. Phys.* **2008**, *10*, 105010. [CrossRef]
92. Krupin, O.; Wong, W.R.; Béland, P.; Adikan, F.R.M.; Berini, P. Long-Range Surface Plasmon-Polariton Waveguide Biosensors for Disease Detection. *J. Light. Technol.* **2016**, *34*, 4673–4681. [CrossRef]
93. Chen, C.; Berini, P. Grating couplers for broadside input and output coupling of long-range surface plasmons. *Opt. Express* **2010**, *18*, 8006–8018. [CrossRef] [PubMed]
94. Homola, J. *Surface Plasmon Resonance Based Sensors*; Springer: Berlin/Heidelberg, Germany, 2006.
95. Golden, J.; Yates, M.D.; Halsted, M.; Tender, L. Application of electrochemical surface plasmon resonance (ESPR) to the study of electroactive microbial biofilms. *Phys. Chem. Chem. Phys.* **2018**, *20*, 25648–25656. [CrossRef] [PubMed]
96. Rusling, J.F. *Biomolecular Films: Design, Function, and Applications*; CRC Press: Boca Raton, FL, USA, 2003.
97. Elgrishi, N.; Rountree, K.J.; McCarthy, B.D.; Rountree, E.S.; Eisenhart, T.T.; Dempsey, J.L. A practical beginner's guide to cyclic voltammetry. *J. Chem. Educ.* **2018**, *95*, 197–206. [CrossRef]
98. Wang, S.; Huang, X.; Shan, X.; Foley, K.J.; Tao, N. Electrochemical surface plasmon resonance: Basic formalism and experimental validation. *Anal. Chem.* **2010**, *82*, 935–941. [CrossRef]
99. Gordon, J.G.; Ernest, S. Surface plasmons as a probe of the electrochemical interface. *Surf. Sci.* **1980**, *101*, 499–506. [CrossRef]
100. Huang, X.; Wang, S.; Shan, X.; Chang, X.; Tao, N. Flow-through electrochemical surface plasmon resonance: Detection of intermediate reaction products. *J. Electroanal. Chem.* **2010**, *649*, 37–41. [CrossRef]
101. Patskovsky, S.; Dallaire, A.M.; Meunier, M. Electrochemical surface plasmon resonance sensing with absorptive redox mediator film. *Sens. Actuators B Chem.* **2016**, *222*, 71–77. [CrossRef]
102. Sannomia, T.; Dermutz, H.; Hafner, C.; Voros, J.; Dahilin, A.B. Electrochemistry on a localized surface plasmon sensor. *Langmuir* **2010**, *26*, 7619–7626. [CrossRef]
103. Ahmed, I.; Shi, L.; Pasanen, H.; Vivo, P.; Maity, P.; Hatamvand, M.; Zhan, Y. There is plenty of room at the top: Generation of hot charge carriers and their applications in perovskite and other semiconductor-based optoelectronic devices. *Light Sci. Appl.* **2021**, *10*, 174. [CrossRef]
104. Reddy, H.; Shalaev, V.M. Plasmonic hot-carriers and their applications: Opinion. *Opt. Mater. Express* **2021**, *11*, 3827–3832. [CrossRef]
105. Tagliabue, G.; Jermyn, A.S.; Sundararaman, R.; Welch, A.J.; DuChene, J.S.; Pala, R.; Davoyan, A.R.; Narang, P.; Atwater, H.A. Quantifying the role of surface plasmon excitation and hot carrier transport in plasmonic devices. *Nat. Commun.* **2018**, *9*, 3394. [CrossRef]
106. Lee, J.; Mubeen, S.; Ji, X.; Stucky, G.D.; Moskovits, M. Plasmonic Photoanodes for Solar Water Splitting with Visible Light. *Nano Lett.* **2012**, *12*, 5014–5019. [CrossRef]
107. DuChene, J.S.; Tagliabue, G.; Welch, A.J.; Cheng, W.; Atwater, H.A. Hot hole collection and photoelectrochemical CO2 reduction with plasmonic Au/p-GaN photocathodes. *Nano Lett.* **2018**, *18*, 2545–2550. [CrossRef]
108. Zhou, L.; Swearer, D.L.; Zhang, C.; Robatjazi, H.; Zhao, H.; Henderson, L.; Dong, L.; Christopher, P.; Carter, E.A.; Nordlander, P.; et al. Quantifying hot carrier and thermal contributions in plasmonic photocatalysis. *Science* **2018**, *362*, 69–72. [CrossRef]
109. Jain, P.K. Taking the Heat Off of Plasmonic Chemistry. *J. Phys. Chem. C* **2019**, *123*, 24347–24351. [CrossRef]
110. Baffou, G.; Bordacchini, I.; Baldi, A.; Quidant, R. Simple experimental procedures to distinguish photothermal from hot-carrier processes in plasmonics. *Light Sci. Appl.* **2020**, *9*, 108. [CrossRef] [PubMed]
111. Cortés, E.; Besteiro, L.V.; Alabastri, A.; Baldi, A.; Tagliabue, A.; Demetriadou, A.; Narang, P. Challenges in Plasmonic Catalysis. *ACS Nano* **2020**, *14*, 16202–16219. [CrossRef] [PubMed]
112. Dubi, Y.; Un, I.W.; Sivan, Y. Thermal effects—An alternative mechanism for plasmon-assisted photocatalysis. *Chem. Sci.* **2020**, *11*, 5017–5027. [CrossRef]
113. Maley, M.; Hill, J.W.; Saha, P.; Walmsley, J.D.; Hill, C.M. The role of heating in the electrochemical response of plasmonic nanostructures under illumination. *J. Phys. Chem. C* **2019**, *123*, 12390–12399. [CrossRef]
114. Bauer, M.; Marienfeld, A.; Aeschlimann, M. Hot electron lifetimes in metals probed by time-resolved two-photon photoemission. *Prog. Surf. Sci.* **2015**, *90*, 319–376. [CrossRef]
115. Kravets, V.G.; Kabashin, A.V.; Barnes, W.L.; Grigorenko, A.N. Plasmonic surface lattice resonances: A review of properties and applications. *Chem. Rev.* **2018**, *118*, 5912–5951. [CrossRef]
116. Deng, S.; Zhang, B.; Choo, P.; Smeets, P.J.M.; Odom, T.W. Plasmonic Photoelectrocatalysis in Copper–Platinum Core–Shell Nanoparticle Lattices. *Nano Lett.* **2021**, *21*, 1523–1529. [CrossRef] [PubMed]
117. Hirbodvash, Z.; Houache, M.S.E.; Krupin, O.; Khodami, M.; Northfield, H.; Olivieri, A.; Baranova, E.A.; Berini, P. Electrochemical Performance of Lithographically-Defined Micro-Electrodes for Integration and Device Applications. *Chemosensors* **2021**, *9*, 277. [CrossRef]

118. Hirbodvash, Z.; Krupin, O.; Northfield, H.; Olivieri, A.; Baranova, E.A.; Berini, P. Infrared surface plasmons on a Au waveguide electrode open new redox channels associated with the transfer of energetic carriers. *Sci. Adv.* **2022**, *8*, eabm9303. [CrossRef] [PubMed]
119. Hirbodvash, Z.; Baranova, E.A.; Berini, P. Real-time convolutional voltammetry enhanced by energetic (hot) electrons and holes on a surface plasmon waveguide electrode. *Anal. Chem.* **2022**, *94*, 13145–13152. [CrossRef] [PubMed]

MDPI AG
Grosspeteranlage 5
4052 Basel
Switzerland
Tel.: +41 61 683 77 34

Chemosensors Editorial Office
E-mail: chemosensors@mdpi.com
www.mdpi.com/journal/chemosensors

www.ingramcontent.com/pod-product-compliance
Lightning Source LLC
Chambersburg PA
CBHW061555210926
51850CB00009B/2111
9783725821792